W9-AAS-587

Beginning SQL Server R Services

Analytics for Data Scientists

Bradley Beard

Apress®

Beginning SQL Server R Services: Analytics for Data Scientists

Bradley Beard
Palm Bay, Florida
USA

ISBN-13 (pbk): 978-1-4842-2297-3 ISBN-13 (electronic): 978-1-4842-2298-0
DOI 10.1007/978-1-4842-2298-0

Library of Congress Control Number: 2016958725

Distributed to the book trade worldwide by Springer Science+Business Media New York, 233 Spring Street, 6th Floor, New York, NY 10013. Phone 1-800-SPRINGER, fax (201) 348-4505, e-mail orders-ny@springer-sbm.com, or visit www.springer.com. Apress Media, LLC is a California LLC and the sole member (owner) is Springer Science + Business Media Finance Inc (SSBM Finance Inc). SSBM Finance Inc is a **Delaware** corporation.

For information on translations, please e-mail rights@apress.com, or visit www.apress.com.

Apress and friends of ED books may be purchased in bulk for academic, corporate, or promotional use. eBook versions and licenses are also available for most titles. For more information, reference our Special Bulk Sales–eBook Licensing web page at www.apress.com/bulk-sales.

Any source code or other supplementary materials referenced by the author in this text are available to readers at www.apress.com. For detailed information about how to locate your book's source code, go to www.apress.com/source-code/. Readers can also access source code at SpringerLink in the Supplementary Material section for each chapter.

Printed on acid-free paper

This book is dedicated to the memory of my late grandmother,
Bessie Dejaynes, who passed away during the writing of this book.
I love you, I miss you, and I will see you again.

Contents at a Glance

Contents

About the Author

Bradley Beard is a software engineer with more than 15 years' experience writing dynamic, interactive web sites using ColdFusion and SQL Server. He graduated from the Florida Institute of Technology in 2007 with a Master of Science in Computer Information Systems, and studied for his undergraduate degrees in CIS and Technology Management at Herzing University. In 2013, he earned the MCSA: SQL Server 2012 certification from Microsoft. In 2016, he earned the MCSE: Business Intelligence certification as well. His continual quest for learning has earned him shelves full of books at home and at work, most of which are about SQL Server, ColdFusion, or general web architectures or frameworks.

He lives in Palm Bay, Florida, with his wife, Jessica, and children, Josh, Kaylee, Matthew, and Emma. He also apparently runs an animal shelter made up of his dogs, Lady and Bella, and cats, Spice, Simba, Mercury, and Dobby. In his free time, he enjoys fishing and spending time with his wife and kids.

Bradley is available for consultation and third-shift remote employment on ColdFusion and SQL Server by contacting him at `bradley.beard@gmail.com`.

About the Technical Reviewer

Kathi Kellenberger, known to the SQL Server community as Aunt Kathi, is an independent SQL Server consultant associated with Linchpin People and a Data Platform MVP. She loves writing about SQL Server and has contributed to more than a dozen books as an author, co-author, or technical reviewer. Kathi enjoys spending free time with family and friends, especially her five grandchildren. When she is not working or involved in a game of hide-and-seek with the kids, you may find her at the local karaoke bar. Kathi's blog is at `www.auntkathisql.com`.

Acknowledgments

Another big thanks to both Jonathan Gennick and Jill Balzano for guiding me through the process of getting this book published. You guys are awesome.

To my first mentor once again, John Wysocki, who is currently enjoying semi-retirement in some lavish resort somewhere: I can't thank you enough for stoking the fire of creativity in me.

To my newest mentor, Suzy Moore, who is hands down the smartest person I know: I can't wait to learn more from you.

To the one and only Chester Flake: Thank you for your guidance on certifications. Anyone needing any sort of certification training needs to go to `www.certificationcamps.com` and sign up today. You won't be sorry.

To my brother Brian, niece Holly, sister-in-law Andie, and Drew, Austin, and Sam: Mom said no fireworks past 10:00 PM!

To my parents Richard and Carolyn, my in-laws Steve and Carey, my out-laws Al and Val, other brothers Joe, Rick, Zimmer, and Dave, and other sisters Morgan, Erika, Jennifer, Kim, and Michelle, and everyone else I forgot.... Well, you know me....

And finally, to my wife, Jessica, and kids—Josh, Kaylee, Matthew, and Emma, who had to deal with me leaving for hours on end while writing this book: Thank you so much.

Oh, yeah.... I can't forget my best fraaaaaaaand Courtney. She's such a little cutie patootie!!

Introduction

In an effort to not sound like a complete Microsoft fan boy, SQL Server 2016 has some seriously cool additions. Not the least of these is the inclusion of a massive data analysis tool widely used throughout the industry. This tool is called, simply, R. Some of you might be asking why Microsoft would possibly include this tool, since it isn't really a database thing as it is an analysis or graphing tool.

The reason, I think, is fairly simple to deduce: Microsoft is expanding their reach. It seems to me that R is a great way to do that. The inclusion of R Tools as a part of the Visual Studio toolset and the SQL Server database instance will most definitely be a game changer for SQL Server development. It used to be where the database developer would have to prepare the data to be consumed by some service for analysis; not anymore. R Tools for Visual Studio (RTVS) allows the user to either prepare their scripts in Visual Studio, or directly in SQL Server Management Studio. Although this isn't recommended by Microsoft, it is still possible.

What We Will Cover

What this book covers is pretty simple and straightforward. We will...

- Set up a new instance on SQL Server 2016
- Set up the necessary R resources to properly create, consume, and execute R
- Briefly review the history, syntax, and functions within R
- Create a custom R solution using R Tools for Visual Studio
- Configure SQL Server Reporting Services
- Install and configure Report Builder
- Create reports in Report Builder based on R code developed in R Tools for Visual Studio
- Consume those reports through Reporting Services

It is important to note a few things at this point. Specifically, ...

- R Tools for Visual Studio is a brand-new release, so the chances of it being buggy are pretty good.
- We are fully installing SQL Server 2016 as a completely new instance because I wanted to be able to show the advantages that a user stands to gain by incorporating R into their workflow, even though they may not be completely sold on the benefits of R.

All the components needed to install and configure what is necessary to run R within SQL Server 2016 are covered on MSDN. As of now, the link for the resources is at `https://msdn.microsoft.com/en-us/library/mt604883.aspx`. Please note that Microsoft may not keep that same link forever, so remember that Google is your friend. I will explain what is needed to get it up and running. It may be up to you to find the necessary tools on Microsoft.com.

Why R?

Good question. Why R? The answer is actually quite simple... I think because Microsoft wants to expand their reach into the data science field with SQL Server, so it makes sense that they would want the best product out there for data analysis, which is arguably R. Microsoft's Business Intelligence offerings are already extremely sharp, so if they could find a way to ...

- Acquire a data analysis product already in heavy usage
- Incorporate that product into their existing database platform
- Offer a GUI which integrates seamlessly into Visual Studio
- Make it all available *for free*

... then the (data analysis) world would be their oyster! To my knowledge, there isn't any other database system that will allow for such complete interaction with R. There are plenty of instances where data is prepared in a database and then imported to R, but this is a totally different level of interaction.

SQL Server 2016 actually runs an instance of the R engine as native to the database engine. Fascinating! For those of you that have worked with SQL Server when they radically changed to SQL Server 2005, you remember the shock of Integration Services and "what happened to my DTS scripts??" This is on that same sort of level; for those intrigued by data analysis or business intelligence, the implications of the addition of R are guaranteed to be long-reaching and will most certainly result in some amazing advancements in data science. I, for one, can't wait to be a part of that.

Also, just to get this out of the way...

```
The MIT License (MIT)
Copyright (c) 2016 Microsoft
Permission is hereby granted, free of charge, to any person obtaining a copy of this
software and associated documentation files (the "Software"), to deal in the Software
without restriction, including without limitation the rights to use, copy, modify, merge,
publish, distribute, sublicense, and/or sell copies of the Software, and to permit persons
to whom the Software is furnished to do so, subject to the following conditions:
The above copyright notice and this permission notice shall be included in all copies or
substantial portions of the Software.
THE SOFTWARE IS PROVIDED "AS IS", WITHOUT WARRANTY OF ANY KIND, EXPRESS OR
IMPLIED, INCLUDING BUT NOT LIMITED TO THE WARRANTIES OF MERCHANTABILITY, FITNESS FOR A
PARTICULAR PURPOSE AND NONINFRINGEMENT. IN NO EVENT SHALL THE
AUTHORS OR COPYRIGHT HOLDERS BE LIABLE FOR ANY CLAIM, DAMAGES OR OTHER
LIABILITY, WHETHER IN AN ACTION OF CONTRACT, TORT OR OTHERWISE, ARISING FROM,
OUT OF OR IN CONNECTION WITH THE SOFTWARE OR THE USE OR OTHER DEALINGS IN THE SOFTWARE.
```

In other words, *thank you, Microsoft, for letting me use a whole lot of your stuff! I really appreciate it.*

If you read my first book, *Practical Maintenance Plans in SQL Server* (Apress, 2016), then you are already familiar with my writing style. I tend to try and keep the mood a bit light and sometimes quirky, without sacrificing technical content. At times, I might divert down a rabbit hole, but I always have a point at the end. Most of the time, it might take me a while to get there, but I do eventually get there in the end.

Right now—before we get into installing and setting up our environment—is a great time for you to take a few minutes to get familiar with R and what it can do.

Once you've done that, let's begin our journey as data scientists!

PART I

Setup and Installation

CHAPTER 1

Setup and Installation of SQL Server 2016

One of the major updates to SQL Server 2016 is the addition of R as an integral part of the database engine. R began in 1993 as a data analysis language developed by Robert Gentleman and Ross Ihaka at the University of Auckland. It started as a language that could rival the S language in statistical analysis and evolved into arguably the most popular language in the world for statistical computing, data analysis, and machine learning.

With the business world making a major shift toward business intelligence and data analysis, the addition of R as an integral part of SQL Server is a smart business move for Microsoft. Not only is Microsoft introducing new functionality into an already widely accepted platform, but they are also leaving the core of the language intact so that current R statisticians can easily move onto the SQL Server platform to enhance their statistical computing methodology. In the end, this enhances visibility for Microsoft in the business intelligence field, and hopefully, leads to even greater acceptance for SQL Server in everyday data analysis operations.

In 2016, Microsoft bought Revolution Analytics, which is built around R and provides both an open source (Revolution R Open) and commercial (Revolution R Enterprise) development platform for R. Heavy integration of R into existing products is now Microsoft's focus, with the obvious choice of SQL Server and, eventually, Azure. This is an exciting release, because it gives Azure hosted services the opportunity to deliver content based on R computations done in an Azure site or database.

Since R has been added as an installation portion of SQL Server 2016, all we need to do is select the option during installation to add it and then run through some minor configuration tasks.

There are certain things that we need to check and install to make sure that R runs, but I will show all that when we get there. For now, download SQL Server 2016, and then follow along with me on how to install it. It is worth noting that your installation screens may vary slightly from mine, depending on the service packs or if Microsoft decides to change the install screens, but I think that the gist of the content will be the same.

B. Beard, *Beginning SQL Server R Services*, DOI 10.1007/978-1-4842-2298-0_1

Planning

First things first though. Once you download SQL Server 2016, you want to plan out the basics, such as the account that you're going to use and where your default file locations are going to be. If you read my last book, then you know that I have a very particular way in which I organize my file system for SQL Server. For this book, I make a separate logical disk (E:\) with the following folder structure:

- E:\SQL Server
    - Backups
    - Data
    - Logs

So, one main folder, SQL Server, and then three folders inside of that folder to hold the different bits as needed. There can be other folders, such as DTSX or Output, which you can use for other things, but for the most part, those three subfolders inside of the main folder work nicely.

▒ **Note** There are other locations that SQL Server wants to place files in during installation; this is fine, since this is how SQL Server wants to categorize the system files to keep everything copacetic. We will have control over our data, logs, and backups in the folders specified earlier.

As far as which account you should use to run the functionality of SQL Server 2016, this should be a no-brainer. It needs to run as if it were a regular database installation, so it needs to have the account assigned that it would normally have. To be clear, assign the same account that you are currently using for whatever version of SQL Server you are running. Most times, this needs to be an administrator in order to install programs.

A quick side-note here: if you haven't read the hardware and software requirements for SQL Server 2016, you probably want to do that. Also, Appendix A covers installation of SQL Server 2016 onto an existing SQL Server 2014 server. If you are running SQL Server 2014 and want to try SQL Server 2016 on the same server, then look to Appendix A for guidance. (But in no case should you ever use a production server to follow along with this book).

Beginning the Installation

Here we go! Double-click the **setup.exe** file in the download folder. You should see what's shown in Figure 1-1.

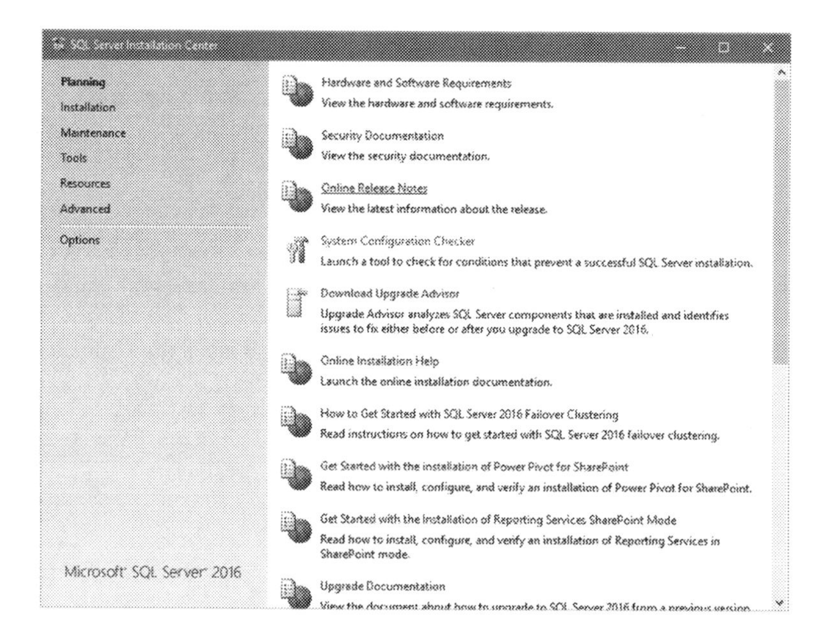

Figure 1-1. *Initial SQL Server 2016 installation screen*

If you see the screen asking to make changes to your system, go ahead and say **Yes**.

Figure 1-1 shows the first screen that you should see when you start installation. This screen should look pretty familiar to you, if you have ever installed SQL Server before. Click the **Installation** link on the left. You should see what is shown in Figure 1-2.

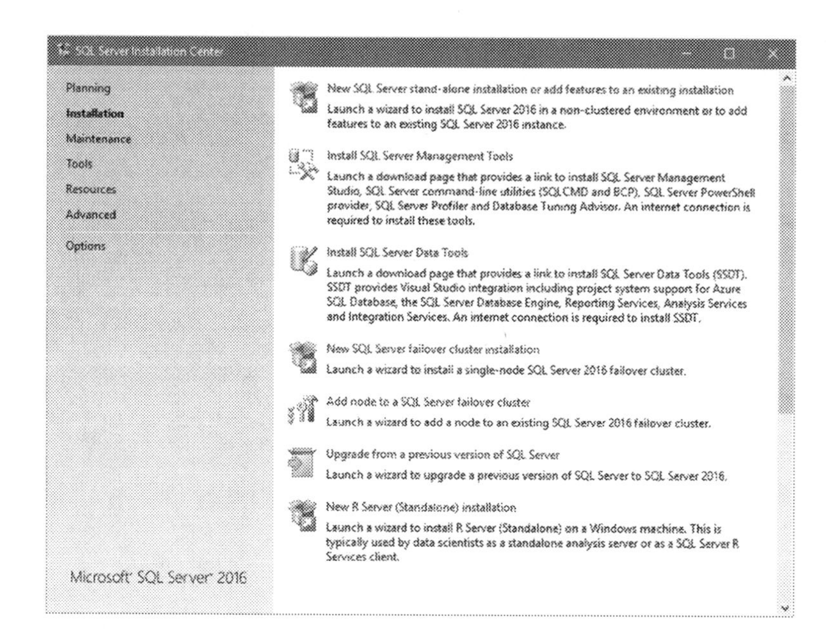

Figure 1-2. *SQL Server 2016 installation options*

Once here, click the top link, titled **New SQL Server stand-alone installation or add features to an existing installation**.

Note that the very bottom option is something new. It says **New R Server (Standalone) installation**. You would select this option if you only wanted to install R Server as either a server (standalone, self-contained data analysis server, in other words) or a client (manipulating data from a remote SQL Server R Services installation). Note that you need the SQL Server 2016 services running as well, so this would be to add R services to an existing SQL Server 2016 installation.

Product Key

Next is to enter your product key. Figure 1-3 shows the screen you see after continuing from Figure 1-2 in the prior section. Here you can specify that you wish to run the free edition, or you may enter a product key in order to run a licensed edition.

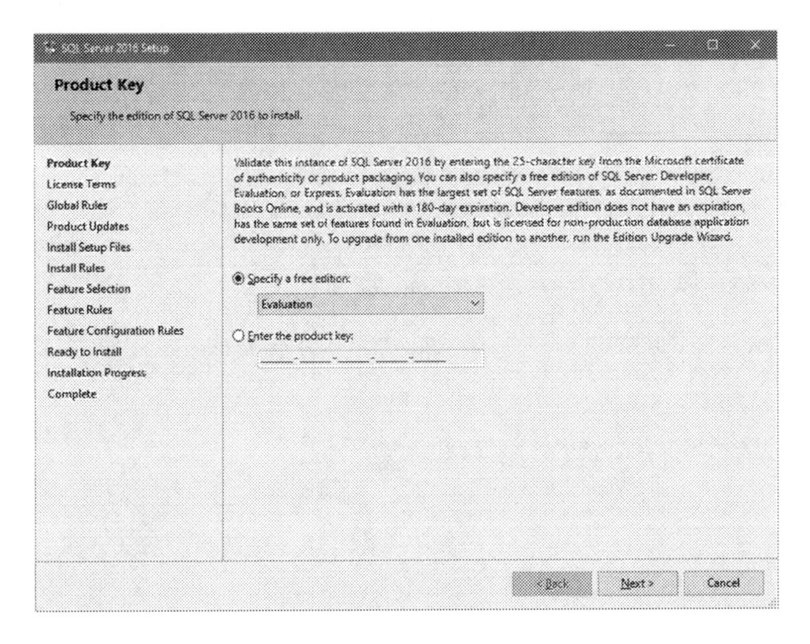

Figure 1-3. *Product Key screen*

SQL Server 2016 can be installed in one of three free editions:

- **Evaluation**: A full set of features; basically, the Enterprise version of SQL Server 2016, but only good for 180-day spans.
- **Developer**: A full set of features, but cannot be used for production database work.
- **Express**: The smallest, bare-bones installation of SQL Server 2016; does not expire and can be used for production use.

If you would like to choose an option other than Evaluation, go right ahead. Just understand the implications of choosing that option; for example, the Express option doesn't support R so I wouldn't choose this option. For what you need here, the Evaluation version is perfect, because you certainly decide within 180 days if this new functionality is something you want to permanently include in your SQL Server installation.

When you have chosen the version you are most comfortable with, click **Next** to continue.

License Terms

The next screen, shown in Figure 1-4, simply asks you to accept the license terms.

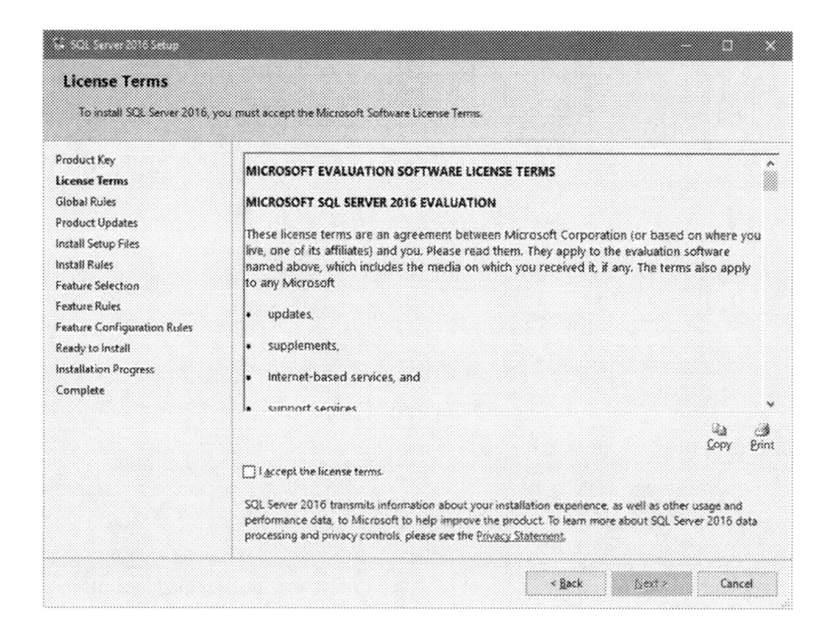

Figure 1-4. *License terms*

I honestly have never read this license all the way through. I can't say that I know anyone who has. Obviously, just click the **I accept the license terms** check box and then click **Next** to move on.

Install Rules

This screen shows you what happens as SQL Server goes through the preliminary steps to check for a clean installation. If you get any errors or warnings, you should look at correcting them so that you can install as cleanly as possible.

My screen flashed a few times and I eventually ended up at the screen shown in Figure 1-5.

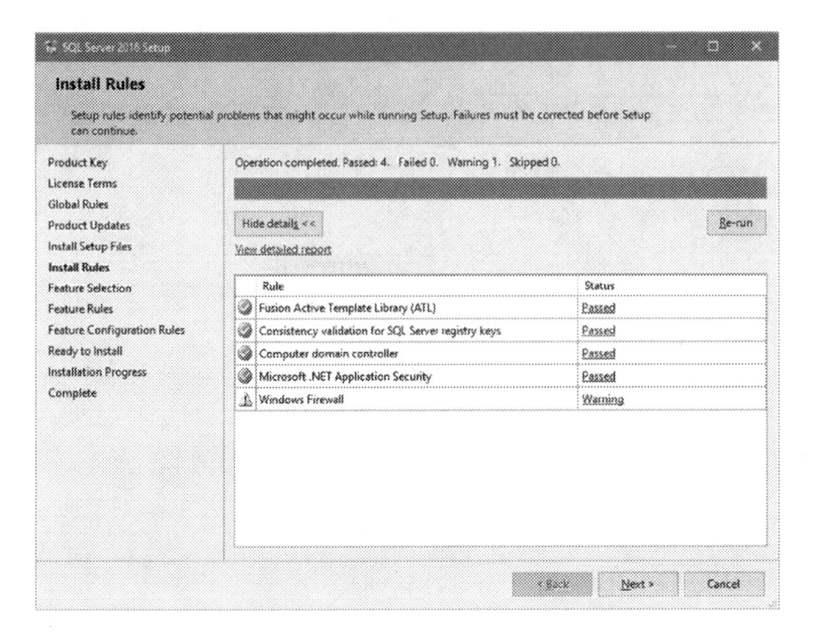

Figure 1-5. *Install Rules*

It's worth noting that there could be an update to SQL Server 2016 that gets downloaded and installed during this step, so if a message comes up with that information, go ahead and install it.

So everything looks good except for my firewall rule. Since no connections come from a network on my laptop, this should be fine, so I'm going to click **Next** to continue.

Feature Selection

Now we get to choosing what we want as part of the actual installation. Figure 1-6 shows the screen we have been waiting for.

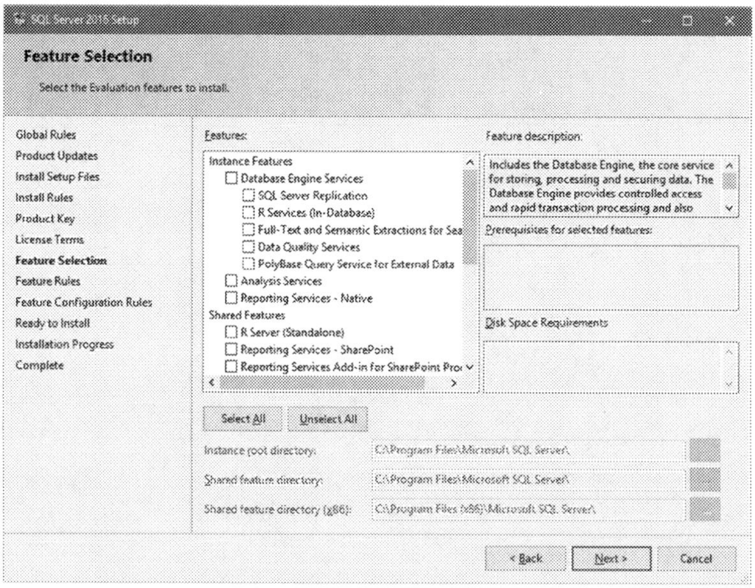

Figure 1-6. *Feature Selection*

At this point, we need to choose just the bare minimum of what we need to test the functionality of R within our database instance. In this book, we get familiar with R and create charts using R Tools for Visual Studio, and then duplicate those results in SSMS, ultimately serving those results in reports through Reporting Services. Because of this, we only install R Services (In-Database) and Reporting Services - Native. This gives us everything we need to really get a feel for R and what it can do for us. It also means that we don't need to install the entirety of SQL Server 2016. Figure 1-7 shows the selected options that you should have at this time.

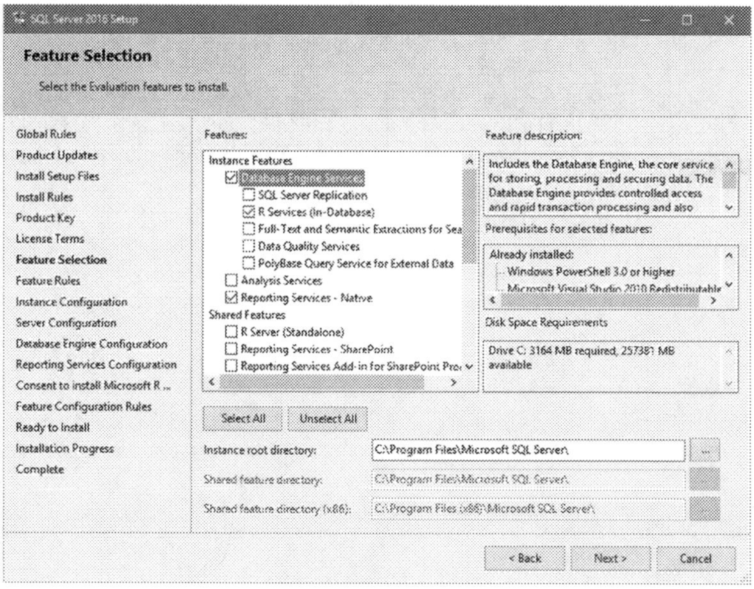

Figure 1-7. *Selected options*

Click **Next** here to move on. It takes a second to think about what it wants to do, but eventually, you see the **Instance Configuration** screen shown in Figure 1-8.

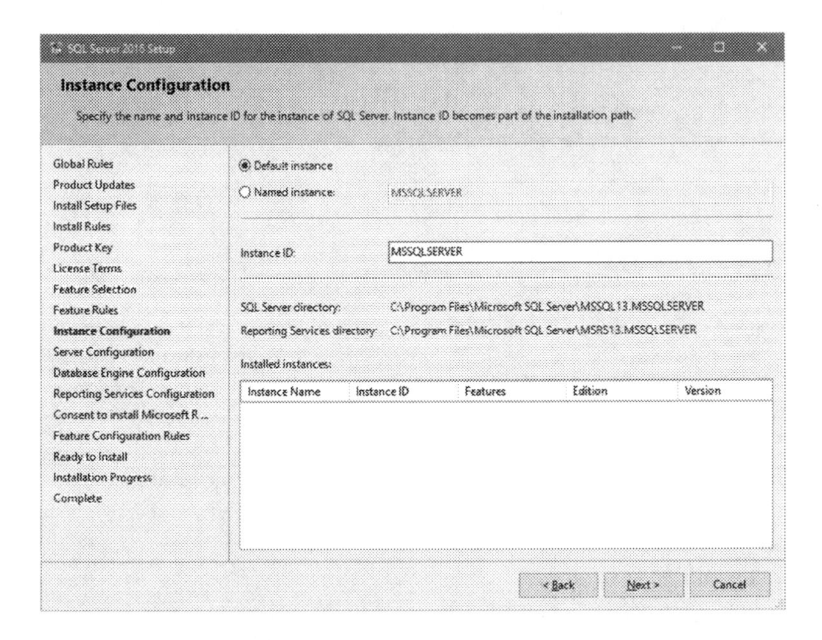

Figure 1-8. *Instance Configuration*

Instance Configuration

Since this is a new installation and there isn't a previous version of SQL Server installed, the option is available for Default Instance. We can certainly do this with no issue, but I usually prefer to name my instances. I leave this up to you, but understand that I will use a Named instance and not a Default instance for the remainder of this book.

At this point, we need to define our new instance. If you look on the **Installed Instances** section, you see that there is nothing there. We choose the **Named Instance** option and call it SQL2016RS for SQL Server 2016 R Services. The Instance ID field should be updated to SQL2016RS as well. Once you do that, you see what is shown in Figure 1-9.

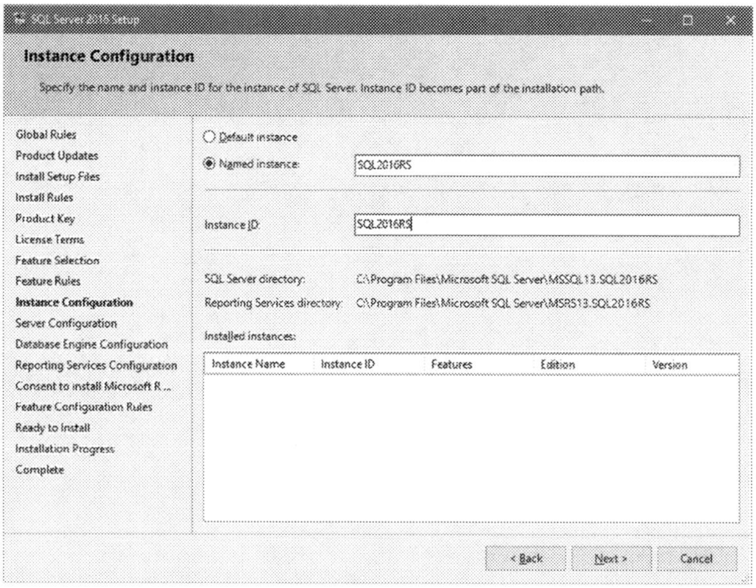

Figure 1-9. *Updated Instance Configuration screen*

Pay attention to the **Named Instance** field, the **Instance ID** field, the **SQL Server directory** location, and the **Reporting Services directory** location listed on this screen. Those need to all have SQL2016RS referenced in them. Once you are satisfied that everything is as it should be, click **Next** to continue.

When you are ready, click **Next** to move on.

Server Configuration

The next screen is where we define the service accounts and startup types for the services. This screen is shown in Figure 1-10.

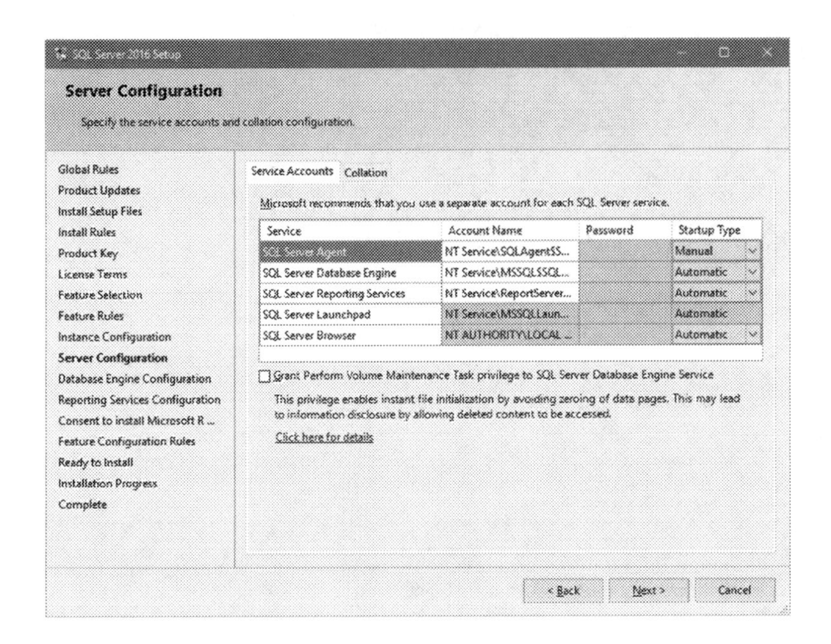

Figure 1-10. *Server Configuration*

The only service to really pay attention to here is the SQL Server Launchpad service. This service handles the execution of R within the database engine, so if R isn't behaving as expected, check this service first.

The SQL Browser service is running under the context of local services, so there isn't a new service account being created. We won't worry about that one, in other words.

These service accounts are the default, but can always be changed to your own service accounts, if you have them. If you don't have your own service accounts, you can keep these suggested service accounts. I know a lot of server administrators that insist on employing the principal of least privilege for services, so if that is the case for your particular environment, then you need to get the service name and login information from the server administrator in order to proceed. Another way you can go about this is to copy these service names and include them in a summary to your system administrator regarding the accounts that were created during installation, so that the system administrator can audit the permissions for this user as needed. It is important to note here that I am referring to a separate individual or entity for "system administrator" that is not a database administrator, but rather the Windows-level administrator. The person in charge of the operating system level, one step up from the Application layer, in other words.

We only want to change a little bit here; specifically, set the SQL Server Agent service **Startup Type** to **Automatic**. That is the only change we need to make. Figure 1-11 shows what you should see at this point.

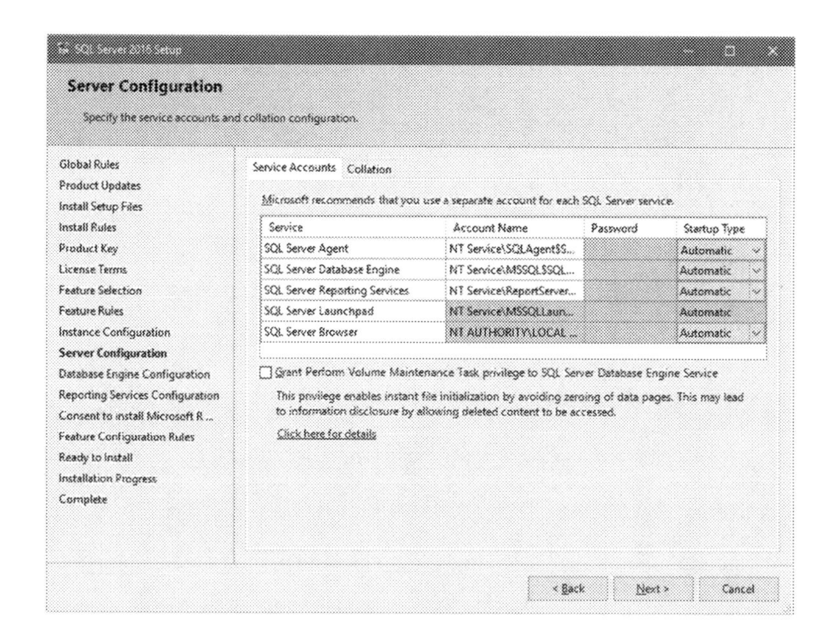

Figure 1-11. *Server Configuration updated*

Notice that we cannot set the password for any of these accounts. This is the same as it has been for every installation of SQL Server that I have ever seen. If you were to change the Account Name box from the default to a custom service account name, then the Password box would become active and accept input. Otherwise, the password is controlled by SQL Server and Windows.

Notice also that there is a new **Grant Perform Volume Maintenance Task privilege to SQL Server Database Engine Service** check box underneath the default services listed. For what we're doing in this book, it's not necessary to check this box. In future installations, or for production environments, it would probably be a good idea to enable this.

At this point, all of our services are configured correctly. Notice that we aren't going to bother with the Collation tab. This should have SQL_Latin1_General_CP1_CI_AS specified in the tab by default. That's it. Go ahead and click **Next** to move on.

Database Engine Configuration

The next screen is Database Engine Configuration, shown in Figure 1-12. Here you set options for the engine in four different tabs, as specified in the subsections to follow.

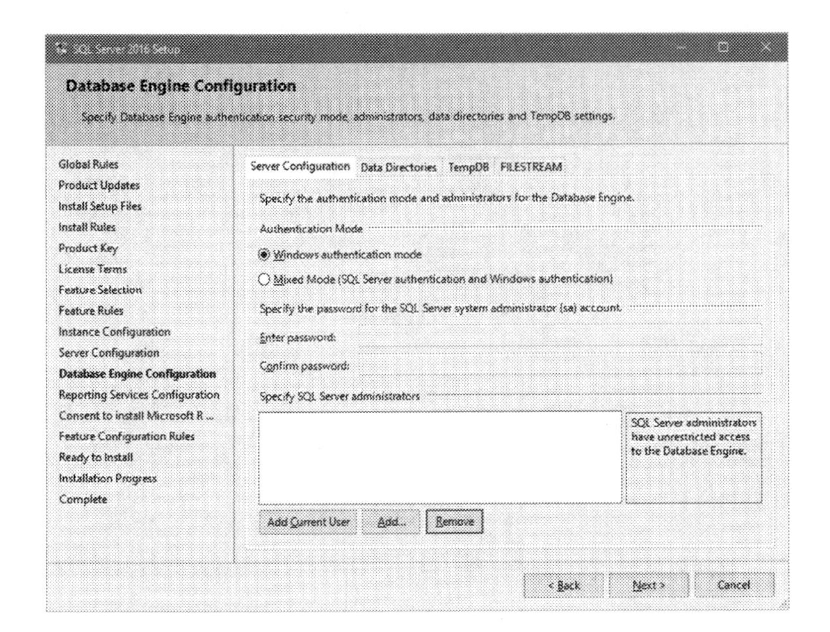

Figure 1-12. *Database Engine Configuration*

Server Configuration

This tab lets you specify the authentication mode and the administrators for this instance of the database engine. Because this is just for testing and evaluation, I am going to add myself in Windows Authentication Mode as the administrator by clicking the **Add Current User** button at the bottom of the screen and selecting **Windows Authentication Mode**. Figure 1-13 shows these options selected.

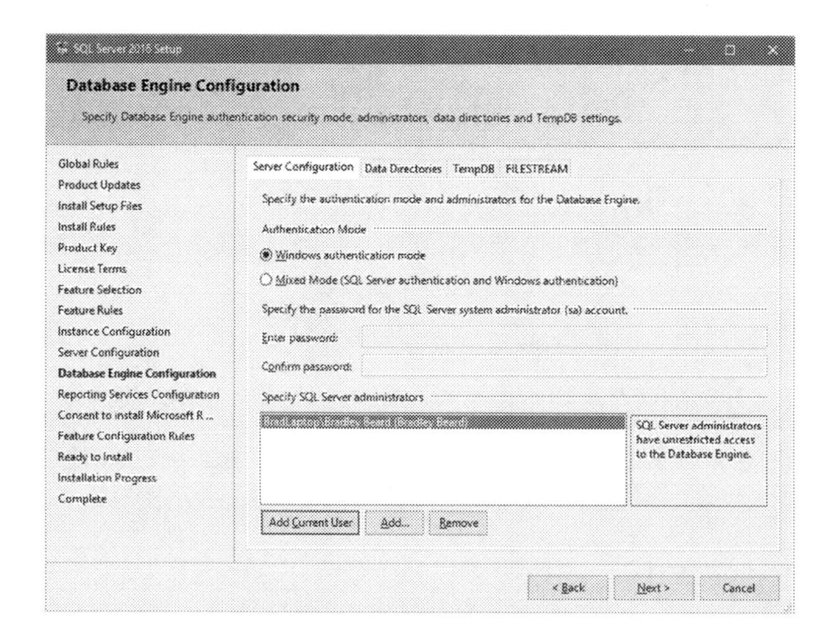

Figure 1-13. *Server Configuration tab with options*

Data Directories

Do you recall how I had my file system set up? The Data Directories section describes the locations of those data directories within the file structure I specified. Figure 1-14 shows what this screen looks like initially. Figure 1-15 shows my selected options. You can leave these however you like, but my personal preference is to not put the files I want in the default labyrinth of folders.

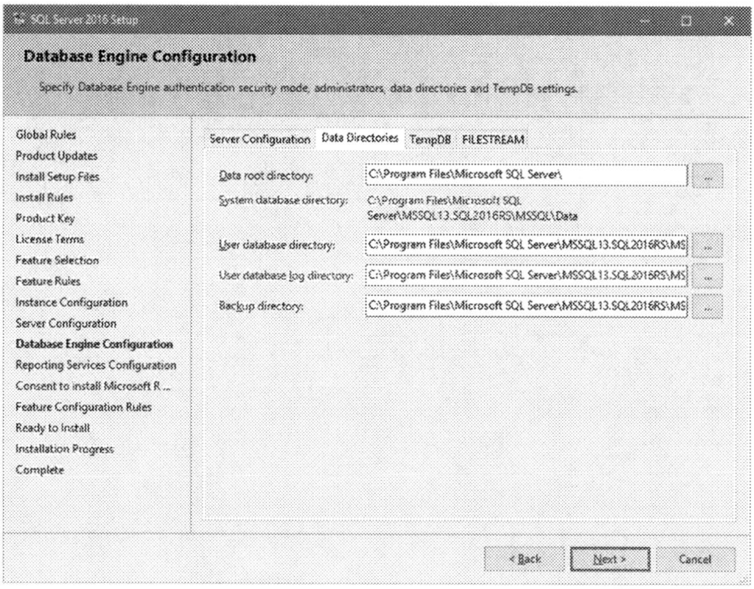

Figure 1-14. *Initial Data Directories tab*

SQL Server 2016 Setup

Database Engine Configuration

Specify Database Engine authentication security mode, administrators, data directories and TempDB settings.

Global Rules
Product Updates
Install Setup Files
Install Rules
Product Key
License Terms
Feature Selection
Feature Rules
Instance Configuration
Server Configuration
Database Engine Configuration
Reporting Services Configuration
Consent to install Microsoft R ...
Feature Configuration Rules
Ready to Install
Installation Progress
Complete

Server Configuration | Data Directories | TempDB | FILESTREAM

Data root directory: C:\Program Files\Microsoft SQL Server\
System database directory: C:\Program Files\Microsoft SQL Server\MSSQL13.SQL2016RS\MSSQL\Data
User database directory: E:\SQL Server\Data
User database log directory: E:\SQL Server\Logs
Backup directory: E:\SQL Server\Backups

< Back | Next > | Cancel

Figure 1-15. *Updated Data Directories tab*

TempDB

Usually, I leave this TempDB option alone. However, in this case, I set the options to mirror the file system that I have enabled. Figure 1-16 shows the default settings and Figure 1-17 shows the updated settings.

SQL Server 2016 Setup

Database Engine Configuration

Specify Database Engine authentication security mode, administrators, data directories and TempDB settings.

Global Rules
Product Updates
Install Setup Files
Install Rules
Product Key
License Terms
Feature Selection
Feature Rules
Instance Configuration
Server Configuration
Database Engine Configuration
Reporting Services Configuration
Consent to install Microsoft R ...
Feature Configuration Rules
Ready to Install
Installation Progress
Complete

Server Configuration | Data Directories | TempDB | FILESTREAM

TempDB data files: tempdb.mdf, tempdb_mssql_#.ndf
Number of files: 2
Initial size (MB): 8 Total initial size (MB): 16
Autogrowth (MB): 64 Total autogrowth (MB): 128
Data directories: C:\Program Files\Microsoft SQL Server\MSSQL13.SQL2016RS\M
Add...
Remove
TempDB log file: templog.ldf
Initial size (MB): 8
Autogrowth (MB): 64
Log directory: C:\Program Files\Microsoft SQL Server\MSSQL13.SQL2016RS\MS

< Back | Next > | Cancel

Figure 1-16. *TempDB default settings*

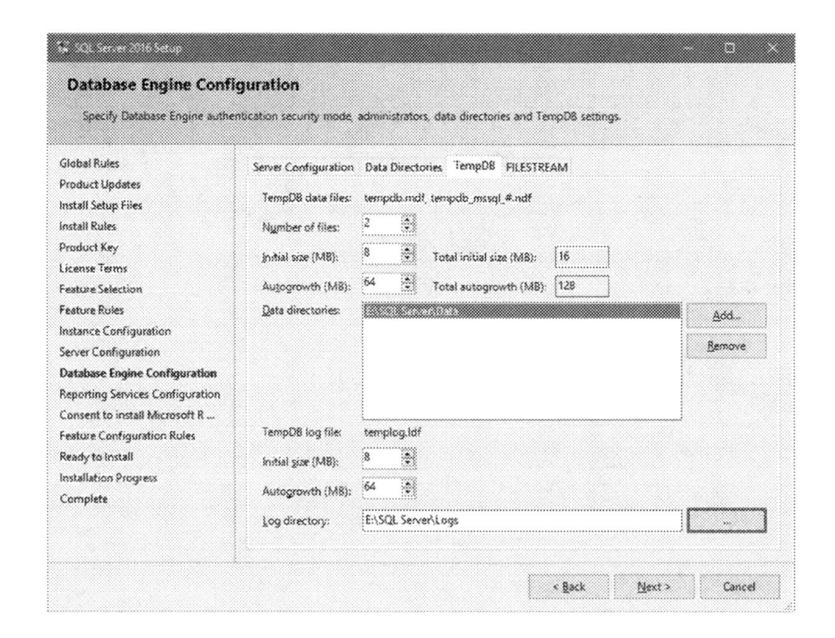

Figure 1-17. *TempDB updated settings*

The changes I made were slight. I first highlighted the existing option in the **Data directories** field and then clicked the **Remove** button. Then I clicked the **Add** button and added **E:\SQL Server\Data** instead. This location was mirrored in the **Log directory** field, so I changed it to **E:\SQL Server\Logs** instead. That's it for this tab.

FILESTREAM

Just leave the FILESTREAM tab alone. We won't be using FILESTREAM for this book.

Once you've got all the other Database Engine Configuration tabs filled in, click **Next**.

Reporting Services Configuration

Now we get to configure Reporting Services. We configure Reporting Services further in Chapter 7 and onward. We install it now using the **Install and configure** option, which is selected by default. Figure 1-18 shows that option.

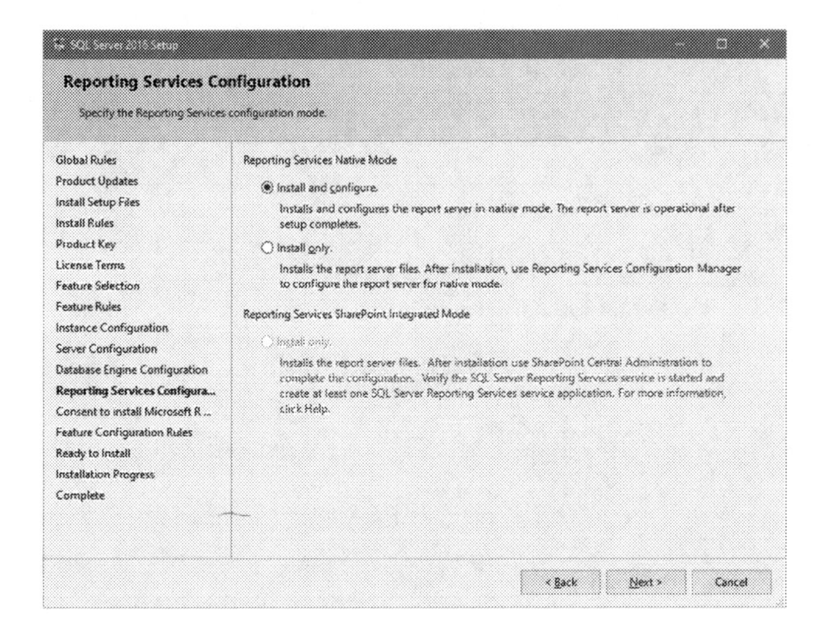

Figure 1-18. *Reporting Services Configuration*

Also note that the **Install only** option is selected for when you are installing Reporting Services SharePoint Integrated Mode. You can leave the option as it is, since there isn't another choice anyway.

One quick thing while we're talking about the configuration of Reporting Services; if you go in later and want to install Reporting Services because you didn't install it with the database engine, you only have the **Install only** option available to you. The reason for this is because Reporting Services Configuration Manager must be used to configure Reporting Services after a database engine instance has been added.

Ensure that the **Install and configure** option is selected on the top, and then click **Next** on the Reporting Services Configuration screen to move on.

Consent to install Microsoft R Open

The addition of R into SQL Server is a major change, as I noted before. In the pre-release editions of SQL Server 2016, it was mandatory to install the components separately in order for R to run correctly. Microsoft updated the installation process gradually through the editions until the final version, which included the full download and installation of the R components. Note that the version of R used in SQL Server 2016 is called out as Microsoft R Open, which is "an enhanced distribution of R made available by Microsoft under the GNU General Public License v2." R, the language, remains the copyright of the R Foundation for Statistical Computing. Microsoft is careful to spell that out exactly in the figure shown in 1-19. This is important, I believe, because Microsoft is making an entirely new software package available to SQL Server; for this reason, we must *consent* to installing Microsoft R Open and therefore accept any patches or updates issued by Microsoft.

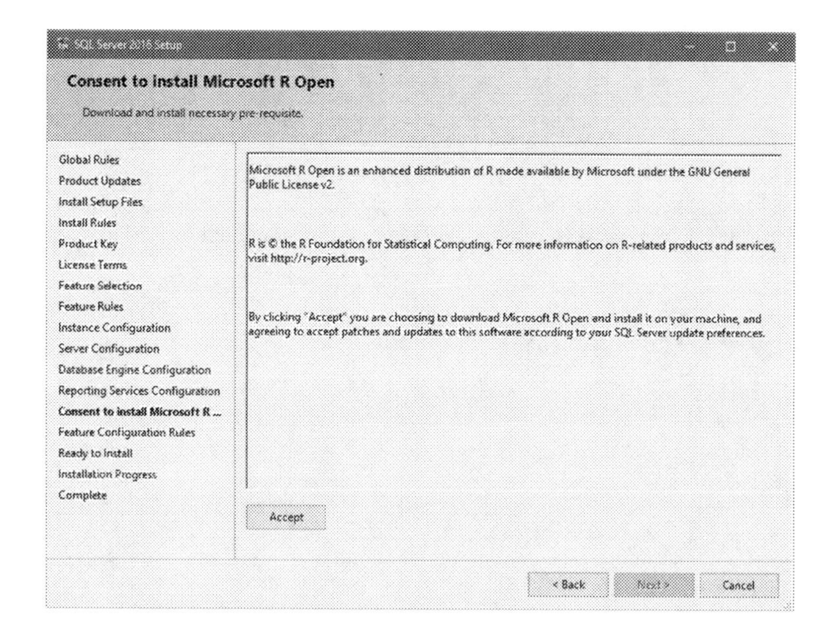

Figure 1-19. *Consent to install Microsoft R Open*

Click the **Accept** button. The **Next** button becomes enabled. Go ahead and click **Next** to move on.

Ready to Install

Figure 1-20 shows the **Ready to Install** screen that you should now see. Read it over and ensure that you match these settings if you're going to do the exercises in this book.

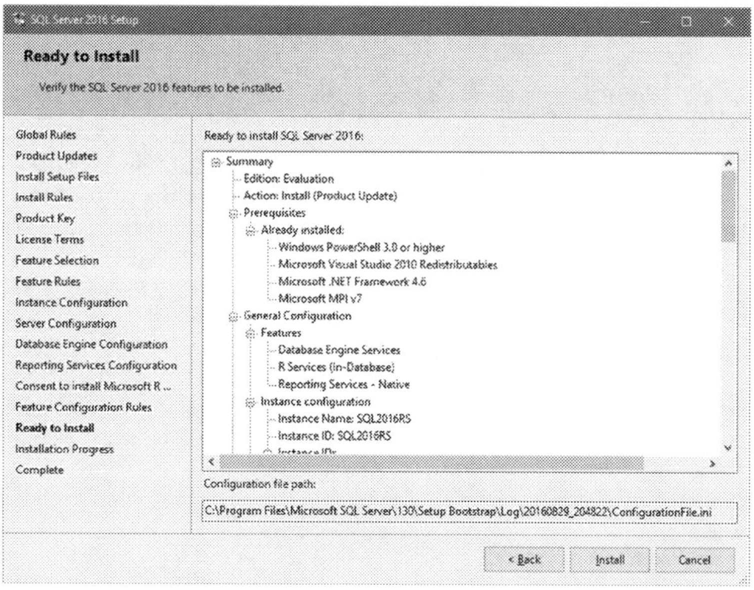

Figure 1-20. *Ready to install*

Once you are ready, click **Install**. Your screen will flash a few times while it is loading and installing what it needs.

Installation Progress

Figure 1-21 shows what you should see while your installation is running. The progress bar shows overall progress, so you can gauge how far along you are and how much time is remaining. Wait patiently. Watch the progress bar. Or go for coffee.

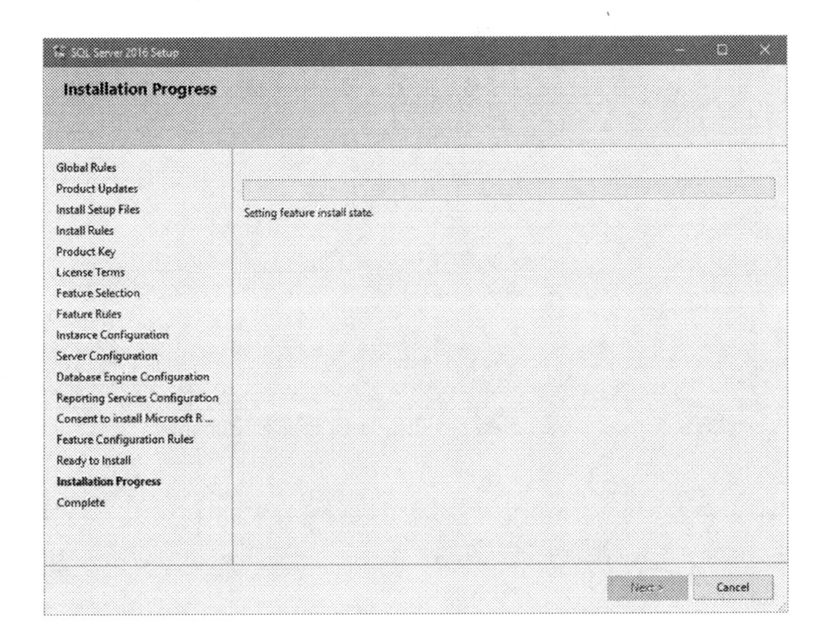

Figure 1-21. *Installation Progress*

Install Complete

The install takes a little while, but eventually finishes with the screen shown in Figure 1-22. My installation took about 10 minutes to complete. Scroll down to see if everything installed correctly, and then click **Close**.

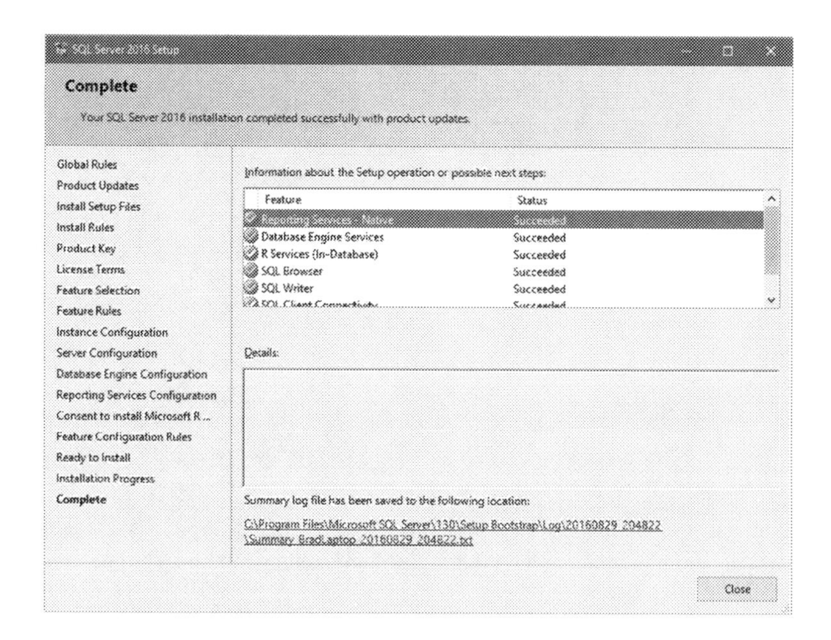

Figure 1-22. *Complete*

Let's take a second to plan out what else we want to do in this chapter. We are already pretty far along so far, but we have a bit more before we can stop and rest.

- What we've done so far
    - Installed the SQL Server 2016 components that we need to test the new R functionality
- What we will do next
    - Verify the SQL Server 2016 services all started correctly
    - Install SQL Server Management Tools

That second bullet there might throw you off a bit. Why do we need to install SQL Server Management Tools? Because, for some strange reason that I'm sure Microsoft can justify, SQL Server Management Studio does not ship as part of the installation for SQL Server. I have never seen this before in any previous installation, so I can assume that this is the new norm for SQL Server.

Services Verification

Let's go make sure that the services that needed to start all started correctly. Start your Services window by pressing your Windows key and then typing **services**. You should see the Services desktop app appear as an option, as shown in Figure 1-23. Go ahead and click the **Services** app to continue. Alternatively, you could also press your Windows key, type **services.msc**, and press **Enter**. This opens the Services window without having to search for it.

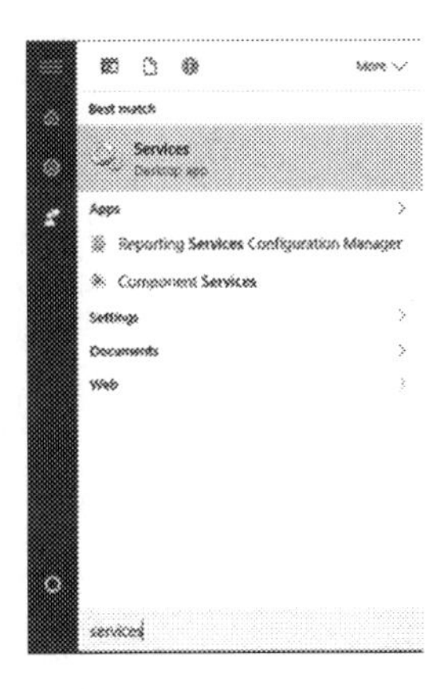

Figure 1-23. *Services location*

The application starts as normal, so scroll down to the SQL services. They should be sorted by Name by default, so to jump to a letter, just type it. You will get there a lot faster than by scrolling. Figure 1-24 shows you the relevant SQL Server services and their current status.

SQL Server (SQL2016RS)	Provides sto...	Running	Automatic	NT Service\MSSQL$SQL2016RS
SQL Server Agent (SQL2016RS)	Executes jo...	Running	Automatic	NT Service\SQLAgent$SQL2016RS
SQL Server Browser	Provides SQ...	Running	Automatic	Local Service
SQL Server CEIP service (SQL2016RS)	CEIP service...	Running	Automatic	NT Service\SQLTELEMETRY$SQL2016RS
SQL Server Launchpad (SQL2016RS)	Service to la...	Running	Automatic	NT Service\MSSQLLaunchpad$SQL2016RS
SQL Server Reporting Services (SQL2016RS)	Manages, e...	Running	Automatic	NT Service\ReportServer$SQL2016RS
SQL Server VSS Writer	Provides th...	Running	Automatic	Local System

Figure 1-24. *Services*

Everything looks good to me. What do you think? All are running as expected and all are set to an **Automatic** startup type.

Congratulations! You have successfully installed SQL Server 2016 and verified that everything is copacetic by confirming that the services all started as expected.

SQL Server Management Tools

Next, we need to install the SQL Server Management Tools. This installation task is part of the SQL Server Installation Center that should still be open on your desktop. Figure 1-25 shows the current state of the SQL Server Installation Center screen.

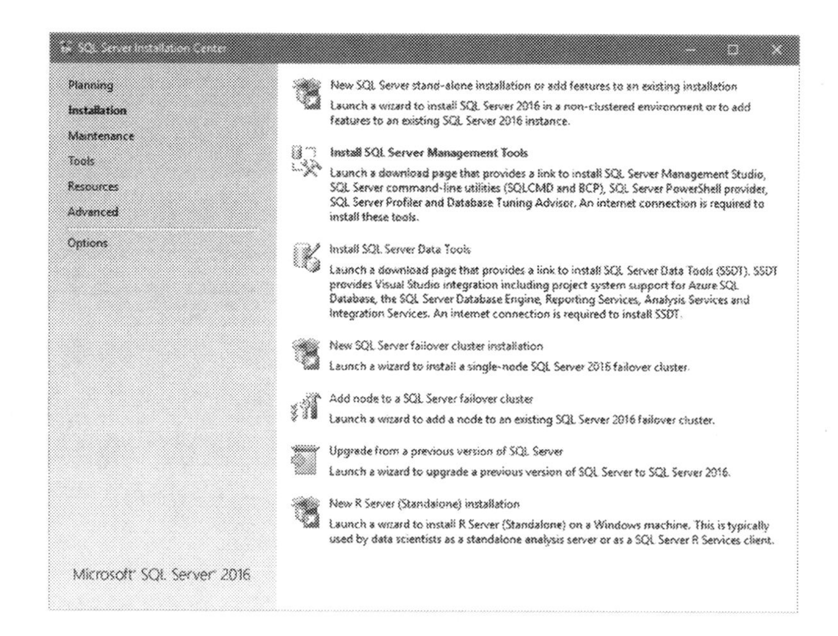

Figure 1-25. *SQL Server Installation Center*

Notice that the link for SQL Server Management Tools has become highlighted since SQL Server 2016 was installed. That's where we want to click now. Figure 1-26 shows the results of this action.

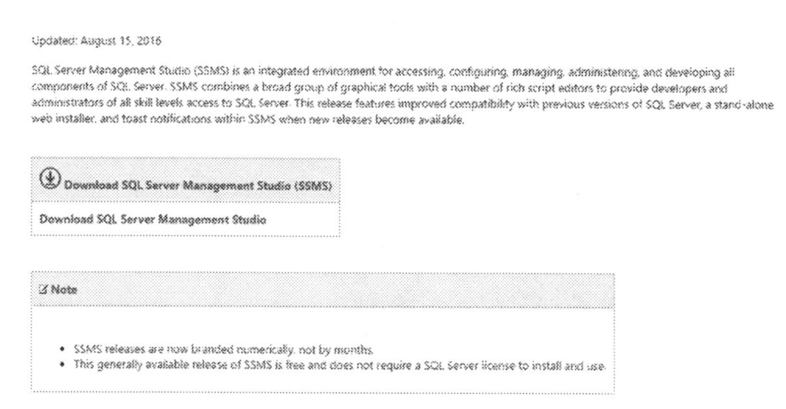

Figure 1-26. *Download SQL Server Management Studio*

When we clicked the **Install SQL Server Management Tools** link on the SQL Server Installation Center, a web page opened on Microsoft.com, where we can download SSMS. Go ahead and click that blue link titled Download SQL Server Management Studio now. The download was 806MB at the time of this writing.

While that is downloading, let's plan out what we want to do next. Essentially, we need to "finish" the installation of SQL Server by verifying that R is correctly installed and communicating normally with the database engine.

You almost forgot this was a book about R, didn't you? I know, this chapter has been a ridiculously long one, but we really did need to get all of the pieces up and running in order to show what R can do in the current context.

Eventually, the download completes. In your browser you should a button similar to what's shown Figure 1-27.

Figure 1-27. *Download complete*

Click this button to begin the install, or go to your Downloads folder and double-click the file named **SSMS-Setup-ENU.exe** to begin the installation of SQL Server Management Studio. Figure 1-28 shows the initial installation screen.

Figure 1-28. *SSMS installation*

▒ **Tip** You might end up with a different Release number than mine, considering that Microsoft may release a newer version in the time between when this was written and when you downloaded SQL Server Management Tools.

Well, that's certainly a new installation interface. I would think this was closer to Visual Studio, but it's not. The instructions are right there for us; click **Install** to begin. Figure 1-29 shows the interface after clicking Install.

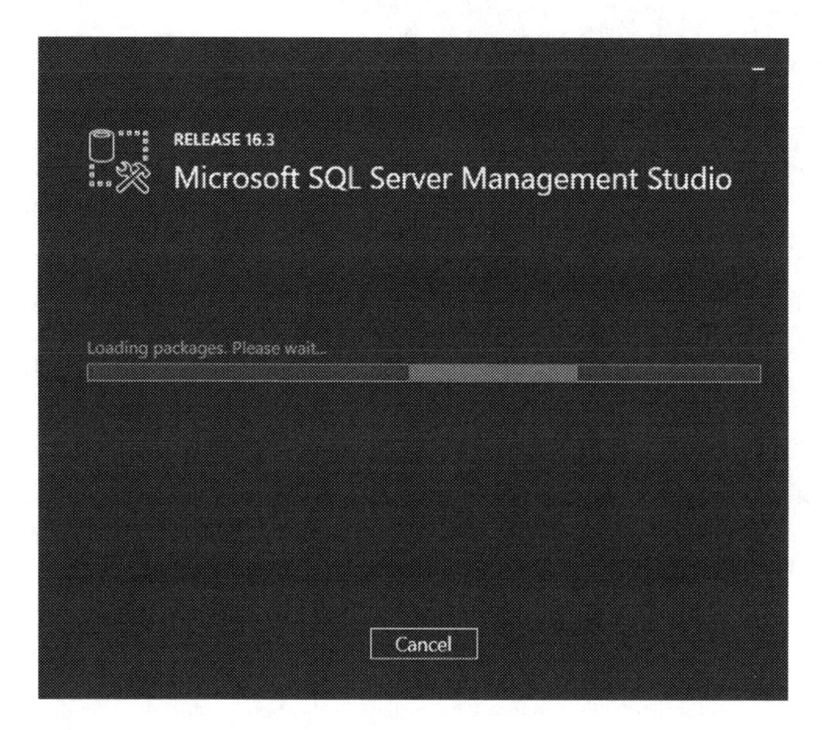

Figure 1-29. *Initial installation interface*

It sits here for a minute or so loading packages, so just let it do its thing and install what it needs. The first major package that gets installed is .NET Framework 4.6.1. It then progresses to other packages, like Visual Studio 2015 Shell. This will be important later on, once we start looking at the different IDEs for developing R projects.

Eventually, the installation completes and we settle on the screen shown in Figure 1-30.

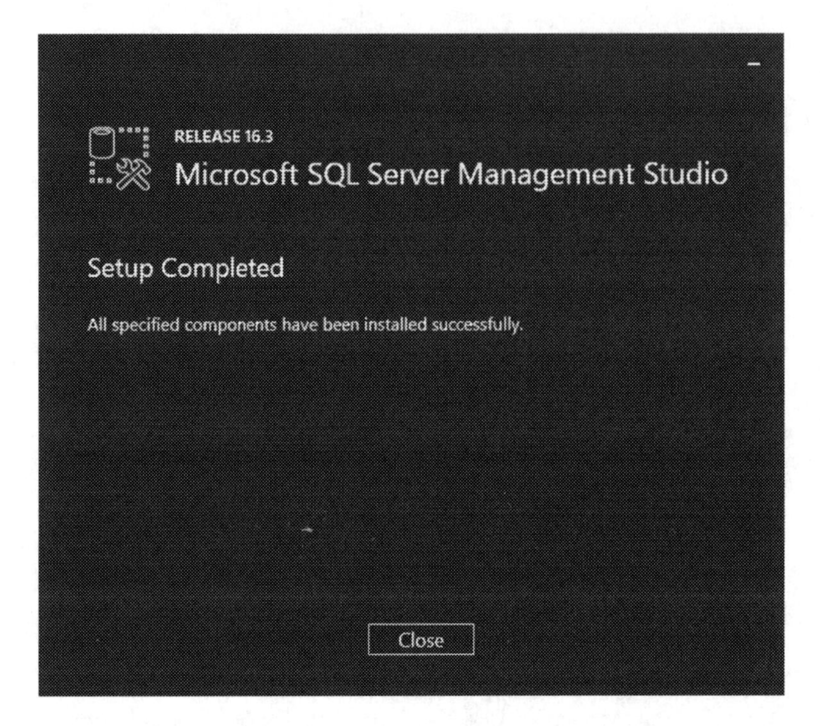

Figure 1-30. *Restart required*

Next, open SQL Server Management Studio and connect to your newly installed instance. Press your Windows key and type **ssms** to show the Microsoft SQL Server Management Studio desktop app. Go ahead and click the app to start the application.

■ **Tip** If you aren't logged in as the local Administrator account, you may need to right-click the desktop app icon that appears in the Start menu and choose Run as administrator instead.

Figure 1-31 shows how the initial interface for SSMS looks in this updated version.

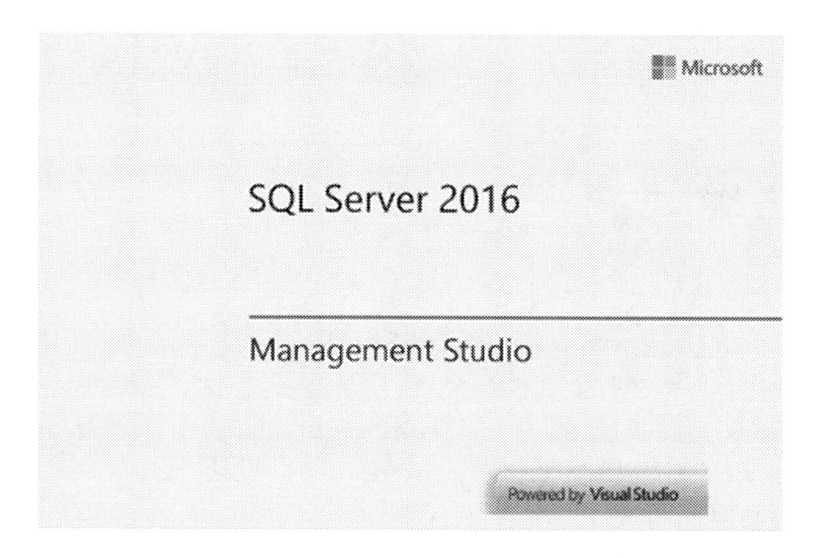

Figure 1-31. *SQL Server Management Studio*

That needs to load the interface and then, finally, we get to the SSMS login screen. Figure 1-32 shows how different this looks from previous versions of SSMS.

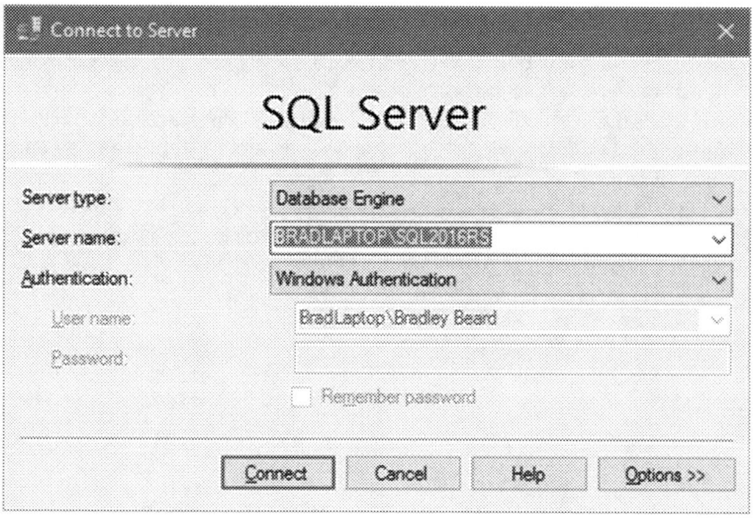

Figure 1-32. *Connect to Server*

Recall that I named the new SQL Server 2016 instance SQL2016RS, so that's the instance I am going to connect to. The format for the Server name field is SERVER\INSTANCE, so that's how I have formatted my connection. You can also pull down the menu and navigate to another instance you may have installed from there. However you are more comfortable is fine, as long as you get there. Click **Connect** to log in to your instance.

The Object Explorer initial screen you should see now resembles Figure 1-33.

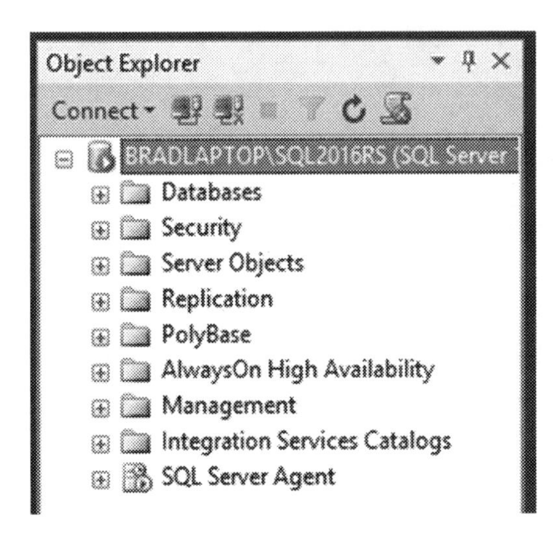

Figure 1-33. *SQL Server Management Studio*

Pay attention to the named instance and the SQL Server version shown. You may need to expand the interface to see the version number, but it will be there unless Microsoft takes it away from the interface. My version is 13.0.1601-5, which is the latest stable release at the time of this writing. As always, you may have a different version of SQL Server 2016. That's fine, considering that Microsoft hasn't deprecated anything major, which would affect the outcome of this book. This also means that we have successfully connected to our new instance and we are ready to get going.

Before we get into anything else, let's turn on the line numbers in SSMS. It's a pet peeve of mine. Go to **Tools** ä **Options**, expand **Text Editor**, and click **All Languages**. Figure 1-34 shows the location of this setting.

Just check that box and click **OK**—and you're good to go.

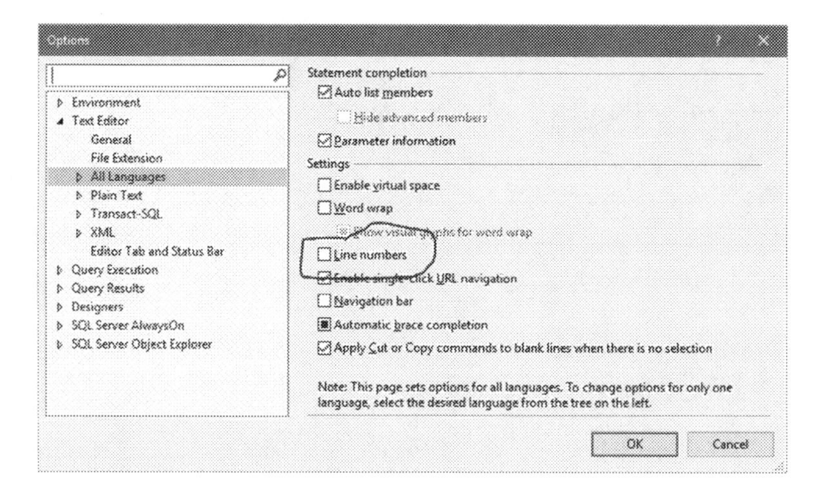

Figure 1-34. *Line numbers*

Microsoft has put out a post-configuration procedure, which we are going to run first. This may possibly be removed and added to the installation in the future, but for now, follow along to complete installation.

Open a New Query window in SQL Server Management Studio and type the following command:

```
exec sp_configure 'external scripts enabled', 1
Reconfigure with override
```

This is the result of executing this code in SSMS:

```
Configuration option 'external scripts enabled' changed from 0 to 1. Run the RECONFIGURE
statement to install.
```

▒ **Note** We already ran the RECONFIGURE statement as part of the query.

Notice that we are executing against the master database and that the query has executed successfully. This means that R scripting is now enabled in SQL Server. This script is essential because, without it, we cannot execute R code within the database engine. The reason for this is because the `sp_execute_external_script` stored procedure is disabled by default; it must be enabled manually.

Next, we need to verify that R is indeed running. To do this, Microsoft says to restart the SQL Server instance and run the following script. Restart the instance first, and then open a **New Query** window and type the following:

```
exec sp_execute_external_script
@language =N'R',
@script=N'OutputDataSet<-InputDataSet',
@input_data_1 =N'select 1 as hello'
with result sets (([hello] int not null));
go
```

Press **F5** to execute the script. The anticipated results are shown in Figure 1-35.

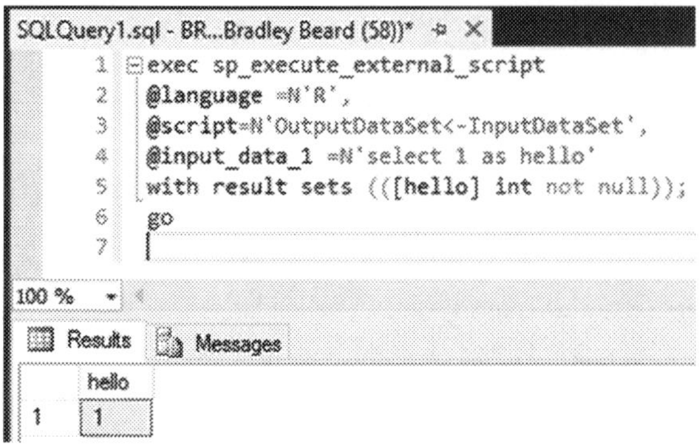

```
SQLQuery1.sql - BR...Bradley Beard (58))*
1  exec sp_execute_external_script
2  @language =N'R',
3  @script=N'OutputDataSet<-InputDataSet',
4  @input_data_1 =N'select 1 as hello'
5  with result sets (([hello] int not null));
6  go
7
100 %
Results  Messages
   hello
1  1
```

Figure 1-35. *R installed and communicating correctly*

Excellent! The successful execution of this simple script shows us that R is alive and well, and communicating normally with the SQL Server instance. We can see from the results in Figure 1-43 that we had a very simple query defined as `select 1 as hello`, which we return as a column named `[hello]` using a data type of, which is `not null`.

For those that haven't yet memorized every single system stored procedure, you won't recognize that `sp_execute_external_script` is a brand-new stored procedure introduced to execute external scripts. This stored procedure can be invoked with:

- `@language`
    - The name of the supported language. Currently, only R is supported.
- `@script`
    - The script that is executed (you can either type it all in to the stored procedure or reference it as a variable).
- `@input_data_1`
    - The SQL query you're using to gather data from the database goes here.
- `@input_data_1_name`
    - The data frame that acts as the result set of the `@input_data_1` query. *This attribute is optional.*
- `@output _data_1_name`
    - The data frame variable in `@script` that holds the output data. *This attribute is optional.*

Believe it or not, there are just a few more things that Microsoft says we need to do, so let's get that out of the way next.

For the Windows users out there, press **Windows key + R**, type **lusrmgr.msc**, and then hit **Enter**. You should see your Local Users and Groups, as shown in Figure 1-36.

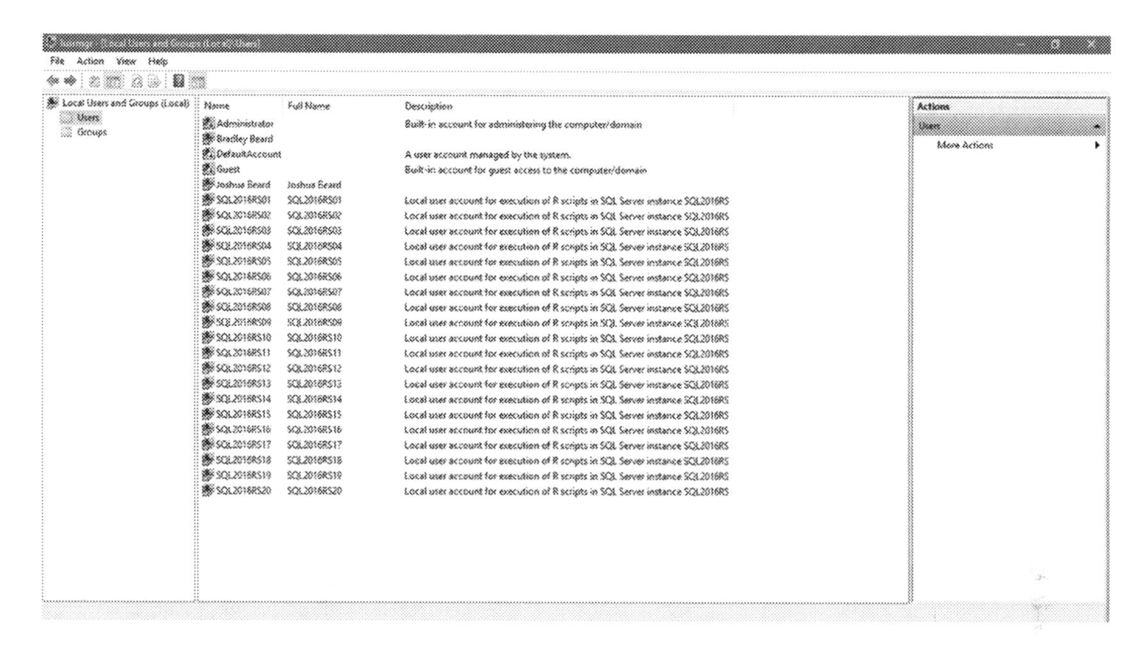

Figure 1-36. Local Users and Groups (Users)

Do you see that? Twenty new user accounts were created during installation and each of them are specifically created to interact with R Services. These 20 accounts are added to a new Group called SQLRUserGroup<instance_name>, where <instance_name> is the name of your instance. So in my case, my group is named **SQLRUserGroupSQL2016RS**. If you followed this naming convention, yours will be named that as well.

Click the **Groups** option on the left of your screen. You should see what's shown in Figure 1-37.

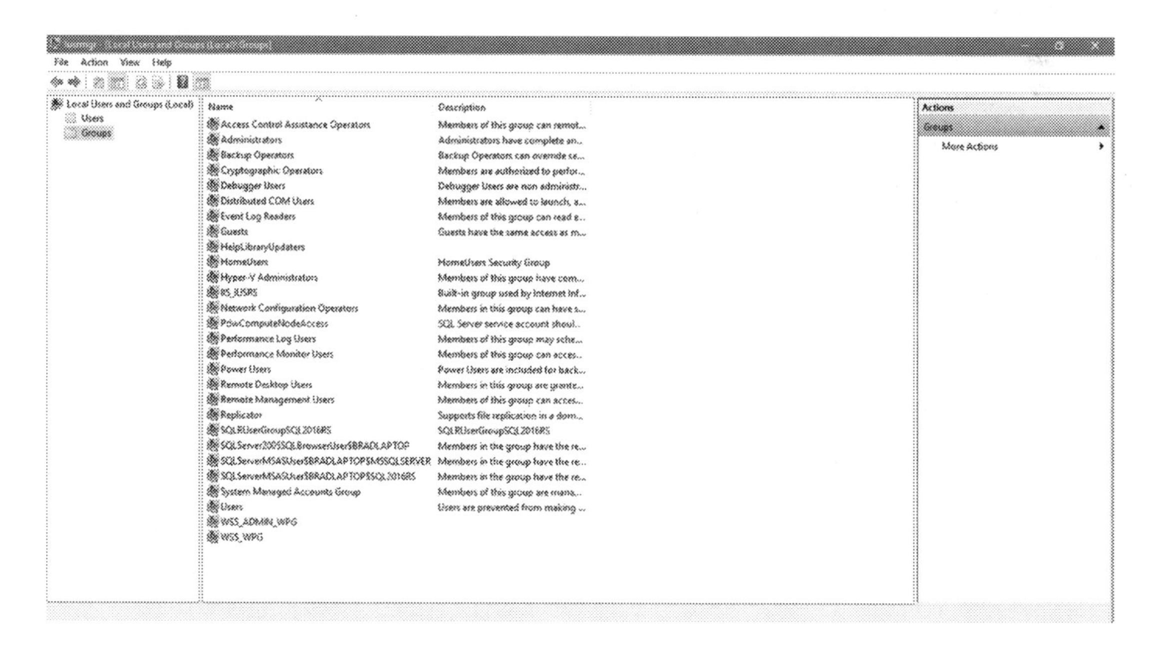

Figure 1-37. Local Users and Groups (Groups)

Double-click the **SQLRUserGroupSQL2016RS** link. You should see what is shown in Figure 1-38.

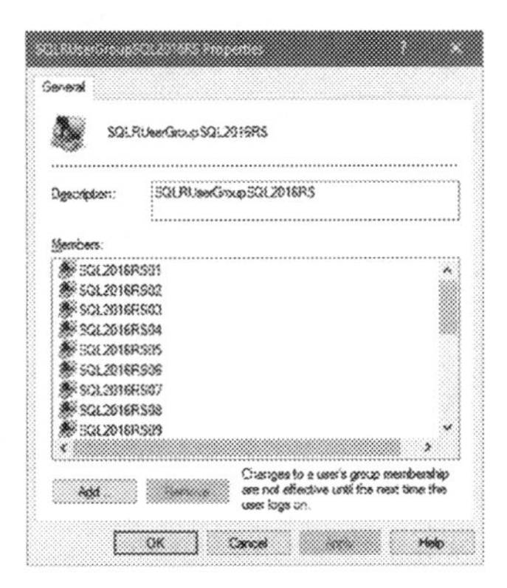

Figure 1-38. *SQLRUserGroupSQL2016RS detail*

The way that this is configured, we can work just fine from this server. We don't need to define anything else or add any other user accounts, since we have already verified that R is working correctly in the database. Microsoft has some pretty good articles about how to configure the database instance to accept R scripts from external developers, which essentially entails adding the SQLRUserGroupSQL2016RS group as a new login in SQL Server. That way, when a user connects to the database instance to run R scripts, one of those 20 new user accounts is used to execute the script through the Launchpad service on behalf of the user. Microsoft has dubbed this *implied authentication*, since a user in the group would then be able to access SQL Server R Services remotely.

Summary

Let's briefly review what we covered in this chapter.

- Installed a full SQL Server 2016 instance
- Installed SQL Server Management Tools
- Configured R after installation
- Verified that R is installed correctly by running the script specified earlier

This was a pretty important first chapter. We covered a lot of ground, so if anything was unclear, now is a great time to go back through and try it again.

Next, we install R Tools for Visual Studio, which we will use to write our R code. This first part was obviously necessary to get the R functionality and the database instance installed at the server level. We now shift our focus to getting the client side set up with our development tools.

CHAPTER 2

Setup and Installation of R Tools for Visual Studio

Now that we've got SQL Server 2016 installed correctly, we need to have some sort of IDE to develop our code in. Microsoft has made Visual Studio 2015 even better by introducing a brand-new R GUI as part of Visual Studio. This is simply named R Tools for Visual Studio.

There are other R GUIs that you can use, if you would like. You are not constrained or forced to use R Tools for Visual Studio at all. If you currently have a favorite IDE, then by all means, continue to use that one.

SQL Server Data Tools

One quick thing before we get started. We will not be using SQL Server Data Tools for this book. Instead, we will use R Tools for Visual Studio. That much should be obvious by the chapter title. The reason is pretty simple, but it took me a minute to figure it out. Let's look at the difference between the two.

- SQL Server Data Tools
  - Develops solutions for Analysis Services, Reporting Services, or Integration Services. Allows for connected database development; that is, the "live" editing of databases.
- R Tools for Visual Studio
  - Specifically designed as an R GUI for current Microsoft users who want to stay with a familiar interface while learning about R. Does not interact with the R instance in SQL Server, so needs to have a separate installation called Microsoft R Open in order to execute any R code.

The main reason that we are using R Tools for Visual Studio (RTVS) and not SQL Server Data Tools (SSDT) for this book is because RTVS can be used to develop R code, and SSDT cannot. SSDT can connect to the database instance and run queries against the database, but it is not an IDE for R. Alternatively, RTVS is an IDE for R, but in the constraints of this book, we are not connecting to the database instance directly.

That should help you figure that out quickly. The main difference to me is that SSDT runs an internal process, while RTVS runs an external process. You may find other differences between the two, so if you do, good work. As I said, though, we use RTVS in this book.

B. Beard, *Beginning SQL Server R Services*, DOI 10.1007/978-1-4842-2298-0_2

Visual Studio

In addition to the installation of Visual Studio, you need to download the extensions for R that go into Visual Studio. That way, RTVS has a version of R to execute code against. To get this done easily and with instructions all in one place, Microsoft has a site that explains R Tools for Visual Studio (RTVS). It is at `https://beta.visualstudio.com/vs/rtvs/`. A portion of this site is shown in Figure 2-1.

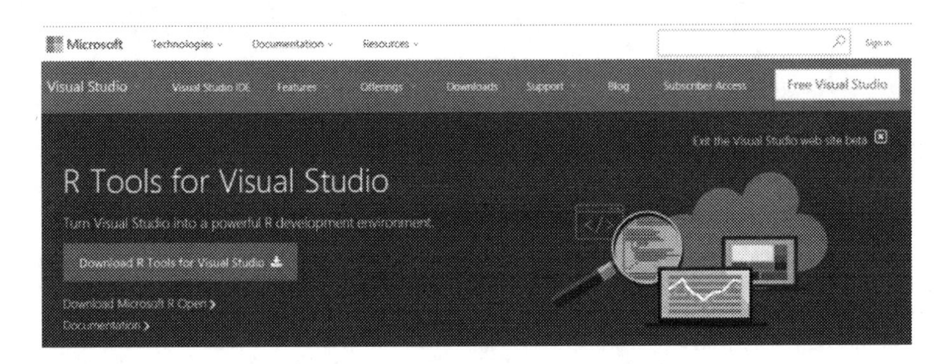

***Figure 2-1.** R Tools for Visual Studio site*

The first step, as I mentioned, is to get Visual Studio. Go ahead and click the **Free Visual Studio** link shown in Figure 2-1. It is the white box in the top-right corner. Another page opens, as shown in Figure 2-2, with a download link for Visual Studio Community on the left of the page.

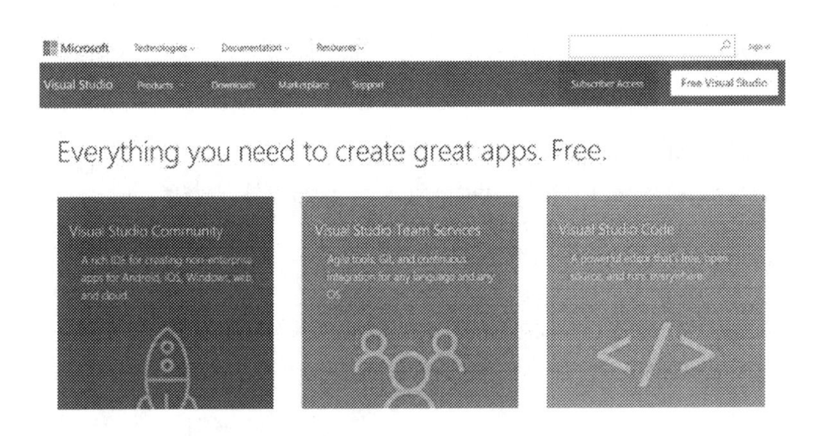

***Figure 2-2.** Download Community 2015*

We can see from the graphic that pretty much anything can be developed by Community 2015. Scroll down and click the **Download** button underneath the purple rocket ship graphic. You should see what is shown in Figure 2-3.

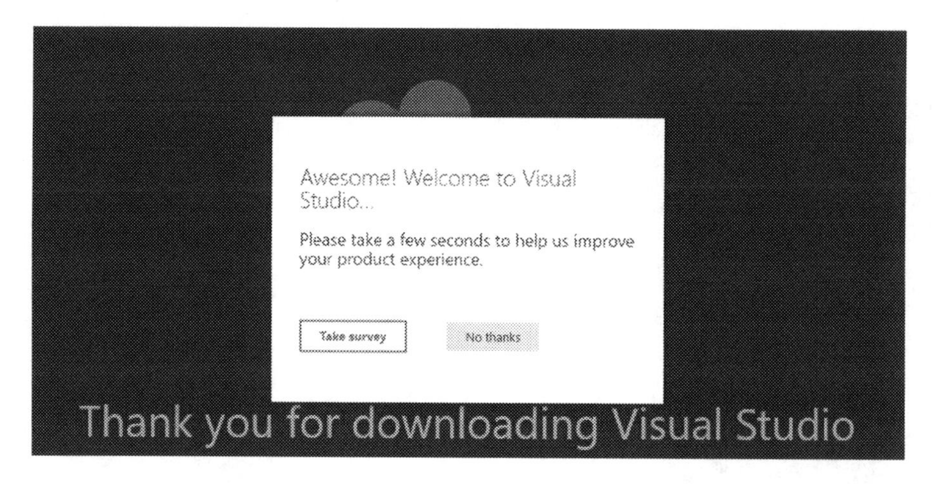

Figure 2-3. *Installer download begins*

If you take a look in the `Downloads` folder, you see that the installer has downloaded. It's a tiny 209KB install, which is very likely to download before you can even open your `Downloads` folder to check for it.

You can go through the survey mentioned in Figure 2-3 if you really want, but I said No thanks. Sorry, Microsoft!

Go to your `Downloads` folder and double-click the file **named vs_community_<random** string that looks like a GUID>.exe. Of course, you have a different random string than I did, so look for the file that begins with **vs_community**. That's the one that you want to double-click. You should then see Figure 2-4 on screen. You are about to begin the install, which consumes several gigabytes of bandwidth.

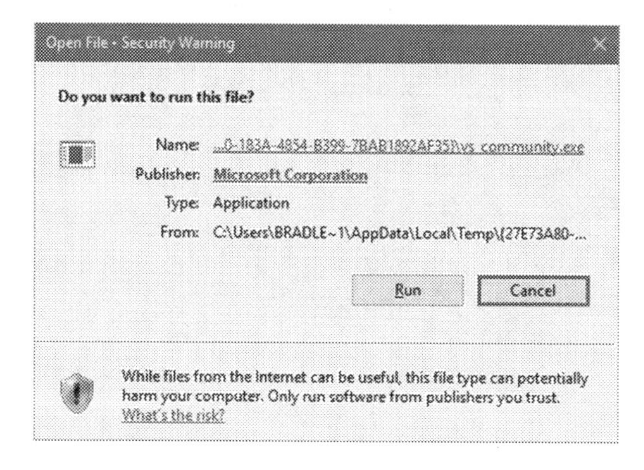

Figure 2-4. *Open File*

Go ahead and click **Run**. You will see Figure 2-5 pop into view.

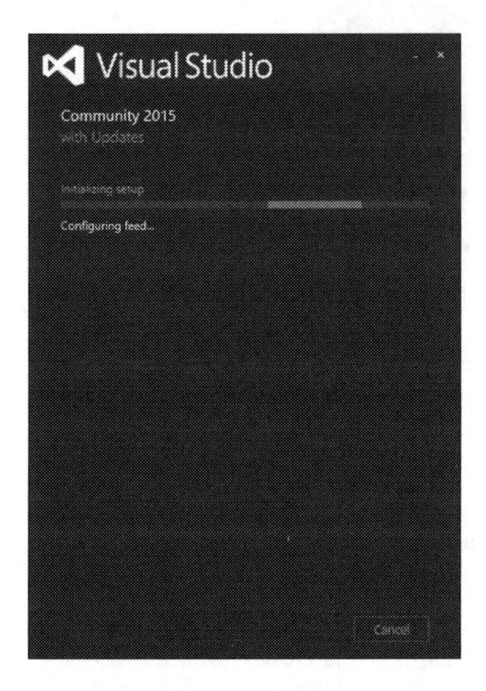

Figure 2-5. *Initializing setup*

Eventually, you get to Figure 2-6, which shows the configuration options for Visual Studio Community 2015.

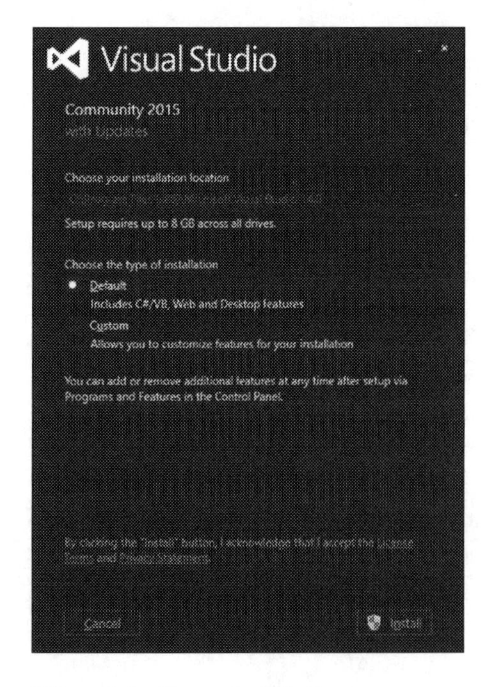

Figure 2-6. *Choose installation type*

From here, we can either choose **Default** or **Custom**. We really only need the basic functionality of Visual Studio, so keep the default option selected and click **Install**.

The screen shown in Figure 2-7 then appears; this begins the actual installation of Visual Studio Community 2015.

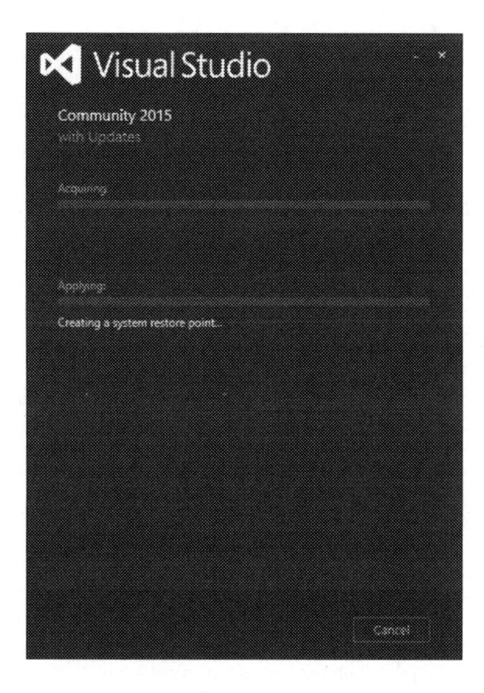

Figure 2-7. *Installation begins*

The installation runs for quite a while until completion. Note that there are probably required updates that need to be downloaded along with the actual application; so be patient and let it do its thing.

You eventually get to see Figure 2-8, which shows that the application has completed installation successfully.

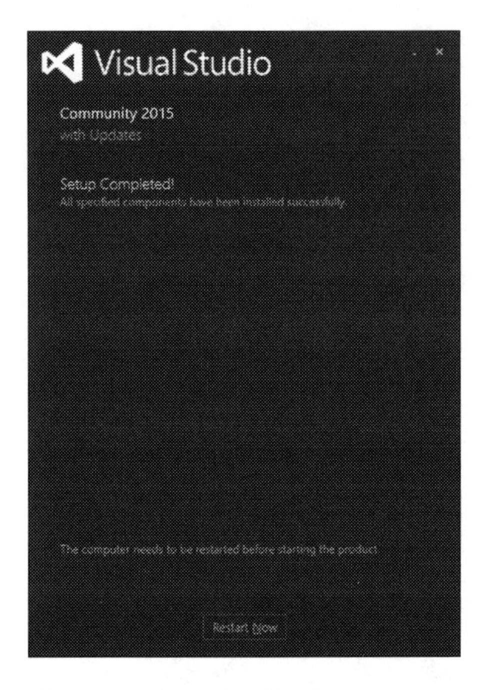

Figure 2-8. *Setup Completed*

Visual Studio is now installed. Next, we need to download R Tools for Visual Studio, but first restart your machine by clicking the **Restart Now** button, or closing and saving everything first and then restarting.

Once you restart, you may see Figure 2-9, which shows a nice little splash screen saying that Visual Studio has been installed.

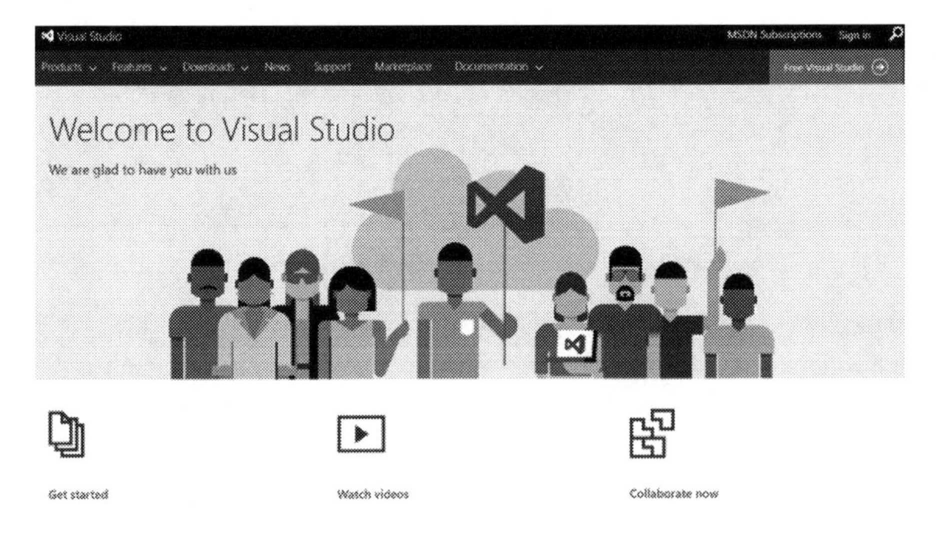

Figure 2-9. *Welcome to Visual Studio*

If you didn't see this screen come up after rebooting, that's probably okay. As long as you successfully installed Visual Studio, you should be good to go. Read further for installing the rest of what we need to start testing R functionality.

Download R Tools for VS

The second step, shown in Figure 2-1, is to Download R Tools for Visual Studio. Clicking the **Download R Tools for Visual Studio** link, also shown in Figure 2-1, starts a download of what I need to install. The downloaded file is named RTVS_`2016-06-23.7.`exe, as shown in Figure 2-10.

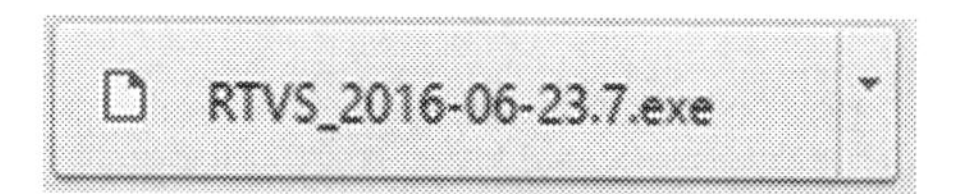

Figure 2-10. *RTVS file name*

You may end up having a different file name than this, but that's okay. Microsoft manages the future releases of this product, so we can assume that it is fine to install whatever application is presented from the link that you have clicked.

Run the executable once it's downloaded. You'll see the Figure 2-11, showing that you are ready to begin installing the R Tools. Click the **Install** button to start the installation process.

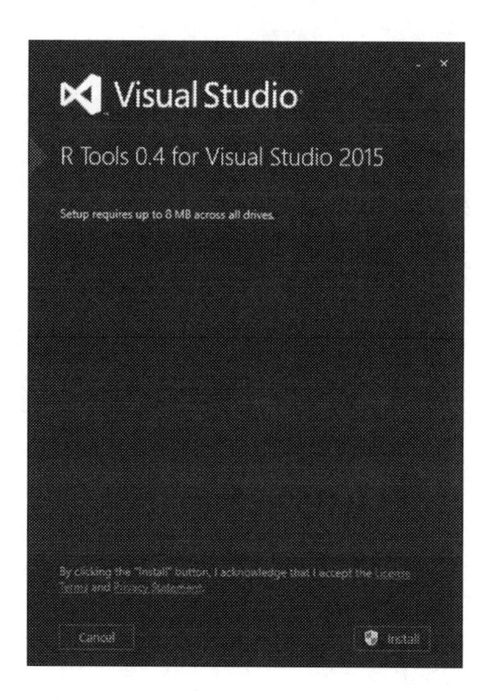

Figure 2-11. *R Tools 0.4 for Visual Studio*

The installer starts doing its thing, eventually showing the completed R Tools installation interface in Figure 2-12.

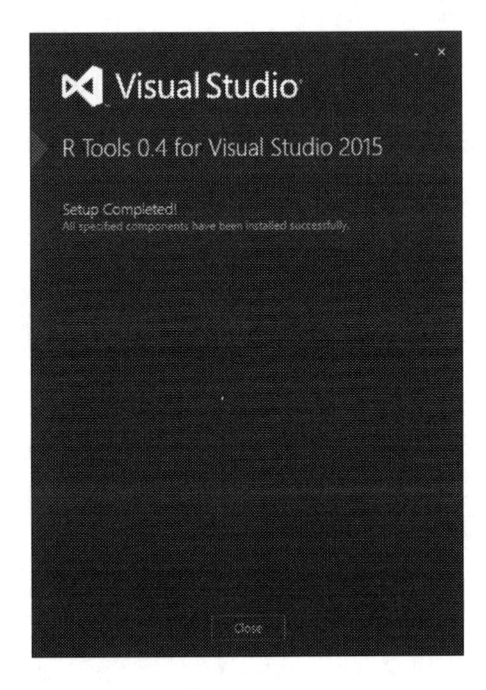

Figure 2-12. *R Tools 0.4 for Visual Studio installation complete*

Clicking the **Close** button, as shown in Figure 2-12, also opens a web page, so let's take a look at that next in Figure 2-13.

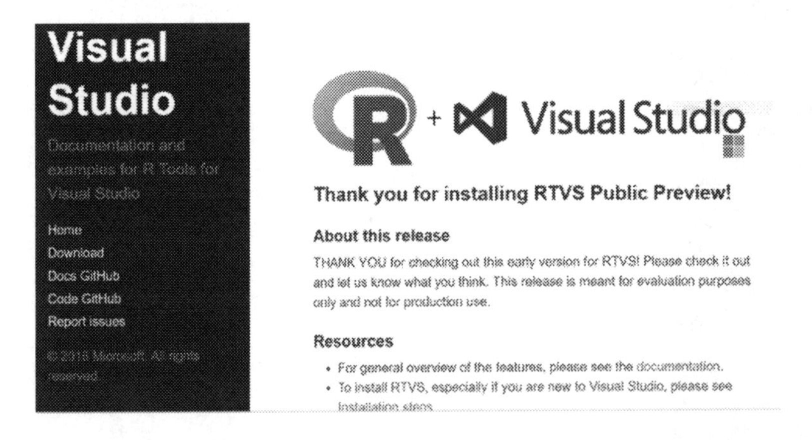

Figure 2-13. *Opened web page*

Isn't that nice! Microsoft gives us a nice page all lined up with resources and everything. This page may change when you set yours up, but it probably will be fairly similar. I strongly urge you to go through the listed resources to gather as much information about R as you can. Once we get into the interface of RTVS,

I show you an even better way for getting documentation on R, but for now, just peruse this site generally so that you are aware of what is available.

Back to the setup screen shown in Figure 2-12, press **Close**. R Tools for Visual Studio has now been successfully installed.

Download Microsoft R Open

The final step to getting the installation all working is to install Microsoft R Open. Go back to the web page provided by Microsoft and shown in Figure 2-1 and click the **Download Microsoft R Open** link. At this point, another web page opens; its URL is `https://mran.revolutionanalytics.com/download/`. Figure 2-14 shows the Microsoft R Open download page that you should see at this point.

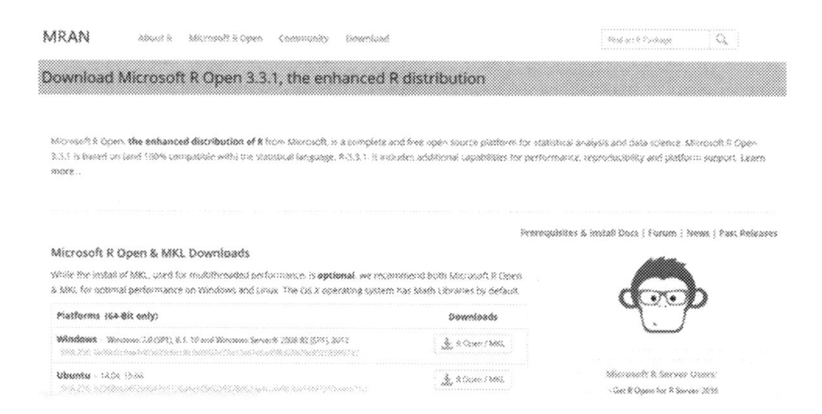

Figure 2-14. *Opened web page*

Notice that the download for Microsoft R Open is available for 64-bit platforms *only*. Microsoft R Open is available for 64-bit versions of Windows (7 SP1, 8.1, 10, Server 2008 R2 SP1, and Server 2012), Ubuntu (14.04 and 15.04), Red Hat Enterprise Linux (7.1 and 6.5), and SUSE Linux Enterprise Server 11.

Go ahead and click the link for your operating system; the download should start. At the time of this writing, the file name is `microsoft-r-open-3.3.1.msi` for the Windows installation and it is 131MB in size. The executable takes a little while to download, so go grab a drink or something while you're waiting.

Has Microsoft R Open finished downloading? Good. Then let's go ahead and install it now. Double-click the executable that you've just downloaded. Figure 2-15 shows what you should see at this point.

Figure 2-15. *Starting Microsoft R Open installation*

Click **Next**. You should be taken to the screen shown in Figure 2-16. This screen provides some detail about the impending installation. Read over the information given, click the **I Acknowledge** check box, and then click **Next** when you are ready.

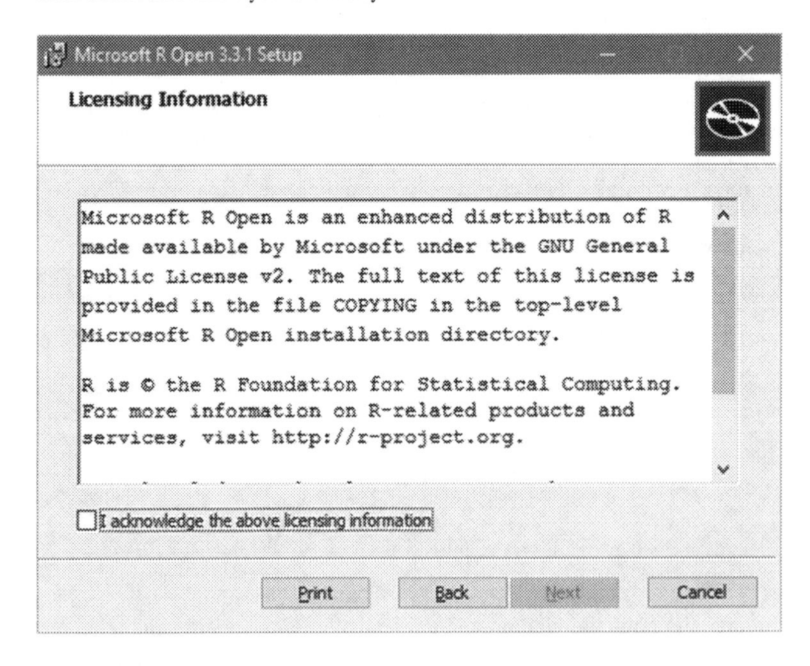

Figure 2-16. *Information*

The next screen, shown in Figure 2-17, shows the Install Math Kernel Library (Intel© MKL) installation option.

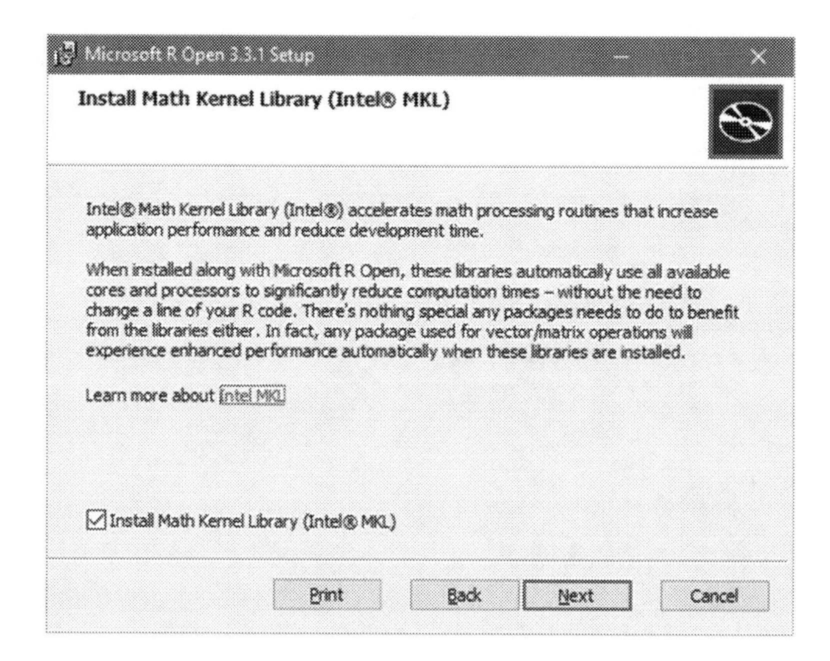

Figure 2-17. *Install Math Kernel Library (Intel© MKL) option*

This is completely your decision to install or not, since it is not a required installation portion, but it is recommended that you install it. The reason it is selected by default is that the addition of the math kernel library allows your R code to use all available resources to generate a result, which means that it gets done a lot faster (depending on the speed of your machine and the available resources). I also recommend leaving this option checked. Click **Next** when you are ready to move on.

Next, we have to accept the license terms for the MKL that we are installing. This is shown in Figure 2-18. At this point, make sure that you click the **I accept** check box, and then click **Next** to move on.

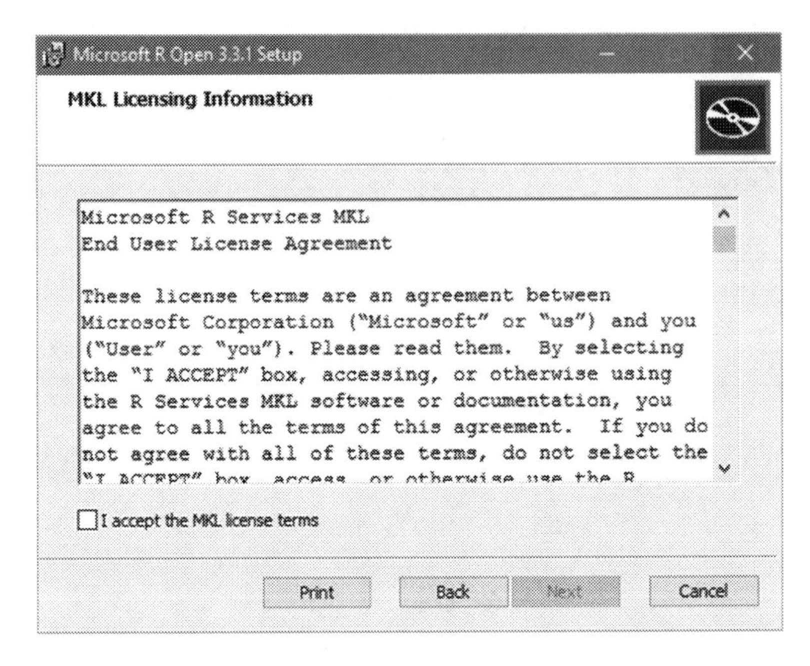

Figure 2-18. *Accept MKL license terms*

Next, the screen in Figure 2-19 shows the destination location information. Click **Next** to accept the target folder.

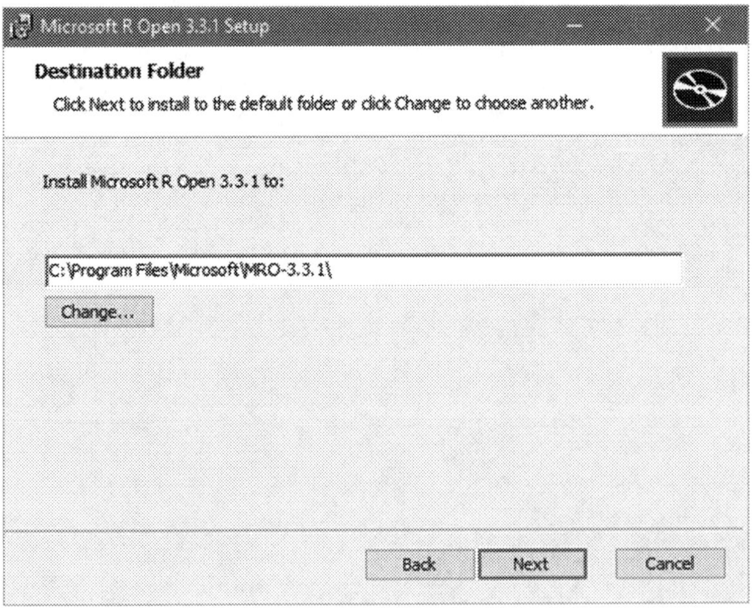

Figure 2-19. *Select Destination Location*

Click **Next** here. The following screen, shown in Figure 2-20, lets us begin the installation; click **Install** when you are ready.

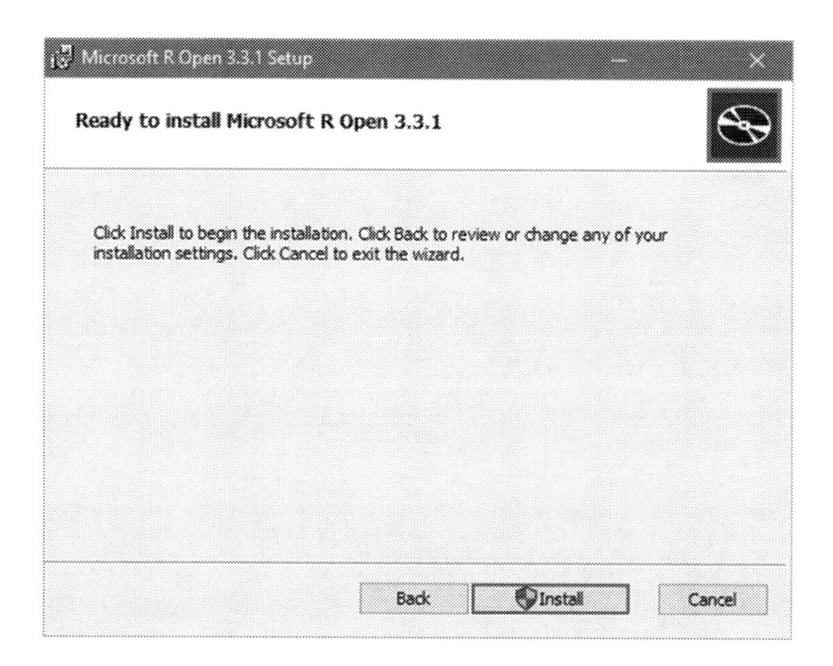

Figure 2-20. *Ready to install*

Figure 2-21 shows the installation running. It doesn't take very long, thankfully.

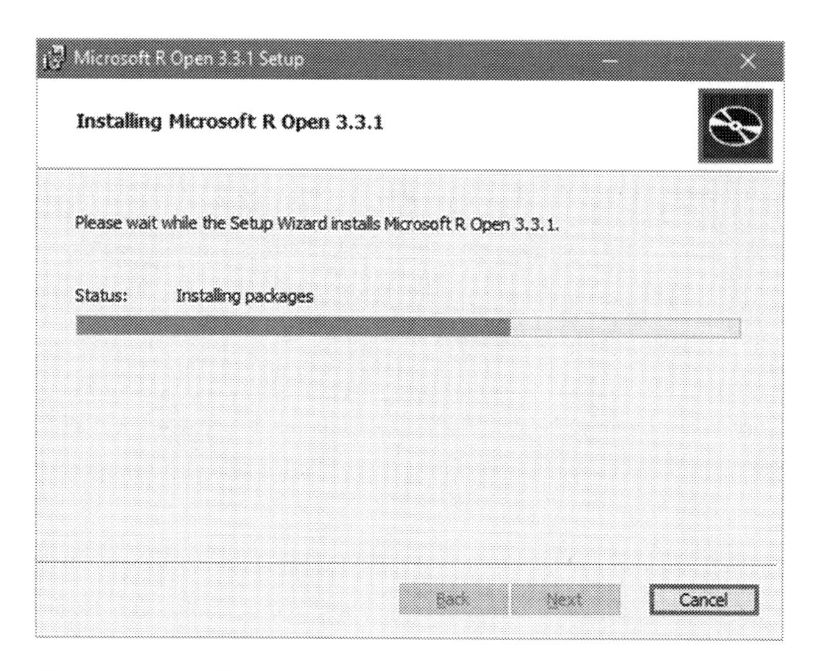

Figure 2-21. *Installing*

Finally, we are done installing Microsoft R Open. Figure 2-22 shows the final screen you should see at this point in the installation.

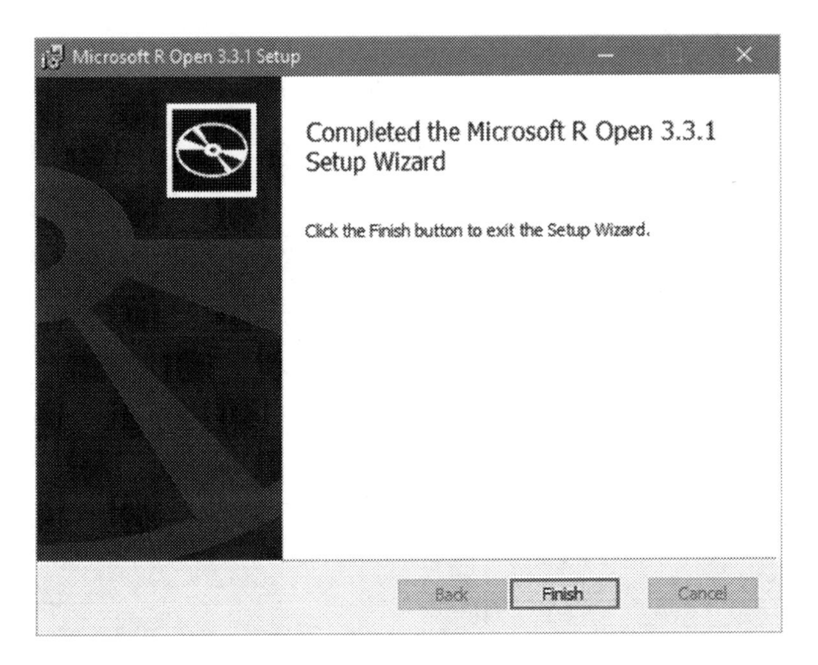

Figure 2-22. *Installing*

Click the **Finish** button, as shown in Figure 2-22, to finish the installation. No reboot is needed at this point. You have all you need to be up and running with R in Visual Studio.

Visual Studio Environment

Now that we have everything installed according to Microsoft, let's take a look around our new development environment and see what is new, changed, or different. For those of you that have used Visual Studio before (and that's probably a large number of you), you see the familiar Visual Studio environment. For those of you that haven't used Visual Studio before, I invite you to spend a little time poking around in the various menu sections to see how comfortable you are with the user interface. Again, if you have a familiar IDE that you would rather use to write R code, by all means, go ahead and use that one instead. I will guide you along a quick introduction to the menus specifically pertaining to the new R functionality in Visual Studio 2015.

Go ahead and start Visual Studio. You might see screens prompting you to connect to developer services. If you see this, go ahead and press **Maybe Later**. Those are expected. You eventually get to the main IDE screen shown in Figure 2-23. Look at the menu bar for the R Tools menu that is just a bit to the right of the center of the menu toolbar.

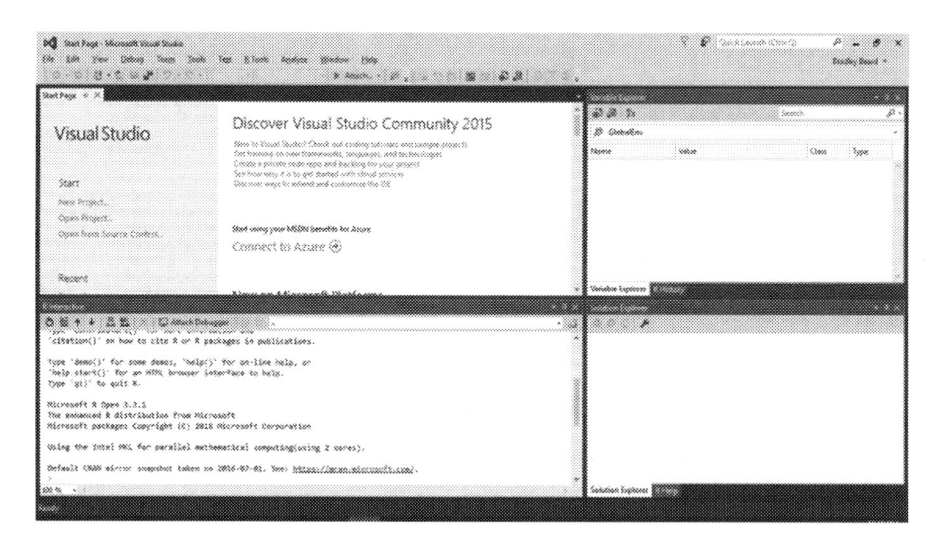

Figure 2-23. *R is installed as part of Visual Studio*

The R Tools menu is home to all the R functionality now available from Visual Studio. Opening this menu shows the options in Figure 2-24.

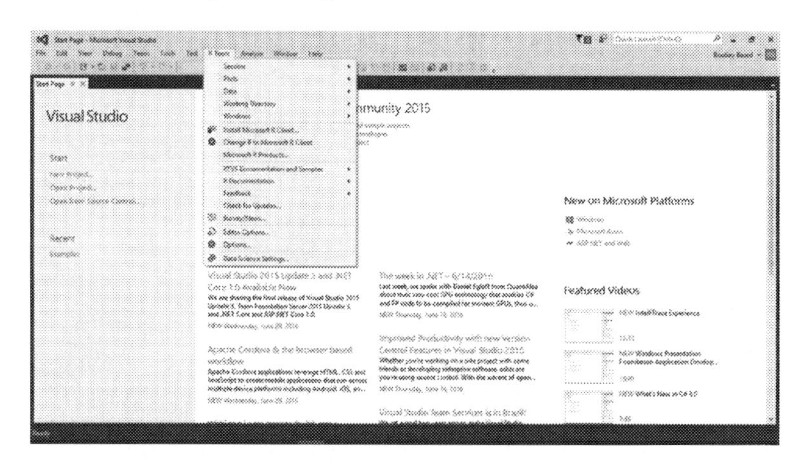

Figure 2-24. *R Tools menu options*

Let's take a look at those menu options. The following subsection gives a brief description of what each menu option does for you and what that menu option allows you to control.

Session

The Session option lets you conduct R sessions. The following are the available options:

- Interrupt R
- Attach Debugger
- Reset

- Load Workspace...
- Save Workspace As...

Plots

The Plots option lets you define how you want your results to be shown. The following are your available options:

- Previous Plot
- Next Plot
- Save as Image
- Save as PDF
- Copy as Bitmap
- Copy as Metafile
- Remove Plot
- Clear All Plots

Clearly, you can see that this is a huge help. The ability to export these plots as the various types mentioned is already a pretty big advantage over some other systems that only generate in Flash or Silverlight, for example.

Data

The Data option lets you define how you want to consume data that already exists. The following are you available options:

- Import Dataset into R Session from Web URL...
- Import Dataset into R Session from Text File...
- Delete All Variables

I wonder if we will get the option to connect to different databases and/or installations in the future...? It would be an interesting development, should Microsoft decide to pursue that path. I think that making more data sources available for consumption is only going to enhance the usability of the product.

Working Directory

The Working Directory option allows you to change your working directory. Your working directory is different from where the installation of R is installed, for example. A working directory means that this is where you save the files that you are working on.

- Set Working Directory to Source File Location
- Set Working Directory to Project Location
- Select Working Directory...

Windows

The Windows menu lets you open new windows to monitor what is going on with your R scripts. The following are the available options:

- Source Editor
- R Interactive
- Help
- History
- Files
- Plots
- Packages
- Variable Explorer

These options replace the regular views that you may have for class view or stack traces, apparently. Since R is a totally different beast, it stands to reason that there are different ways of viewing the data and the development tasks.

Install Microsoft R Client...

You select the Install Microsoft R Client option when you want to install the R client as a download from Microsoft. We don't want to do this, so just leave this alone. I explain the different components to Microsoft's R products shortly.

Change R to Microsoft R Client

You select Change R to Microsoft R Client when you havea local installation of R that you want to convert to an instance of Microsoft R Client. The implication is that Microsoft R Open is installed, since that is a prerequisite to running R Client. In the instance, we also don't want to change to an R client. This implies that we are connecting to a remote database instance, which we are not. We are connecting locally.

Microsoft R Products...

The Microsoft R Products menu option is essentially a shortcut to a URL. Selecting the option takes your web browser to `https://www.microsoft.com/en-us/cloud-platform/r-server`, where you can view Microsoft's product options for R Server.

RTVS Documentation and Samples

The RTVS Documentation and Samples option takes you to documentation and samples relating specifically to R Tools for Visual Studio. Two sub options are available:

- **Documentation**: This option takes you to a web site (`http://microsoft.github.io/RTVS-docs`) that goes through the documentation for the tool.
- **Samples**: This option takes you to a web site (`http://microsoft.github.io/RTVS-docs/samples.html`) that lets you see a rather large sample of R scripts available. We will come back to this section in a little while.

R Documentation

The R Documentation menu item provides easy access to documentation on the R language itself. There are four options available here:

- **Intro to R**: Clicking this option takes you to `https://cran.r-project.org/doc/manuals/r-release/R-intro.html`. I thought this was an interesting addition. Microsoft linked the entire R documentation set right into Visual Studio. All you have to do is click the link—and you're there.
- **Task Views**: This link goes to `https://cran.r-project.org/web/views/`, which shows a list of task views that can be downloaded and installed. A task view, in this context, is a group of libraries that work together for a common purpose. There are a lot of them here, so go explore them when you get a chance.
- **Data Import/Export**: This link goes to `https://cran.r-project.org/doc/manuals/r-release/R-data.html`, which is another documentation set. Light reading, no big deal.
- **Writing R Extensions**: Finally, this link goes to `https://cran.r-project.org/doc/manuals/r-release/R-exts.html`, which is yet another extensive documentation set.

There is a lot of information available on R from these menu options. Take advantage of what's available anytime you have questions about the language. I highly recommend reading through the documentation, at least briefly, so you can get familiar with how R works syntactically. That makes the later part of this chapter much easier to work with.

Feedback

Select the **Feedback** menu option to rate the product. The following are the available options:

- Report issue on GitHub
- Send Smile via E-mail
- Send Frown via E-mail

Any feedback that you choose to send gets routed eventually to Microsoft.

Check for Updates

Select **Check for Updates** to check for and download updates to R Tools for Visual Studio.

Survey/News

Choose the **Survey/News** option to be taken to `http://rtvs.azurewebsites.net/news/`, where Microsoft posts news about R. There is also the Survey option, which at this time is "experimental."

Editor Options

Choose **Editor Options** to get into customizing the interface options. This is the same action as **Tools** ä **Options** in the regular menu, except this menu confines the options by default to the R context menu. There are various options, including how IntelliSense operates, formatting, and the general sort, such as line numbers and word-wrap features.

Options

This option lets you customize the environment options. This is different than the Editor Options menu because the Editor Options menu only dealt with the options available from the Editor context. The Options menu lets you deal with the actual environment options for R, such as debugging, CRAN mirror location, general help settings, and the R installation location.

Data Science Settings

Selecting this option opens up a modal dialog window shown in Figure 2-25. You are given the option to reconfigure Visual Studio for use by so-called data scientists. You'll get a window layout and some keyboard shortcuts that Microsoft deems useful to those doing data analysis in R.

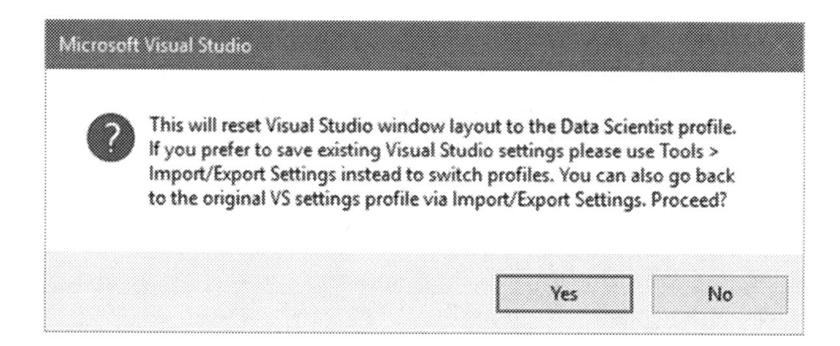

Figure 2-25. *Reset options*

Go ahead and click **Yes**. Another screen might open, as you can see in Figure 2-26.

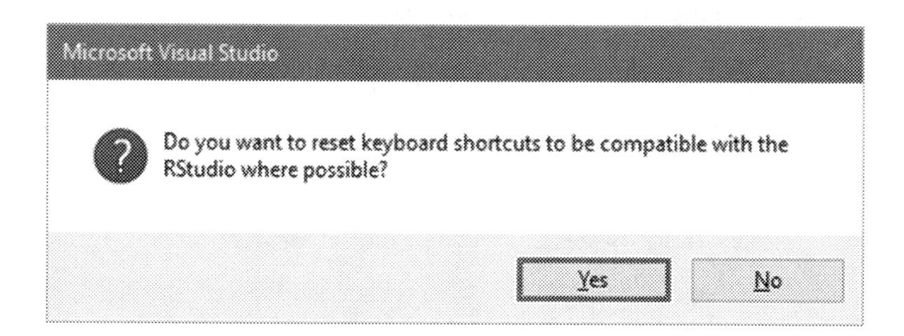

Figure 2-26. *Reset keyboard options*

If you would like to reset your keyboard shortcuts while using RTVS, click **Yes**; otherwise, click **No**. We can now see an updated interface. Take a look at Figure 2-27 for a look at the changes.

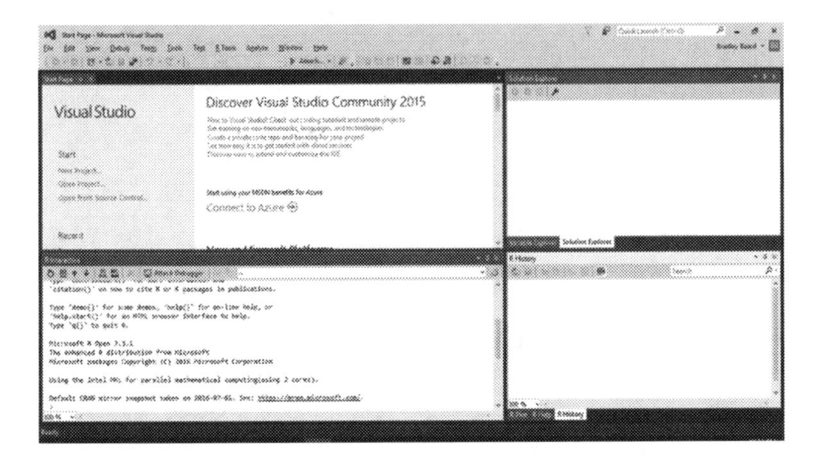

Figure 2-27. *Updated interface*

You can see that I have my interface laid out so that R Interactive is the main window in the bottom-left corner. On the right, I have Variable Explorer and Solution Explorer at the top. R Plot, R Help, and R History are all laid out nice and neat at the bottom. This is proof of concept that R is installed and working correctly.

Let's briefly discuss what Microsoft makes available in their R product line.

- **Microsoft R Server**: An enterprise-class platform running Hadoop, Teradata DB, or even Linux to provide powerful interactions with data. Uses ScaleR technology for parallelization.
- **Microsoft R Client**: A free data science tool that works with Microsoft R Open. R Client uses remote data and processes the operations locally. Uses ScaleR technology for parallelization.
- **Microsoft R Open**: Microsoft's "version" of R. Called MRO for short, it is fully compatible with R code in every way. It does not reference the proprietary ScaleR technology though.
- **SQL Server R Services**: Integrates R into SQL Server's database engine natively. ScaleR is available in Enterprise Edition only though.

Additionally, R Server is a server-class analysis platform and R Client is a client-based tool. MRO provides the interface to R from a client. So, for example, in a typical data analysis environment, there would be an R Server running an instance of SQL Server R Services, Hadoop, Teradata DB, or Linux, with one or more R Client connections running MRO to interact with the data in the R Server or in the SQL Server R Services instance. The clients could either consume the resources on their local workstations using the ScaleR functions, or do the computing on the R Server using SQL Server R Services or one of the other analysis tools (Hadoop, Teradata DB) on either Windows or Linux servers.

For the examples in this book, the R Server instance is our SQL Server R Services database instance and the R Client is our installation of RTVS with MRO.

Exploring Samples

Let's go back to the Samples link in the **RTVS Documentation and Samples** menu item. That link is `http://microsoft.github.io/RTVS-docs/samples.html`. On that page, there is a download of a `.zip` file containing examples that we will use to get familiar with the new R environment. Unzip that `.zip` file to a location that you can access, and then navigate to `RTVS-docs-master/examples` and double-click **README.MD**. This opens this doc in RTVS. Figure 2-28 shows this document once opened.

```
1   ### RTVS examples
2
3   Here are two sets of examples to get you started with R Tools
4   for Visual Studio.
5
6   **A First Look at R** shows you how to open and run an R script in
7   the interactive terminal.
8
9   **MRS and Machine Learning** shows you how to use R and Microsoft R Server
10  to create machine learning models and handle large data sets.
11
12  The README files in each directory give more detail about each example.
```

Figure 2-28. *Readme file*

We need to get a little bit familiar with R as a language before we attempt to get into any sort of development activity, so let's step through some of the examples given in *A First Look at R*. In the tutorial featured later in this book, we deal with the R Server aspect a lot more, since we will directly interface with SQL Server R Services to create a report with embedded information.

There is an awful lot of information about R on the internet, so if you already know about it, then you can consider this a refresher course. If not, no worries. I'm not going to get into the complete history of R and I'm not going to make this a comprehensive guide to all of R's functionality. Instead, I highlight the basics—and we can go from there. I think that will be enough to whet the proverbial whistle and get our minds keen on the practicality of using R for serious data analysis.

A First Look at R

Navigate to `RTVS-docs-master\examples\A first look at R` and double-click the **README.MD** file in that directory. Figure 2-29 shows what you should see at this point.

```
README.md
1   ### A first look at R
2
3   These scripts let you take R for a test drive.
4
5   #### 1 Getting Started with R.R
6   The comments in this script give helpful information about R and how to use it.
7   To get the most out of it, set the cursor at the top of the script and press
8   CTRL-Enter to scroll through it line by line. This will let you see the output of
9   each command. Getting Started shows you some R fundamentals, like installing
10  packages, loading data, plotting it and fitting a model.
11
12  #### 2 Introduction to ggplot2.R
13  ggplot2 is an R graphing package known for its visually appealing plots
14  and simple syntax.
15  Execute this script line by line by pressing CTRL-Enter to see several options
16  for visualizing earthquake data from Fiji.
17
```

Figure 2-29. *A first look at R README*

That document tells us that we get to "take R for a test drive" by running R scripts provided by Microsoft in the `.zip` file that we downloaded earlier. Navigate back to the `RTVS-docs-master\examples\A first look at R` directory and double-click **1-Getting_Started_with_R.R**. The script is opened in RTVS, as shown in Figure 2-30.

```
1-Getting_Started_with_R.R    README.md
1   ## Getting Started with R
2
3   ### Some Brief History
4
5   # R followed S. The S language was conceived by John Chambers, Rick Becker,
6   # Trevor Hastie, Allan Wilks and others at Bell Labs in the mid 1970s.
7   # S was made publically available in the early 1980's. R, which is modeled
8   # closely on S, was developed by Robert Gentleman and Ross Ihaka in the early
9   # 1990's while they were both faculty members at the University of Auckland.
10  # R was established as an open source project (www.r-project.org) in 1995.
11  # Since 1997 the R project has been managed by the R Core Group.
12  # When AT&T spun of Bell Labs in 1996, S was no longer freely available.
13  # S-PLUS is a commercial implementation of the S language developed by the
14  # Insightful corporation which is now sold by TIBCO software Inc.
15
16  # The R Core Group: http://www.r-project.org/contributors.html
17  # Download R: http://cran.r-project.org/
```

Figure 2-30. *Getting Started with R script*

At this point, all that we are going to do is step into this R script and execute some portions to get an idea of how R is laid out syntactically and how it might compare to other languages. We pretty much go line by line through the *Getting Started with R* script so that we can really understand what this introduction is getting across.

It is worthwhile to read through the first 75 lines of comments, as this sets up your basis as a new R user, or refreshes your memory if you're a legacy R user. Either way, there is something for everyone here, so be sure you read it thoroughly, particularly the R Resources and R Blogs sections. The Help section is always good, so don't skip over that one either.

Line 76 is the first executable R script. That line is very simply `installed.packages()`. This simple line lets us see what packages are already installed; so highlight line 76 and press **Ctrl+Enter** to execute it. Note that your R Interactive window (which should still be up) starts loading up a lot of information, as shown in Figure 2-31.

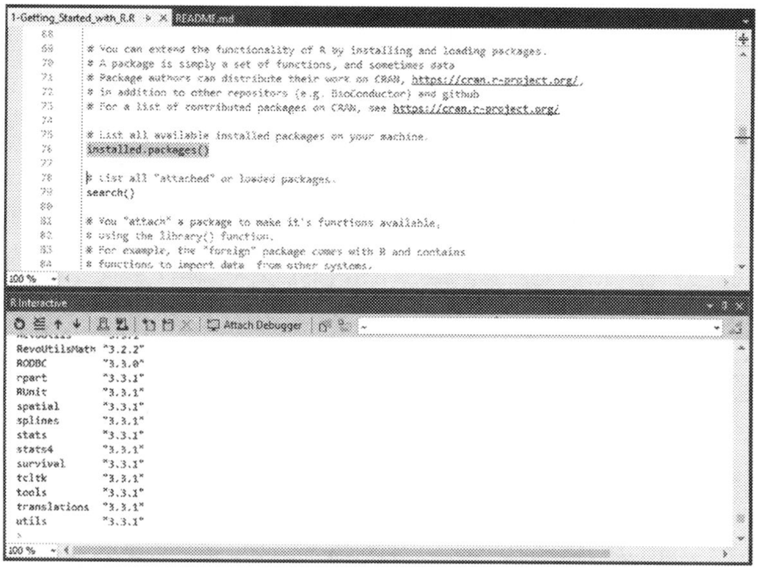

Figure 2-31. *R Interactive window*

You can scroll up in the R Interactive window to see exactly what happened, but it's strictly informational, at this point. It can be useful to peruse this generated content to ensure that you have the latest version of an installed package, for example, but for general use, it's good to know that it's there.

■ **Note** there is now a value in the R History window as well. This is very handy in case we ever need to re-execute a line of code. All you need to do is highlight the code you want to re-execute and press **Enter**. This moves the code from the R History window to the R Interactive window. From here, press **Ctrl+Enter** to execute the line of code as normal.

Highlight line 79, which reads `search()`, and execute it. This gives us a listing of the currently loaded packages for this R session. Next, we attach a package using the `library()` function, which is how R makes functionality particular to a specific package available to the session.

Skip down to line 85, which reads `library(foreign)`. This means that we include the `foreign` library functionality in the current session. Highlight line 85 and press **Ctrl+Enter** to execute it. As soon as you see the caret at the bottom of the R Interactive window turn back to the greater than sign (>), then you know that the code has completed executing. Figure 2-32 shows what you should see at this point.

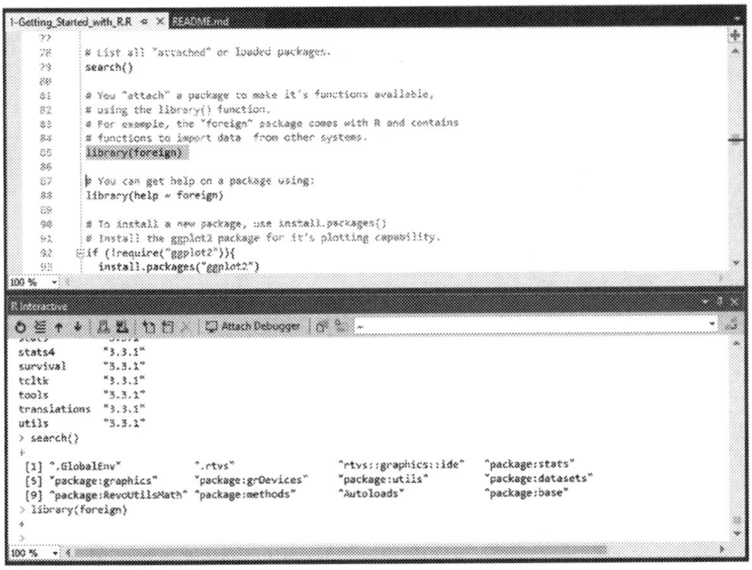

Figure 2-32. *Line 85 execution*

This isn't in the R script we are working with, but if you go back now and execute line 79 again, you should see what is shown in Figure 2-33.

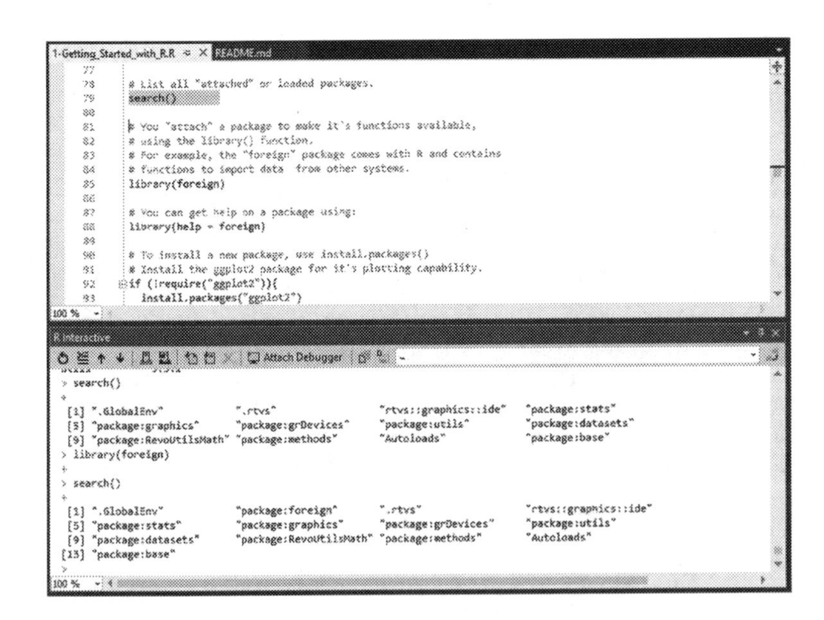

Figure 2-33. *Line 79 execution showing foreign library*

Note how R Interactive shows the listed items? It is done in groups of four, with the starting n-index number being shown as the leftmost column, followed by four packages. The next line begins with the n+4 index, and then lists another four packages, and so on. Now see how the first execution of `search()` showed

12 packages, but we can now see 13 packages returned in the newly returned `search()` command. We can see the addition of the `foreign` package as the reason for this, so that is our proof that the package was successfully added to our current R session.

When packages are added to the R mirrors, they always include a Help section. You can reference this Help section by highlighting line 88 and pressing **Ctrl+Enter**. This opens the help documentation as another page within the top frame in RTVS. Figure 2-34 shows this result.

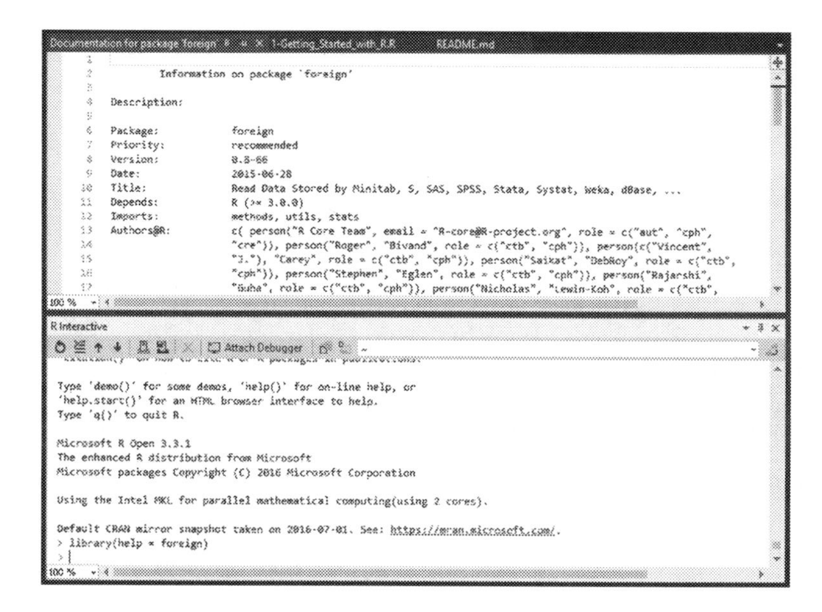

Figure 2-34. *Help documentation for foreign library*

You can close that documentation. Next, skip down to line 90. We are going to install `ggplot2`, which is probably the most popular and robust charting package available for R. Highlight line 92 through 94 and press **Ctrl+Enter** to execute. Figure 2-35 shows the result of this.

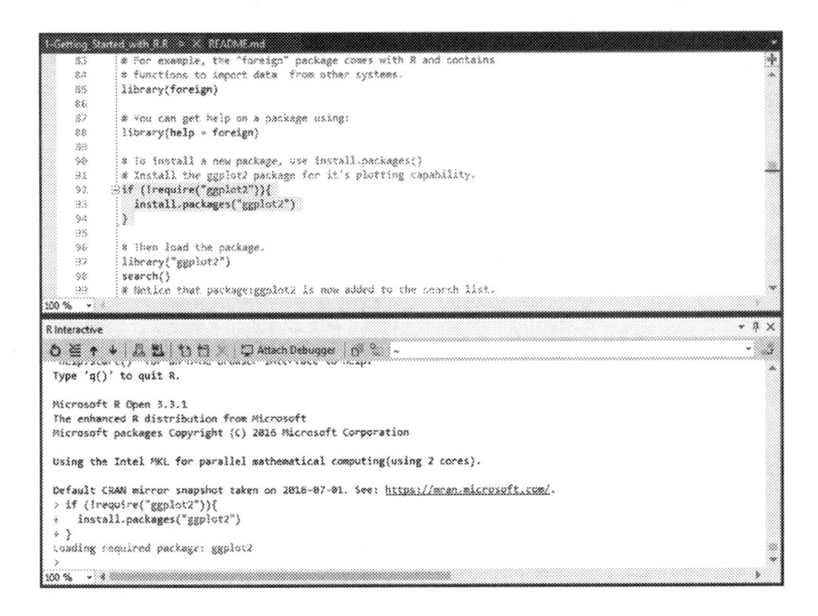

Figure 2-35. *Loading ggplot2 package*

Now that `ggplot2` is loaded, we need to run the `library()` function in order to load it into the current R session. This is shown in line 97; highlight this line and execute it, and then execute line 98 as well. Line 98 says `search()` shows the currently loaded packages. Notice that `ggplot2` is now added to the list of currently installed packages for this session.

Next, we look at a simple regression example, as shown in the script. First, I need to point out that the `ggplot2` package comes preloaded with quite a few sets of data that is to be used to test the functionality of the package. This data is accessed as shown on line 105, using the syntax `data(package = "ggplot2")$results`. This syntax says that we want to run the `data()` function against the `ggplot2` package and return the subset of the output referenced as `results` to the screen, as shown in Figure 2-36.

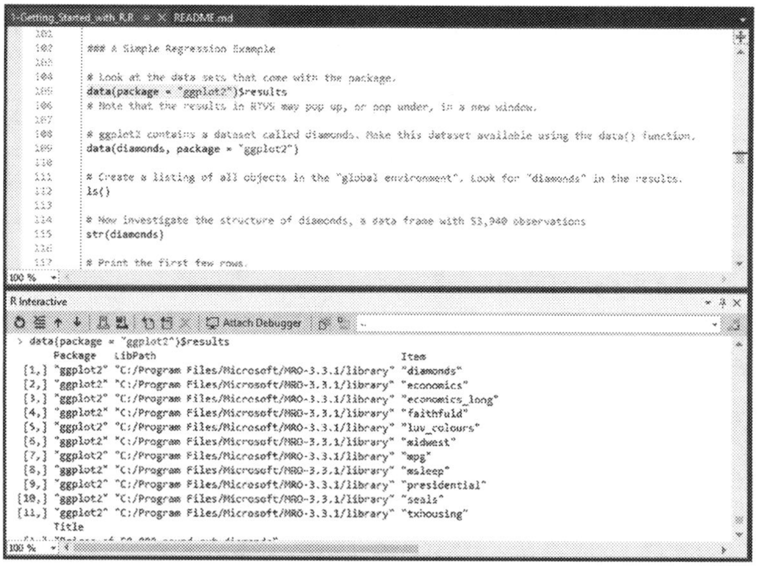

Figure 2-36. *ggplot2 results*

Consequently, we could also execute the command `data(package = "ggplot2")` to see a listing of the datasets included in the package. Figure 2-37 shows this result.

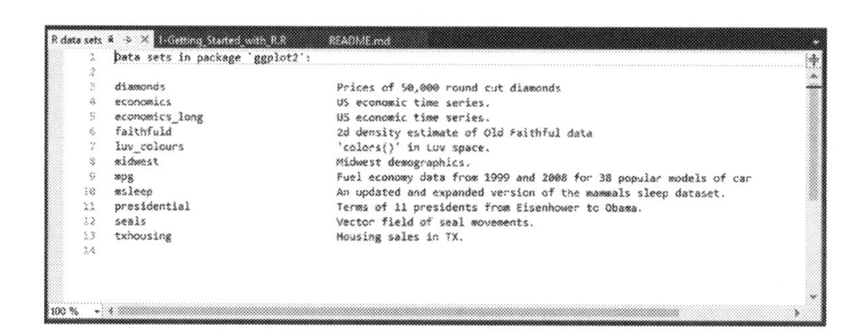

Figure 2-37. *ggplot2 Datasets*

Go ahead and close that window, but keep your R script open. We've got the executed result of `data(package = "ggplot2")$results` shown in the R Interactive window, so skip down to line 109 next. This line says `data(diamonds, package = "ggplot2")`. The syntax of this says that we want to run the `data()` command against the `diamonds` dataset in the `ggplot2` package. Highlight that line and press **Ctrl+Enter** to execute it. There isn't any huge change or anything here; all that happened was that the `diamonds` dataset was just made available for analysis. The way you know that it was just loaded is to check your Variable Explorer window. Figure 2-38 shows what the Variable Explorer window should look like at this point.

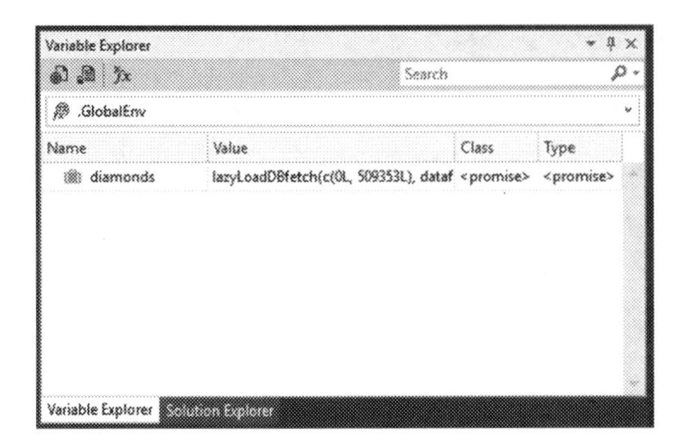

Figure 2-38. *diamonds dataset loaded*

So there is our `diamonds` dataset, loaded and ready to go. Go down to line 112 next, which says `ls()`. That line by itself, without any arguments, only returns the datasets or functions defined by the user in the current session. Running this simple line of code in this instance only results in the output of a single word: *diamonds*. The reason for this is that this is the only dataset loaded for this session. If you were to execute this line inside of a function with no arguments, you would be able to see the local variables for that particular function. As you can tell, this can be a useful debugging tool.

Now go down and execute line 115, which says `str(diamonds)`. This command allows us to examine the structure of the dataset passed in as an argument. Figure 2-39 shows the structure displayed in RTVS.

```
1-Getting_Started_with_R.R    README.md
107
108   # ggplot2 contains a dataset called diamonds. Make this dataset available using the data() function.
109   data(diamonds, package = "ggplot2")
110
111   # Create a listing of all objects in the "global environment". Look for "diamonds" in the results.
112   ls()
113
114   # Now investigate the structure of diamonds, a data frame with 53,940 observations
115   str(diamonds)
116
117   # Print the first few rows.
118   head(diamonds)
119
120   # Print the last 6 lines.
121   tail(diamonds)
122
123   # Find out what kind of object it is.
```

```
R Interactive
[1] "diamonds"
> str(diamonds)
Classes `tbl_df`, `tbl' and 'data.frame':   53940 obs. of  10 variables:
 $ carat  : num  0.23 0.21 0.23 0.29 0.31 0.24 0.24 0.26 0.22 0.23 ...
 $ cut    : Ord.factor w/ 5 levels "Fair"<"Good"<..: 5 4 2 4 2 3 3 3 1 3 ...
 $ color  : Ord.factor w/ 7 levels "D"<"E"<"F"<"G"<..: 2 2 2 6 7 7 6 5 2 5 ...
 $ clarity: Ord.factor w/ 8 levels "I1"<"SI2"<"SI1"<..: 2 3 5 4 2 6 7 3 4 5 ...
 $ depth  : num  61.5 59.8 56.9 62.4 63.3 62.8 62.3 61.9 65.1 59.4 ...
 $ table  : num  55 61 65 58 58 57 57 55 61 61 ...
 $ price  : int  326 326 327 334 335 336 336 337 337 338 ...
 $ x      : num  3.95 3.89 4.05 4.2 4.34 3.94 3.95 4.07 3.87 4 ...
 $ y      : num  3.98 3.84 4.07 4.23 4.35 3.96 3.98 4.11 3.78 4.05 ...
 $ z      : num  2.43 2.31 2.31 2.63 2.75 2.48 2.47 2.53 2.49 2.39 ...
>
```

Figure 2-39. *str(diamonds) output*

Briefly look at the returned data there; you can see that there is a lot of information but it appears to be truncated. Have no fear; take a look at your Variable Explorer again. You see that there is a very slight change: there is now an indicator next to diamonds, as opposed to just a table graphic. Click this indicator; you should see what is shown in Figure 2-40.

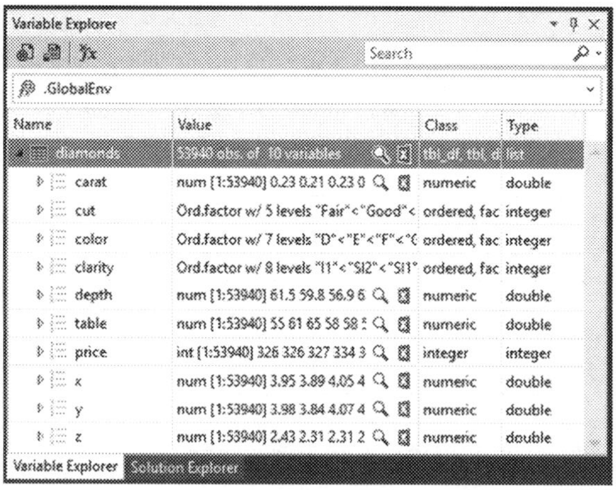

***Figure 2-40.** diamonds detail information*

This is much easier for me to read and decipher. Across the top, just above the blue bar shown in Figure 2-40, the columns are defined as Name, Value, Class, and Type. Go ahead and browse around in there for a minute and get familiar with how it looks. This is a very cool feature that allows the introspection of data before it is really analyzed.

Go to line 118 next. This line says `head(diamonds)`, which simply instructs R to output the first six rows of data in the dataset. Conversely, line 121 says `tail(diamonds)`, which, as you've probably guessed, outputs the last six rows of data in the dataset. The results of these two executions are shown in Figure 2-41.

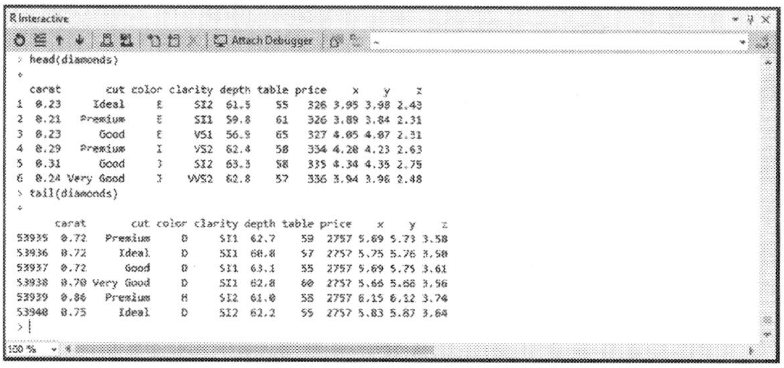

***Figure 2-41.** head() and tail() demonstrated*

Line 127 is next, which says `dim(diamonds)`. The `dim()` function allows us to see the dimension of the object passed as an argument, which would be `diamonds` in this case. Executing this line simply shows what we have already gathered by examining the Variable Explorer earlier: there is one dataset returned with 53940 objects of 10 variables.

Next, we move on to plotting in R. There are three main packages used for charting in R: `base`, `lattice`, and `ggplot2`. This is also referenced on line 136.

```
Lines 133 and 134 say the following:diamondSample <- diamonds[sample(nrow(diamonds), 5000),]
dim(diamondSample)
```

Let's break the syntax of this apart before anything else.

- **diamondSample**: Declares an object for storage of the results of the command.
- `<-` : An assignment operator that declares that the result of the command(s) on the right are to be assigned to the object on the left.
- **diamonds**: References the `diamonds` dataset.
- **[sample(nrow(diamonds), 5000),]**: draw a random sample of 5000 data points from the `diamonds` dataset
- **dim(diamondSample)**: show the dimension of the `diamondSample` object

As expected, the result of the `dim(diamondSample)` command was the output `[1] 5000    10`, which means that one dataset and 5000 objects in 10 variables were returned.

Next, we go to line 140, which reads: `theme_set(theme_gray(base_size = 18))`. This means that we set the font size to 18pt and use the gray theme. Go ahead and highlight and execute that line, then skip down to lines 143 and 144. Those lines show the following code:

```
ggplot(diamondSample, aes(x = carat, y = price)) +
  geom_point(colour = "blue")
```

First off, note that this is on two different lines, but it is actually one long command. The presence of the plus sign indicates a line continuation character, outside of usage inside a function or argument.

Let's take a look at the syntax of this command before we move on.

- **ggplot**: Defines the function to be executed.
- **diamondSample**: References the `diamondSample` object as the active dataset.
- **aes**: Defines the aesthetic qualities of the chart; in this example, the x axis is the `carat` set and the y axis is the `price` set.
- **geom_point**: Customizes the points once they are charted.

So this means that we will generate a chart. Highlight these two rows and press **Ctrl+Enter** to execute them. Figure 2-42 shows the result of this action.

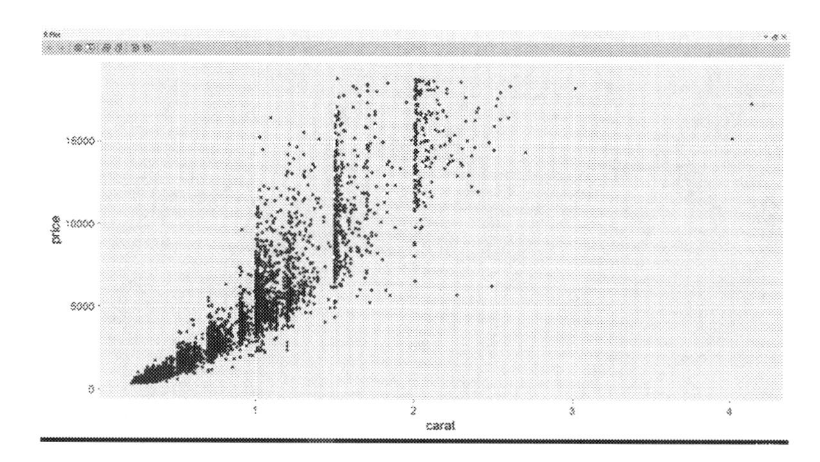

Figure 2-42. *Our first chart!*

Notice that the R Plot window came into focus here? That's pretty handy. Notice also that the x axis is `carat` and the y axis is `price`, as we defined in the `aes()` command.

Go down to lines 147 through 149 next. You'll see there is a very slight addition of `scale_x_log10()`. This gives us a nice little logarithmic scale on the x axis (which is also stated in line 146) so go ahead and highlight and execute lines 146 through 149 now. Figure 2-43 shows what you should see now.

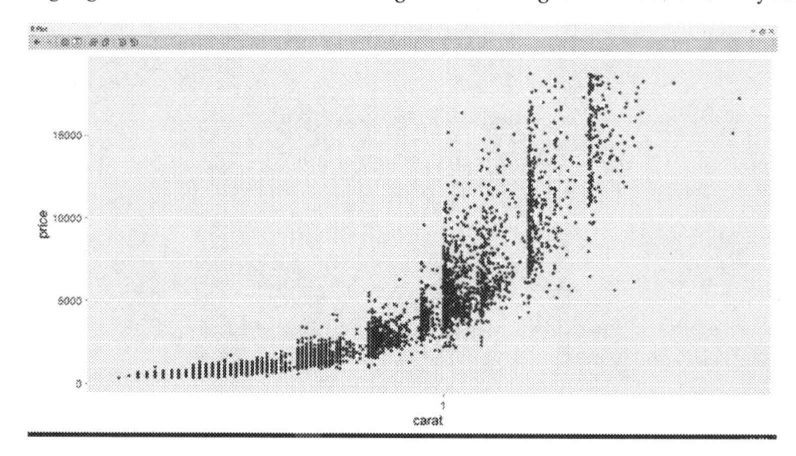

Figure 2-43. *Logarithmic scale addition*

That's starting to make a bit more sense now. Next, another line is added to let us add another logarithmic scale on the y axis. This is on lines 152 through 155; so highlight these lines and press **Ctrl+Enter** to execute them. Figure 2-44 shows what you should see now.

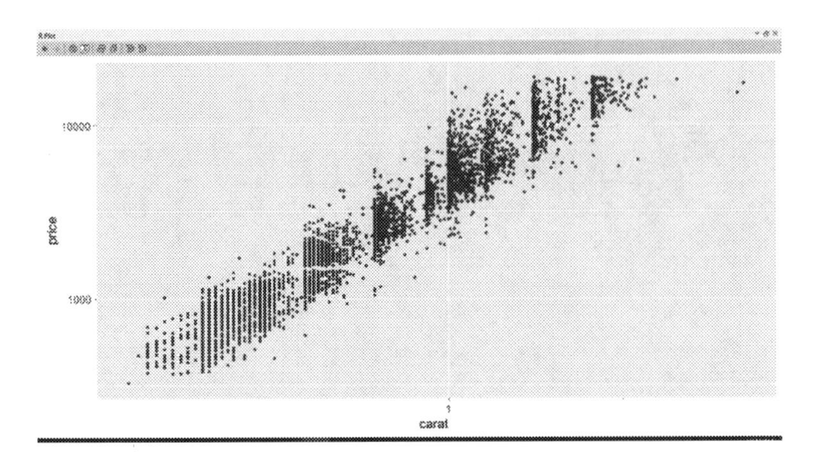

Figure 2-44. *Logarithmic scale on x and y axes*

The chart is even better now, I think.

Next, we look at linear regression in R within the confines of the loaded dataset. Linear regression is basically the way that statisticians define the relationship between a scalar variable and one or many explanatory variables. *Simple linear regression* is when there is only one explanatory variable, which is essentially what we will be dealing with in this example.

In R, we must first build a model of the data before we can access that model. To do that, take a look at line 163, which says `model <- lm(log(price) ~ log(carat) , data = diamondSample)`. Again, let's look at the syntax of this before we move on.

- **model**: Defines the object that inherits what is defined in the command.
- <- : An assignment operator that declares that the result of the command(s) on the right are to be assigned to the object on the left.
- **lm**: Defines a linear model.
- **log(price)**: A logarithmic scale of the `price` data (also the scalar variable).
- ~ : Separates the scalar variable and the explanatory variable(s). Scalar variables are shown to the left of this symbol and explanatory variables are shown to the right.
- **log(carat)**: A logarithmic scale of the `carat` data (also the explanatory variable).
- **,data = diamondSample**: Defines the data object to be used (`diamondSample`, in this case).

Highlight line 163 and execute it. Notice that there is no discernible action taken in the IDE; so let's check the Variable Explorer again. There is now another data object in there referenced as `model`. Figure 2-45 shows what you should be looking at now.

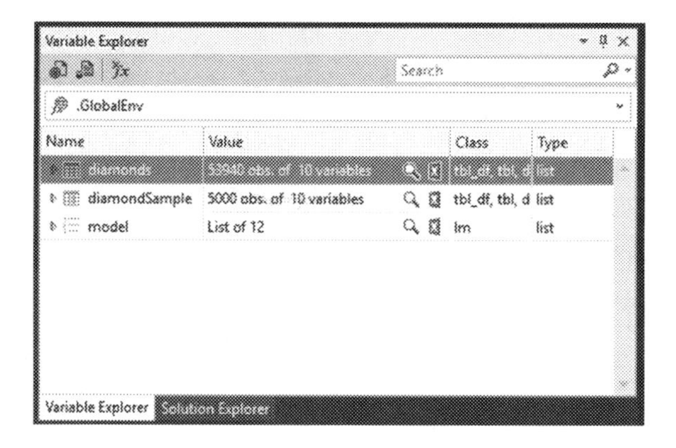

Figure 2-45. *model dataset*

Line 166 simply says `summary(model)`. Go ahead and execute that line. A lot of information is shown in the R Interactive window now. Take a look at Figure 2-46.

```
1-Getting_Started_with_R.R    README.md
160
161   # Build the model. log of price explained by log of carat. This illustrates how linear regression works.
        Later we fit a model that includes the remaining variables
162
163   model <- lm(log(price) ~ log(carat) , data = diamondSample)
164
165   # Look at the results.
166   summary(model)
167   # R-squared = 0.9334, i.e. model explains 93.3% of variance

R Interactive
Attach Debugger
> summary(model)
+

Call:
lm(formula = log(price) ~ log(carat), data = diamondSample)

Residuals:
     Min       1Q   Median       3Q      Max
-1.23362 -0.17217 -0.00395  0.17259  1.14454

Coefficients:
            Estimate Std. Error t value Pr(>|t|)
(Intercept) 8.453487   0.004448  1900.6   <2e-16 ***
log(carat)  1.673569   0.006321   264.8   <2e-16 ***
---
Signif. codes:  0 '***' 0.001 '**' 0.01 '*' 0.05 '.' 0.1 ' ' 1

Residual standard error: 0.2613 on 4998 degrees of freedom
Multiple R-squared:  0.9334,    Adjusted R-squared:  0.9334
F-statistic: 7.01e+04 on 1 and 4998 DF,  p-value: < 2.2e-16

>
```

Figure 2-46. *summary(model) information*

Take a look at what is returned by this action. You have a lot of really useful information returned about the model here, so understand that this is a great place to get meaningful statistical information about a model.

Head on down to lines 170 through 172, which show the following code:

```
coef(model)
coef(model)[1]
exp(coef(model)[1])
```

Line 169 says that we extract the model coefficients. So let's highlight those lines and execute them now. Figure 2-47 shows what you should see in your R Interactive window now.

```
> coef(model)
+ coef(model)[1]
+ exp(coef(model)[1])
(Intercept)  log(carat)
   8.453407    1.673569
(Intercept)
   8.453407
(Intercept)
   4691.029
```

Figure 2-47. *model coefficients*

Now that we've got that, let's take a look at the next bit of code on lines 175 through 179. This code is defined as follows:

```
ggplot(diamondSample, aes(x = carat, y = price)) +
  geom_point(colour = "blue") +
  geom_smooth(method = "lm", colour = "red", size = 2) +
  scale_x_log10() +
  scale_y_log10()
```

We've already stepped through the syntax of this command, but I see that there is a line that says `geom_smooth(method = "lm", colour = "red", size = 2) +` that we didn't define before. The addition of this line of code adds a thin red trend line to the chart. Highlight these lines and execute them now. Figure 2-48 shows what you should see as a result.

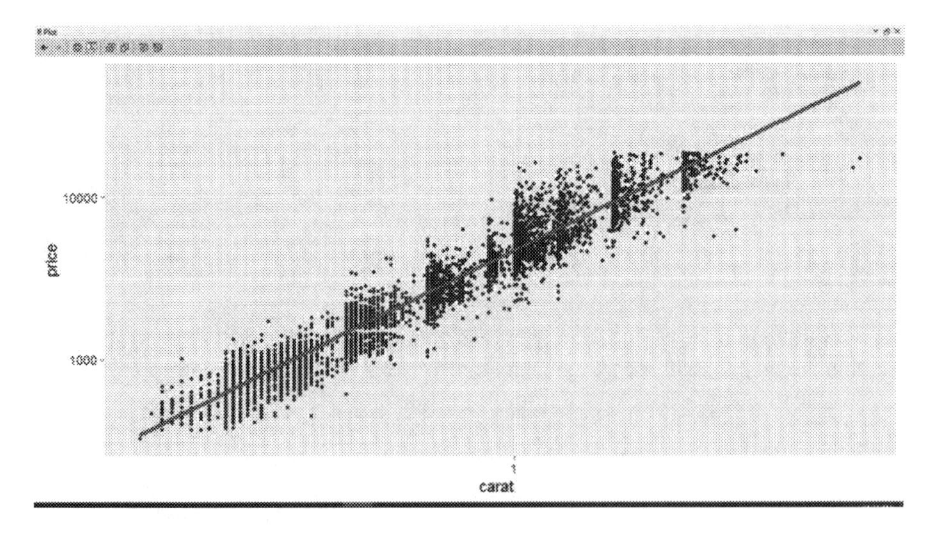

Figure 2-48. *Charted result*

That is one meaningful chart! The charts are getting progressively better and more interesting, as you can see.

Skip down to line 202 now. This line says `model <- lm(log(price) ~ log(carat) + ., data = diamondSample)` and it is closely related to the model we built before, but this time, we declared `log(price) ~ log(carat) + .`, which means that we want to model the log of the price column against all the other columns in the dataset. In other words, this is true linear regression and not simple linear regression.

Highlight line 202 and execute it. Notice that Variable Explorer now shows **List of 13** where it once showed **List of 12**; this means that we have added the model to the list and it is now available in the dataset. You can always double-click **model** to view the data contained within, if you would like.

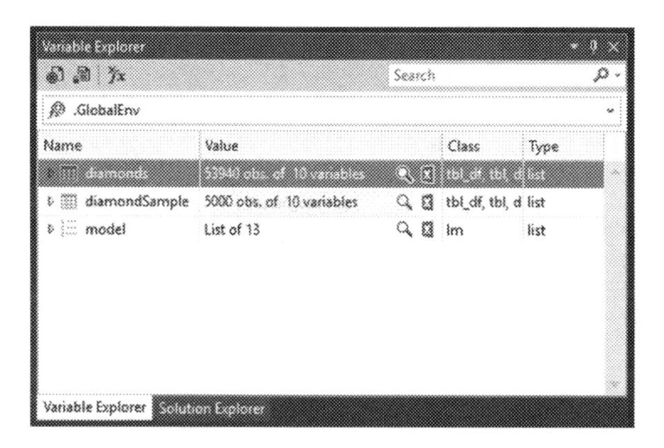

Figure 2-49. *model value increased*

Line 204 shows the summary of the model; so highlight this line and execute it. Figure 2-50 shows the result of this action.

```
Call:
lm(formula = log(price) ~ log(carat) + ., data = diamondSample)

Residuals:
     Min       1Q   Median       3Q      Max 
-0.59504 -0.08640  0.00063  0.08060  1.32134 

Coefficients:
              Estimate Std. Error t value Pr(>|t|)    
(Intercept)  7.867350   0.224753  35.004  < 2e-16 ***
log(carat)   1.737683   0.026715  65.046  < 2e-16 ***
carat       -0.067950   0.019644  -3.459 0.000547 ***
cut.L        0.115715   0.008827  13.109  < 2e-16 ***
cut.Q       -0.034995   0.007016  -4.988 6.32e-07 ***
cut.C        0.011090   0.006078   1.824 0.068141 .  
cut^4       -0.005822   0.004834  -1.205 0.228412    
color.L     -0.437917   0.006594 -66.409  < 2e-16 ***
color.Q     -0.104306   0.006029 -17.300  < 2e-16 ***
color.C     -0.008947   0.005581  -1.603 0.108988    
color^4      0.009815   0.005115   1.919 0.055041 .  
color^5     -0.003020   0.004867  -0.620 0.534972    
color^6     -0.002552   0.004471  -0.571 0.568176    
clarity.L    0.886421   0.011428  77.564  < 2e-16 ***
clarity.Q   -0.250352   0.010607 -23.602  < 2e-16 ***
clarity.C    0.121507   0.009075  13.390  < 2e-16 ***
clarity^4   -0.075253   0.007287 -10.327  < 2e-16 ***
clarity^5    0.025475   0.005997   4.248 2.19e-05 ***
clarity^6   -0.011436   0.005232  -2.186 0.028883 *  
clarity^7    0.030808   0.004602   6.694 2.40e-11 ***
depth        0.001640   0.002087   0.786 0.431986    
table       -0.001954   0.001148  -1.703 0.088674 .  
x            0.093449   0.020742   4.505 6.78e-06 ***
y           -0.003096   0.002834  -1.092 0.274785    
z            0.022161   0.017820   1.244 0.213690    
---
Signif. codes:  0 `***' 0.001 `**' 0.01 `*' 0.05 `.' 0.1 ` ' 1

Residual standard error: 0.1314 on 4975 degrees of freedom
Multiple R-squared:  0.9833,	Adjusted R-squared:  0.9832 
F-statistic: 1.217e+04 on 24 and 4975 DF,  p-value: < 2.2e-16
```

Figure 2-50. *summary(model) information*

This is significant (bad pun) because the R-squared value is 98%, so that means that 98% of the data fits closely to the regression. Not bad!

Lines 211 through 214 show that we are going to now create a data frame, which is how R structures data within its models. These lines are defined as follows:

```
predicted_values <- data.frame(
  actual = diamonds$price,
  predicted = exp(predict(model, diamonds))
)
```

So, again, let's step through this to get a feel for what the syntax is doing here.

- **predicted_values**: The object that holds the result of the command.
- **data.frame**: The way that R handles structured data, similar to a table.
- **actual = diamonds$price**: Sets a column variable named `actual` equal to the value represented by the `price` values in the `diamonds` dataset.
- **predicted = exp(predict(model, diamonds))**: Sets a column variable named `predicted` equal to the exponential value of the predicted values in the `model` linear model in the `diamonds` dataset.

Execute lines 211 through 214. You'll see that nothing happens again, which is expected. Look at your Variable Explorer again; you'll see that there is now another dataset available called `predicted_values`, which we just added. Figure 2-51 shows what you should now see.

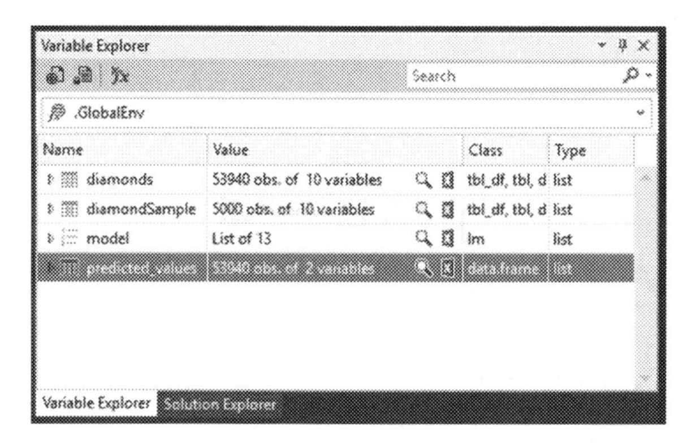

Figure 2-51. *predicted_values dataset*

Line 217 is next, which says `head(predicted_values)`. We saw this before, so recall that the `head()` command allows us to see the first six rows of data. Execute this. You should see what is shown in Figure 2-52.

```
> head(predicted_values)
+
  actual predicted
1    326  284.2928
2    326  272.5036
3    327  375.3411
4    334  418.4202
5    335  288.7658
6    336  303.4273
```

Figure 2-52. *head(predicted_values)*

Next is the big show. Lines 220 through 224 contain the ggplot command that charts all this out for us. So far, we've laid out the data.frame object with the actual and predicted columns. Now we see what that data looks like visualized. The R code for this command is defined as follows:

```
ggplot(predicted_values, aes(x = actual, y = predicted)) +
  geom_point(colour = "blue", alpha = 0.01) +
  geom_smooth(colour = "red") +
  coord_equal(ylim = c(0, 20000)) +
  ggtitle("Linear model of diamonds data")
```

I'm pretty sure that we can read that and decipher what the syntax says, but basically, we are running the ggplot command against the predicted_values dataset. We use the aesthetic value for the x axis to be the actual data and the y axis to be the predicted data. The data points are blue with a varying alpha (transparency), depending on the data value, with a red trend line, a y-limit of 20000 (which forces scale), and a title.

Figure 2-53 shows what you should see after you execute lines 220 through 224.

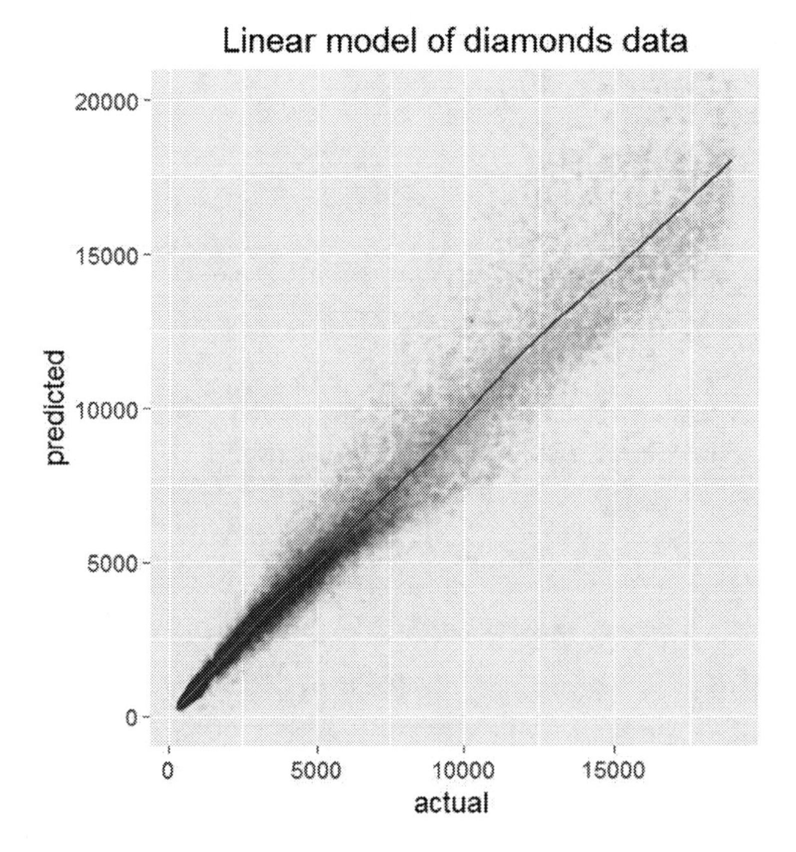

Figure 2-53. *ggplot of predicted_values*

Really, really useful chart there, don't you think? If you got this far with no problems, excellent work! You've actually learned quite a bit about the basics of R syntax now and you have seen real-world application of the most commonly used R plotting package, ggplot2.

Summary

Let's review what we've done in this chapter.

- Read the documentation for RTVS at `http://microsoft.github.io/RTVS-docs`
- Went through the samples at `http://microsoft.github.io/RTVS-docs/samples.html`
- Reviewed the R documentation at `https://cran.r-project.org/doc/manuals/r-release/R-intro.html`
- Reviewed the task views at `https://cran.r-project.org/web/views/`
- Reviewed how data import/export works in R at `https://cran.r-project.org/doc/manuals/r-release/R-data.html`
- Learned about R extensions at `https://cran.r-project.org/doc/manuals/r-release/R-exts.html`

Do I expect you to be an expert in R right now? No. Do I expect you to pore over the preceding URLs in great detail? No. What I do expect is a minimum of the following:

- Visual Studio is installed correctly.
- R Tools for Visual Studio is installed correctly.
- SQL Server 2016 is installed correctly and is communicating with the R engine.
- You have at least briefly gone over the documentation listed earlier in order to get familiar with R syntax.

The installation of these tools may be different with subsequent service packs. Either way, there should be a clear path to get to the solutions I just outlined though. At a minimum, these four things listed should be met before you move forward with this. If any of them aren't functioning 100%, go back and work out the issue. For example, the script I gave earlier in Chapter 1 showed that SQL Server was working with the R engine. Figure 2-29 showed that Visual Studio is configured for R as well. Therefore, we should all be on the same page at this point.

In the next chapter, we go over our project scenario definition. That's where we clearly define what we want the project to accomplish and the steps we're going to take to get there. After that, we begin to develop our solution and implement it using SQL Server 2016 and R Tools for Visual Studio.

CHAPTER 3

Project Scenario Definition

Before we get started with this chapter, I want to take a second and explain that this chapter deals with the program management side of development, and not the actual development side. Sometimes, dealing with the management side can be perhaps the most frustrating part of your job, but it is still necessary. For those that don't have a lot of experience in this area, I included this chapter. Hopefully, it will help you to understand how essential it is to have a plan moving forward in your next project.

The first thing that we want to do is make sure that everything is working together as it should be. Recall from Chapter 1 that R is set up correctly from SQL Server's point of view. Chapter 2 shows us that RTVS is also set up correctly. That means that the software portion of what we're doing is good to go. What's left?

This chapter focuses more on the management side of a program, as opposed to the development side. The reason for this is to prepare you for the eventuality that you will one day have to manage a program, if you haven't already. While this is in no way meant to be everything you need to know about technical program management, it does give you a pretty good primer into the world of managing expectations from a technical point of view. I've found that the best managers are those that can put on different hats as the situation demands; they can be either technical or managerial, as the situation demands.

Whenever you start a new project, it is always best to have a good idea about what you want to do and how to get where you want to go. Just as with starting a road trip, you have to be sure that you have a map and directions for how to get to your destination; otherwise, you will just drive around in circles and never really get anywhere. This is exactly the analogy to describe a poorly managed project; lots of hours are expended, lots of money is spent, but the project is never *really* finished. This is the hallmark for a poorly managed program. In my professional life, I've seen a lot of projects fail because of improper or incomplete planning, and I can tell you that it's very easy to fall victim to the biggest obstacle facing new development, and that is the element of *scope creep*.

Scope Creep

What exactly is scope creep? It's when the requirements of the project are not firmly adhered to. Plain and simple. There really is no point in going through all the time, trouble, energy, and effort of defining project requirements if they aren't going to be followed. In this case, it would be easier if development started, and then ad hoc requests came from the customer as to the intended development for that day or week. The flip side of this is to have a firm set of requirements that are agreed on by the customer and followed religiously by the development team. Hence, no misinterpretation from anyone, because everything is very clearly laid out and agreed upon from the beginning. We get into this a bit later in this chapter.

My favorite example of scope creep is to describe it using a rock. Pretend that you are on one side of a fence and your customer is on the other. Your customer shouts over the fence that they would like a painted rock. You respond that you are very good at painting rocks and that you will get it to them ASAP. So, you look around and find a decently sized rock, and then find your most popular color of paint. You are very good at this after all, so you're assuming what your customer wants. The finished rock looks fantastic, so you throw

B. Beard, *Beginning SQL Server R Services*, DOI 10.1007/978-1-4842-2298-0_3

it over the fence to your customer. They immediately throw it back, because it is not the right color. When was the color defined? Well, it wasn't. You assumed that your color was what the customer wanted; but turns out, it wasn't right at all. So instead of learning your lesson, you decide to paint the rock a different color and then toss it back over the fence. The customer immediately tosses it back, because it is again the wrong color. Now you're getting frustrated because you know this is the correct color—it has to be! Finally, you ask the customer what color they would like; they answer "blue." Not a problem, because you've got blue. So, you paint the rock blue and then toss it over the fence. Guess what. They throw it right back over, saying it isn't the right shade of blue. This goes on and on until you finally get the right shade of blue. But they throw the rock back over again! Why? This time, they're complaining that the shape of the rock isn't correct, and actually, the rock is also the wrong kind of rock all together.

So the project started out as an assumption from "paint this rock," and then into "use this particular shade of paint for this particular shape of rock on this particular kind of rock." See that? It's very frustrating, I can tell you. Unfortunately, a lot of projects are managed exactly like this. And if you're one of the unfortunate souls on the development side and not the managerial side, you get the most amount of grief because you constantly deliver things that may make perfect sense technically, but zero sense aesthetically to the customer. It becomes particularly painful when it becomes obvious that the development team is the last in the line of those taking orders to fulfill customer requirements, so ultimately, the on-time and under-budget status of a project is the responsibility of the development team. For this reason, it is important to stick to the agreed upon requirements and to not deviate.

I've found that the best way to manage any large task is to break it into smaller, more manageable tasks. This is commonly known as modularization, and is a fantastic way to handle just about anything. Splitting a task into phases is what makes sense to me in this aspect, so let's concentrate on that for a minute.

Project Definition Phases

Instead of going through all the trouble of scope creep, it's best to have a very structured idea of what is expected, when it is expected, and how it is expected. There are clearly defined phases for project development, which I discuss shortly. Trust me; it would serve you very well to adopt these phases, or something very similar, in your next project, if you aren't already using something to manage the project. The need for very specific goal posts is very important, no matter the size of your project. Please understand that this is in no way the only way to handle project management. I'm sure that there will be disagreements about the methodology and the necessary steps involved. But what I really want you to take from this if you've never had to deal with this before, is that a management plan is an absolute necessity when planning a project.

Phase I: Requirements Gathering

This phase consists of both software and hardware definitions. This is perhaps the most important phase in the entire process, because if this is wrong or incomplete, it affects all phases after it. For that reason, we need to be absolutely sure that our definitions are complete and agreed upon by all parties and stakeholders. This phase requires a set of documentation to be created by the development team, delivered by the development team to the customer, and signed and dated by the customer. At this point, the documentation set is then considered to be the official scope of the project.

What needs to take place in this requirements gathering phase? The following is a list of what needs to occur.

- Development team needs to produce a *software requirements document*.
    - A problem or issue needs to be clearly defined.
        - Get as much detail as possible from the customer, including how this problem affects their day-to-day work or how much more expedient they could work with the addition of a certain feature.

- A solution to the stated problem needs to be broadly defined.
    - This means that it is not answered in the minutia just yet. Offer many different solutions, if possible, so the customer is aware of their varying choices.
- The software language and technology needs to be determined.
    - Details are provided on the languages and technologies used to develop the interface or application.
- The medium needs to be determined.
    - The medium in which it will be delivered; usually, either web-based or on the desktop.
- A structured *software design document* needs to be produced by the development team.
    - This document provides details on the interface presented to the customer. It needs to be pointed out to the customer that this interface will more than likely change slightly over time, depending on current web or desktop trends and options available in the language you're working in. The large majority of this document will consist of mock screen shots and report scenarios.
- A structured *database requirements document* needs to be produced by the development team.
    - This document provides details about the database tables and scripts that will be created to present the interface, and ultimately, the entire solution to the customer. Notice that this document encompasses the design and the requirements of the database. The reason for this is because the customer, for the most part, is not bothered with the mundane details of the database such as data type or length of the field. While this can certainly be presented to the customer in a different document, it is probably best to keep this information at more of an arm's length away, so as not to overload the customer with meaningless information (from their point of view). What is the overall purpose of this document? Much more important than you may realize at first. This document can be used to provide a mapping of the database for future interaction with different, remote databases, for example. For this reason, it is imperative that this be the last document created, in case something changes in the database and the documentation does not reflect this change.

This list represents what I consider the minimum amount of documentation to deliver as part of a formal request. There could be more, but this is a good start.

Once this phase is complete and a documentation set has been produced, a formal acceptance and signature is required. It needs to be clearly related to the customer that there are no changes of any kind after the documents are signed and development begins. If there is a change that needs to be made, you have a choice to make. You can either begin a spiral development process, or implement a *software change request system*. I highly recommend a software change request system. What are the differences?

Spiral Development Process

Spiral development is a particularly heinous form of development that occurs when scope creep is actually allowed and encouraged on a project. The horrors! Can you see the implications of allowing too much scope creep? That's correct; *the project is never truly done*. The project constantly stays in the definition and development phases, and never progresses to completion.

For those of us who get paid by the project, it is entirely unacceptable to encourage scope creep. Think about it like this: if you estimate a project will take 10 hours and you charge $500 for the project, you're making $50 an hour. There is no guarantee that you will be done in 10 hours, though. Let's say it takes you 20 hours; you're now at $25 an hour. Even worse, it takes you 100 hours. Wow! You're making $5 an hour.

Obviously, this is not how you want to work. Instead, you should just deliver what was agreed upon, and any changes are considered "over and above" what was estimated and paid for, and subject to a new agreement. I can tell you from experience that it keeps everyone happy when there is a very clear definition of what is expected to be done.

Another danger in the spiral development process is like I said: the project is never truly done. It just continues on and on because the customer is getting more than what they paid for, and the developer is oftentimes too timid to bring up that something is, in fact, outside the agreed-upon scope of the project. Does this work for you? Would you like a reputation as someone who takes the time to go through all the motions of correct management techniques, and then cave to the customers' unsubstantiated request for change in the middle of a project? I don't think you would. I think we all would like the reputation of the competent developer that has the integrity to call the customer out on possible or overt deviations or expectation of deviation from the contract. In this case, there needs to be a clearly defined remediation clause in a contract that deals specifically with the steps necessary to enforce the rigidity of the original contract and the scope therein.

Software Change Request Process

Formal change requests are what I recommend when making changes to a project and its scope. The reason is very simple: the project stays structured! The initial project is completed and is labeled as v1.0, right? Any change you make after this would be v1.1 for a small change, or maybe 2.0 for a major change. The *software change request process* has different steps that need to be followed in a workflow to facilitate a complete change request. It isn't a customer sending an e-mail; it is a full process that has sign-offs at each step of the workflow.

Request Submission

In this step, any user can request a change to the software. Requests need to be complete and organized though, and not just vague statements like "I would like a rock painted blue." Request submission is generally a fairly complex step, with the user being forced to provide a large amount of information pertaining to the request. Be sure to gather not just the customer information, but the specific task needing to be done.

Administrator Approval

The administrator then receives a notification that a change has been requested. At this point, the admin can either accept or reject the change. There is also the case that the administrator could contact the requestor for more information, then reject the request so that the originator can expound on the original request. This is very common, since nine times out of ten, regular users don't understand the complexity requirements needed to properly fulfill their request.

Design

Once you have an accepted request, the first step is to design the solution. Sometimes, literally drawing out the solution is best. It gives you a good idea of what to expect for layout and the space necessary to fulfill the solution. The customer expects to see something like screen mockups in Photoshop or something similar, more than likely in a PDF. This isn't always a requirement, but if the user wants to be kept aware of the progress of the request, then it is a good idea. Either way, whether the customer is involved or not, it's always worth taking a few minutes to sketch out a quick picture and perhaps a short pseudo code example of what is going on.

Code

Next is the fun part: writing the code. This could be in whatever language is necessary to get the job done, and whatever is used to deliver solutions in your application. In some cases, you have multiple languages. In web development, this is very common. I often use a pretty steady mixture of HTML, ColdFusion, CSS, and jQuery in my day-to-day application building. Some days are heavier in certain languages than others, but that's just about the gist of what I do. This is the second-longest activity in the process though; the longest is ... the documentation.

Document

Easily the most overlooked part of development is the documentation of the changes made. Make sure that this step is followed religiously! This is different from the inline documentation that all good developers do regularly. Those are fantastic too, don't get me wrong. This documentation, specifically, is for the user. It consists of two documents, but they really only see one of them. One of the documents needs to define the request and the steps taken to implement the request. The reason for this is so that the change can be essentially rewound, if necessary. The second document is the test script for the user to test the change. This needs to be fairly simple and straightforward; something along the lines of "Click the OK button." It needs to be spelled out exactly what is expected to happen when an action occurs, and what the failure condition is as well.

Unit Test

This means testing the solution by itself, in a modular setting. Does it do what is required by itself, or is interaction with another unit or module necessary? It should be as tightly contained as possible, in other words. If it requires other things to run, you maybe want to reconsider this particular technique. Is it possible that it can be rewritten to be cleaner and more modular using native code? These are questions that must be answered objectively in this phase.

Regression Test

This is where you test the solution as part of an over-arching solution that encompasses the entire application. In other words, how does this solution work with the current application? Does it "mesh" or is it inconsistent and in need of further work? Most times, it seamlessly integrates, but other times—not so much. For this reason, it is best to look at the change as part of the application as a whole, instead of as the modular way in the unit test. Consider the success and failure paths of the change (if there are separate paths). Do they make sense, considering the paths of the rest of the system? How well does it do what it is billed to do?

Acceptance Test

This is after the solution has passed all the other tests and documentation is complete. This is the only step that the requesting user should see. Everything else should be a black box to that user, lest you fall into the spiral development trap. Don't do it! This step is very simply a single line stating that the customer agrees to the change and it was delivered as expected. A signature and the current date is all that is needed to complete this step.

Installation

After the solution has been accepted by the customer, it needs to be installed. This is sort of obvious, but still a necessary part of the process. This needs to be documented as well, although much more leniently. Make sure you take a backup of the database immediately before the installation. I recommend a server image if you're running a virtual instance, or a system image with backup if you're on a regular server. The reason for this should be clear; to recover quickly in case everything goes pear-shaped. Once everything loads correctly, delete any backups to retain space on your system.

Archiving

Finally, after installation, the solution needs to be archived. This entails two things: first, the solution should be zipped with all documentation and added to a source repository of some kind. I'll leave it to you to decide which to use. Second, the physical record of the request on the web site or bulletin board should be removed and added to an archived or completed section for later review, if necessary.

Please note that the preceding workflow is not all encompassing, but rather is intended to get you started with understanding exactly what a software change request workflow looks like.

So now you understand what scope creep and spiral development mean and how they are both detrimental to a project. You also understand the software change process and how it is beneficial to just about any system. What's next? Let's continue to define the separate phases of the project definition, and then discuss some more factors contributing to the success and sometimes to the failure of the project.

Phase II: Initial Interface Design

This phase allows the developer to create an initial user interface, if applicable. Sometimes, projects don't have an interface, because they happen "under the covers." If that's the case with your project, just skip to the next phase. If that's not the case, and you have to create an interface, then this is the place for you.

Initial interface design is fairly self-explanatory: you want to create an initial interface based on the requirements documents and by using any other information gathered from the customer. Note that *this does **not** involve going back to the customer for ideas or suggestions.* Basically, they had their chance to suggest implementation changes before, and it is simply too late to implement them now, without falling into the dreaded *spiral development cycle*. Remember how you got the customer to sign off on the scope of the project? Just like they aren't allowed to come to you, you technically shouldn't go back to them. If you discover an authentic issue with the way that the interface needs to be implemented, or an issue that is preventing you from completing the job, then you obviously want to interact with the customer. The point to this is "training the customer" to understand that you work strictly from the requirements document and nothing more. That way, when you deliver what was requested, you know that it is technically and functionally 100% correct, according to the requirements document.

We have already gathered our requirements, which I lay out in the example download for this book. The download contains our requirements documents and initial dataset. Make sure that you have these following two documents, or you will have a really hard time following along; they are available from this book's catalog page on the publisher's web site:

- `Software Requirements Document.docx`
- `Weather_Sample.csv`

For reference, the `Weather_Sample.csv` file is the same document provided in the download available from Microsoft that we downloaded earlier. We use the same example in this book by permission.

With that in mind, let's take a look at the reports that we have agreed to provide and plan out a solution.

Loading the R Solution

Recall that back in Chapter 2, we downloaded a `.zip` file from `https://microsoft.github.io/RTVS-docs/samples.html`. It contains everything that we need for what we will ultimately build. Extract that `.zip` file somewhere easy for you to access, because we're going to load the solution in it now.

Start Visual Studio and go to **File** ä **Open** ä **Project/Solution...**. Navigate to where you saved and unzipped the `.zip` file. Follow the folders down to `RTVS-docs-master/examples/Examples.sln`. Select it and click **Open**, as shown in Figure 3-1.

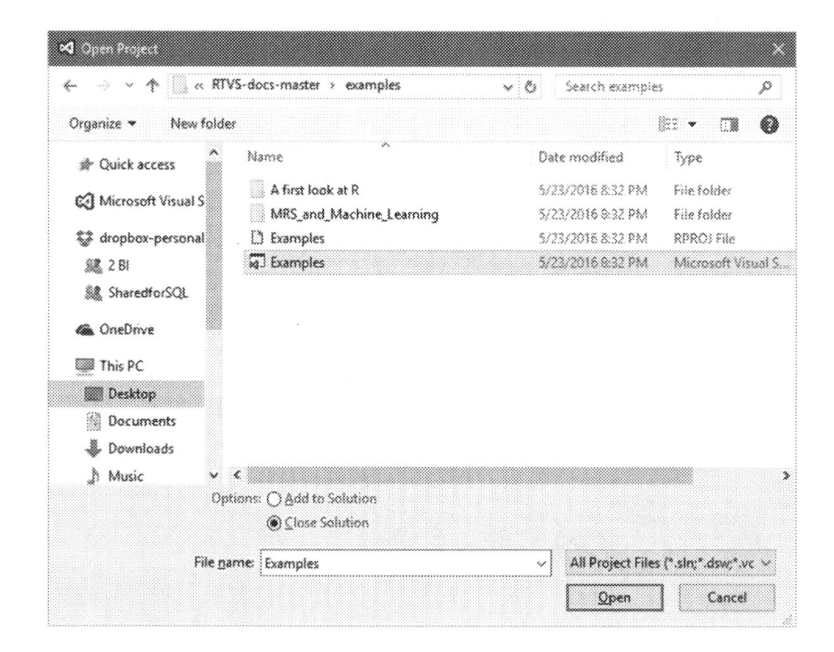

Figure 3-1. *Open Project*

It takes a second to open the project, and then loads the solution file into Visual Studio.

Notice in your Solution Explorer window (if it's not up, press **Ctrl+Alt+L** to make it appear) that there is a set of files, as shown in Figure 3-2.

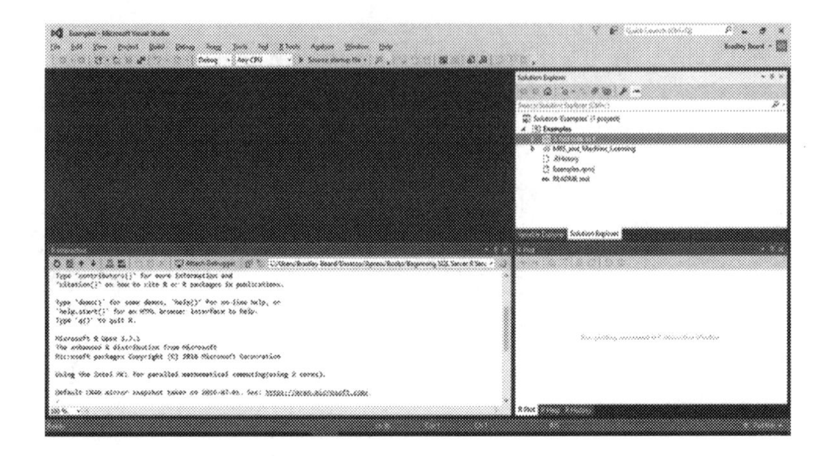

Figure 3-2. *Solution Explorer files*

Double-click the bottom **README** file; the one not inside either of the folders. Figure 3-3 should be what you see at this point.

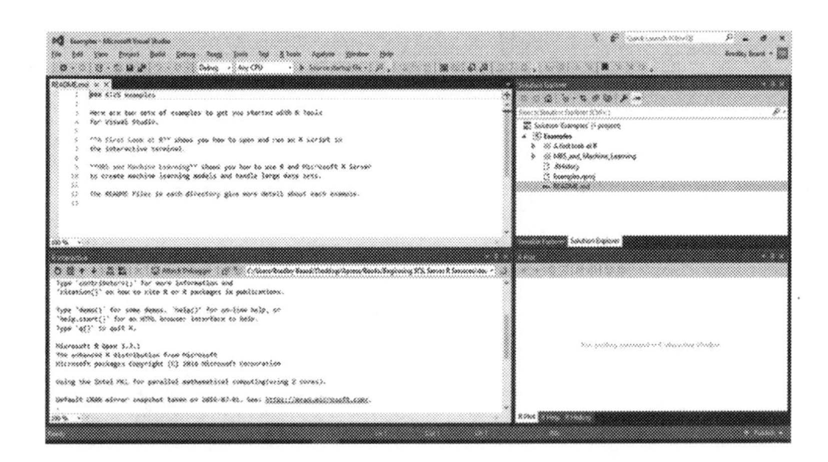

Figure 3-3. *README file*

Does that look familiar? We saw this in Chapter 2, Figure 2-30, before we started stepping through the R examples. Note that there are also more README files inside of the directories for even more guidance from the R gurus at Microsoft.

Expand the **A first look at R** folder in Solution Explorer and then click the **Getting_Started_with_R.R** file in the right pane. You should see something similar to what is shown in Figure 3-4.

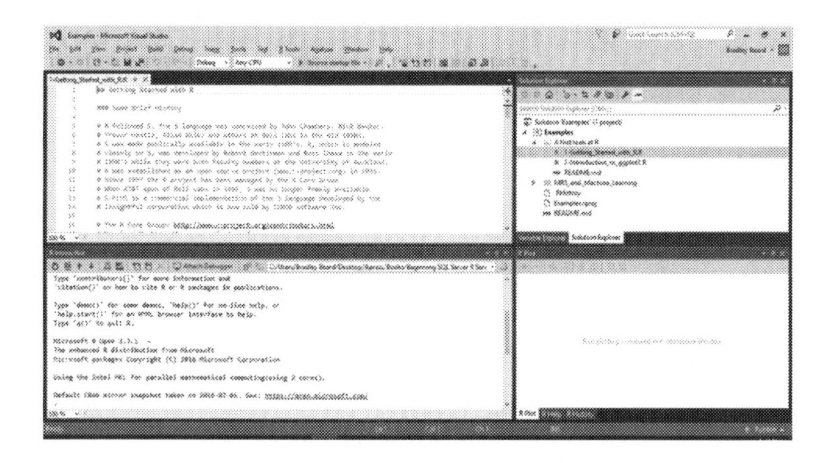

Figure 3-4. *Getting Started with R initial screen*

Just to be clear, the top-left pane is the opened file and the bottom-left pane is the R Interactive window (basically, the results of the R operations). At the top right corner of the screen are the Solution Explorer and Variable Explorer panes, and at the bottom right are the R Plot, R Help, and R History panes. You can rearrange your windows however you like so that it is comfortable for you, as long as you have them readily available in case you need them later.

I want to go over a couple of things before we end this chapter. First, let me pose two questions for you. Which version of R is running in the database? Which version of R is running in Visual Studio? I had to stop and think about this for a while, and I finally reached the conclusion that I could not tell from SQL Server what version of R is installed. I could easily tell what version of R was running from Visual Studio though, because it was prominently displayed in the R startup messages from within RTVS, as shown in Figure 3-5.

```
R version 3.3.1 (2016-06-21) -- "Bug in Your Hair"
Copyright (C) 2016 The R Foundation for Statistical Computing
Platform: x86_64-w64-mingw32/x64 (64-bit)

R is free software and comes with ABSOLUTELY NO WARRANTY.
You are welcome to redistribute it under certain conditions.
Type 'license()' or 'licence()' for distribution details.

R is a collaborative project with many contributors.
Type 'contributors()' for more information and
'citation()' on how to cite R or R packages in publications.

Type 'demo()' for some demos, 'help()' for on-line help, or
'help.start()' for an HTML browser interface to help.
Type 'q()' to quit R.
```

Figure 3-5. *R version in RTVS*

How interesting! What if I were running two different versions of R? Certainly that would cause problems, particularly if I write code in RTVS and then want the same result from SSMS. Wouldn't this give me different results? Logic says that it more than likely would. So, obviously, we want to mitigate this as much as possible.

Recall the stored procedure that we use to execute R in SQL Server? Here is that procedure once again:

```
exec sp_execute_external_script
@language =N'R',
@script=N'OutputDataSet<-InputDataSet',
@input_data_1 =N'select 1 as hello'
with result sets (([hello] int not null));
go
```

Let's rewrite this procedure to get the version of R that is executing this code. Our rewrite is a simple little script that you may want to keep in your bag of tricks, just in case it is needed later on. You may want to switch to **Results** to **Text** by pressing **Ctrl+T** while in the New Query Window. Here is the script:

```
exec sp_execute_external_script
@language =N'R',
@script=N'OutputDataSet<-InputDataSet;
message (R.Version()$version.string);'
with result sets (([Version] varchar));
go
```

I'm sure we're all familiar with generic scripting practices, so we can see that I am using R's `message()` function to return the version of R by invoking `R.Version()$version.string`.

The result of this operation is shown in Listing 3-1.

Listing 3-1. R Version in SQL Server 2016

```
Version
-------

(0 row(s) affected)

STDERR message(s) from external script:
R version 3.2.2 (2015-08-14)
```

We have just verified that Visual Studio is running R version 3.3.1 and SQL Server is running R version 3.2.2. The reason for this is because SQL Server R Services is the server version, and RTVS is the client version. What we want to do is point RTVS to R version 3.2.2, that way we get the same results from RTVS to SQL Server. To do this, open Visual Studio, navigate to the **R Tools** menu, and then click **Options**. Figure 3-6 shows this menu location.

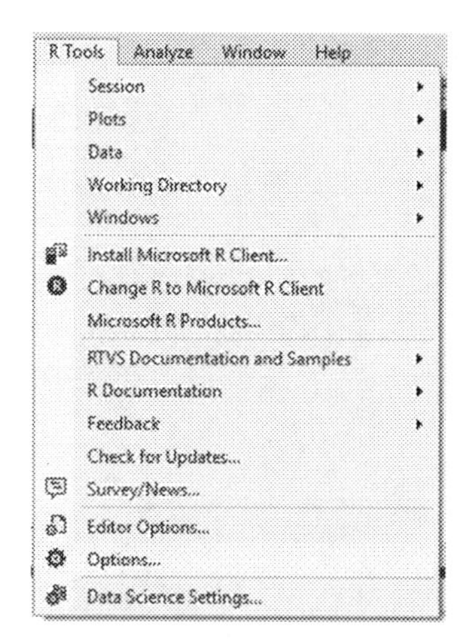

Figure 3-6. *Options menu*

A screen opens, as shown in Figure 3-7.

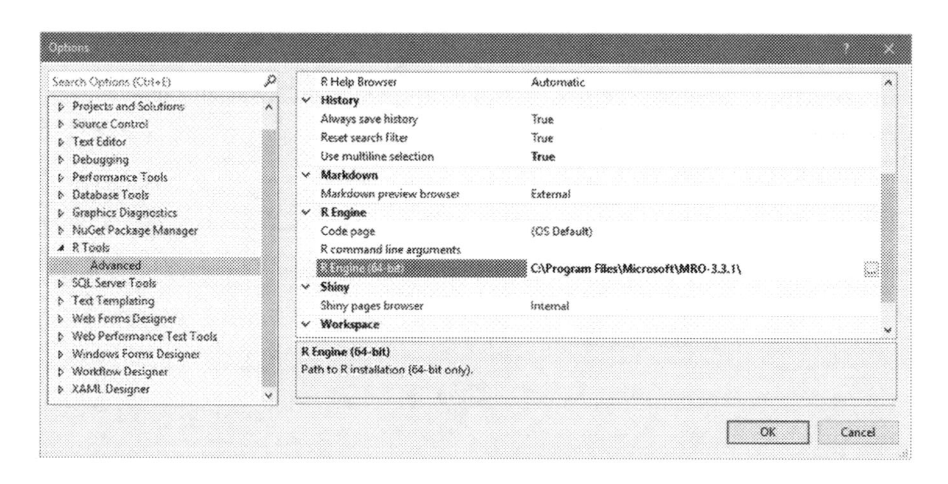

Figure 3-7. *R Tools Advanced options*

In the screen shown in Figure 3-7, the `R Engine (64-bit)` option is selected, and the value of this selection is `C:\Program Files\Microsoft\MRO-3.3.1\`. This needs to be changed to `C:\Program Files\Microsoft SQL Server\MSSQL13.SQL2016RS\R_SERVICES`, since that is the location where SQL Server is referencing the in-database R library. Note that your folder location may be different from mine, so find your `R_SERVICES` folder location and use that instead. Once you replace this folder location, click **OK**. You are prompted to restart Visual Studio, so go ahead and close down Visual Studio and open it again. Once it opens again, notice that the R Interactive window holds some pretty important information for us. Listing 3-2 shows this information.

Listing 3-2. Updated R Information

```
R version 3.2.2 (2015-08-14) -- "Fire Safety"
Copyright (C) 2015 The R Foundation for Statistical Computing
Platform: x86_64-w64-mingw32/x64 (64-bit)

R is free software and comes with ABSOLUTELY NO WARRANTY.
You are welcome to redistribute it under certain conditions.
Type 'license()' or 'licence()' for distribution details.

R is a collaborative project with many contributors.
Type 'contributors()' for more information and
'citation()' on how to cite R or R packages in publications.

Type 'demo()' for some demos, 'help()' for on-line help, or
'help.start()' for an HTML browser interface to help.
Type 'q()' to quit R.

Microsoft R Server version 8.0 (64-bit):
Microsoft packages Copyright (C) 2016 Microsoft Corporation

Type 'readme()' for release notes.
```

This proves that we have now successfully synced our versions of SQL Server and Visual Studio. The next step is to update the packages, which we can easily do via R Package Manager. Recall from Chapter 2 that we need to install two packages (and their dependents): `ggplot2` and `data.table`.

Open the R Package Manager in Visual Studio by going to **R Tools** ä **Windows** ä **Packages**. Figure 3-8 shows the R Package Manager once opened.

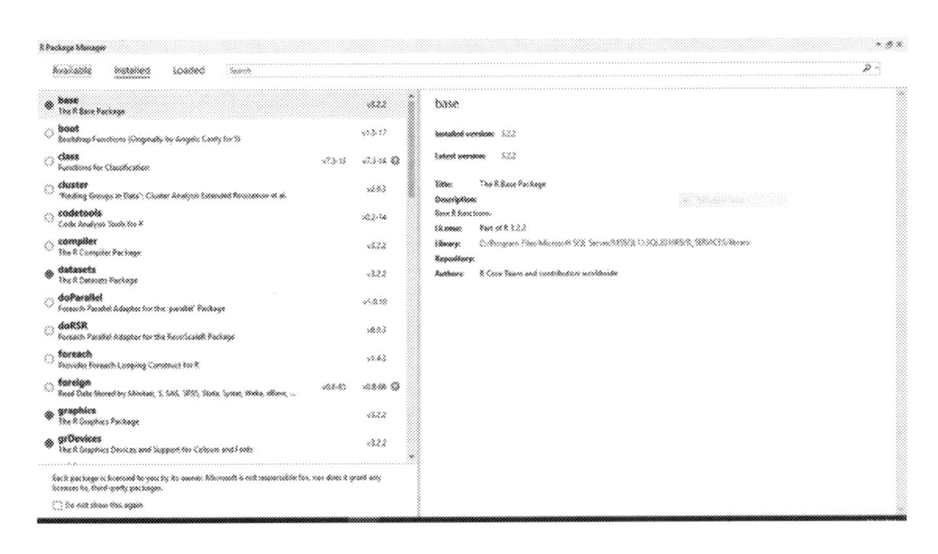

Figure 3-8. *R Package Manager*

In the top pane, click the **Available** line, to show what is seen in Figure 3-9.

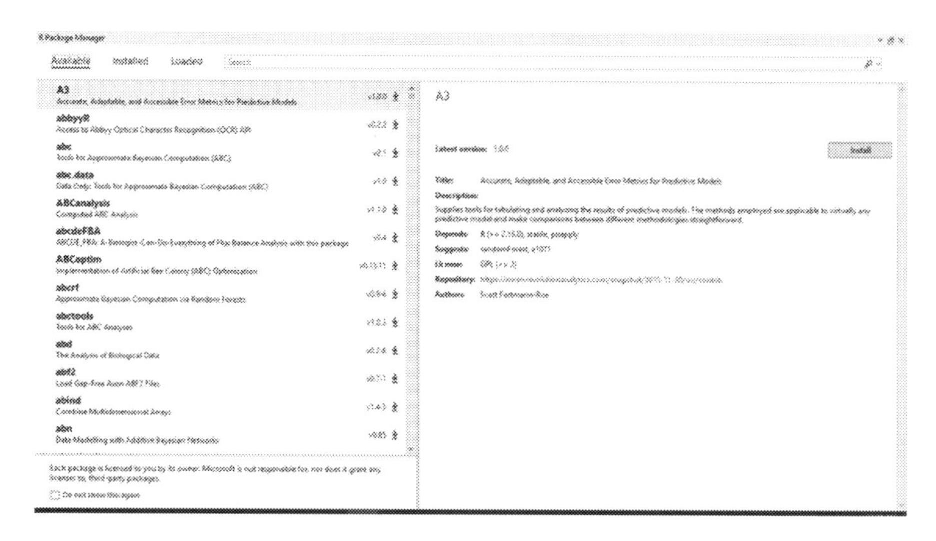

Figure 3-9. *Available R packages*

All we need to do is search for ggplot2; so type it in the Search box at the top of the window. The ggplot2 package appears in the left pane. Figure 3-10 shows this result.

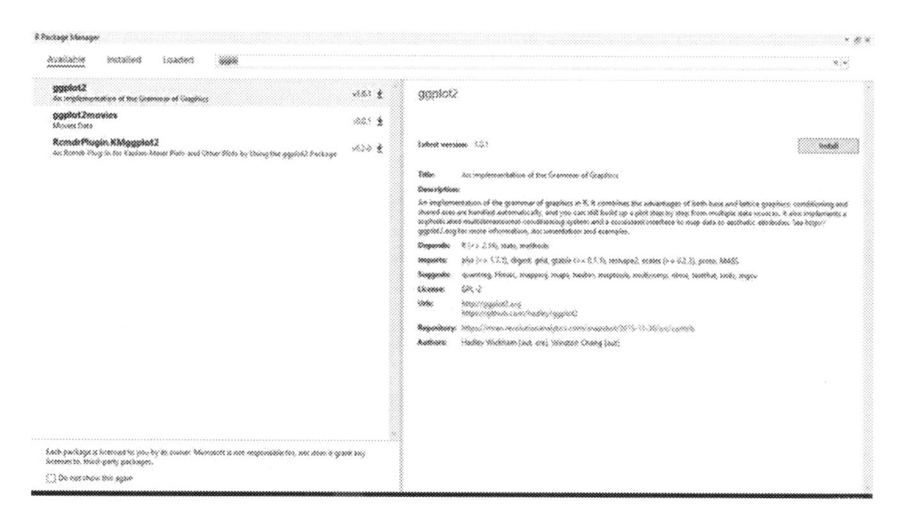

Figure 3-10. *Finding ggplot2*

Next, we just need to click the **Install** button in the ggplot2 pane on the right to install the package. The installation is done in the background here, but you can always check the R Interactive window to see the current status of the installation.

Once the package has finished installing, you see what is shown in Figure 3-11.

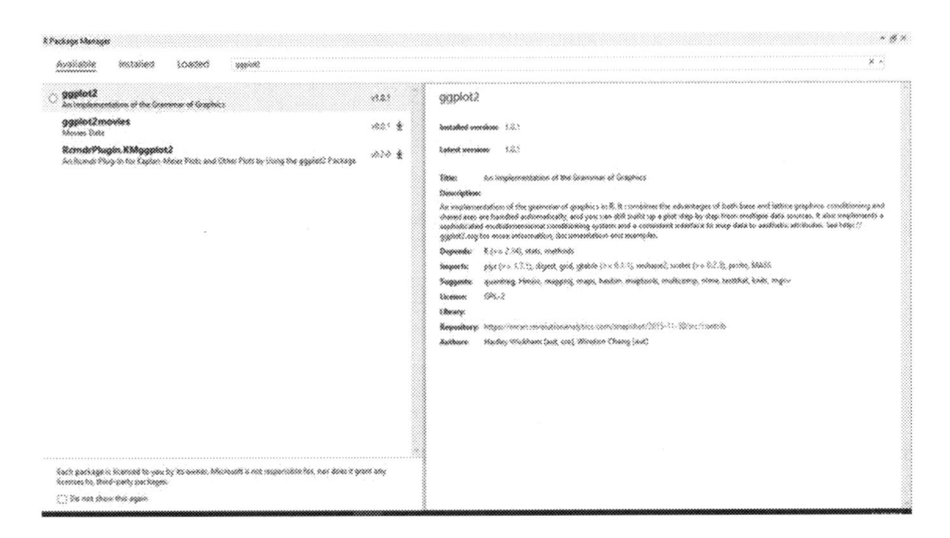

Figure 3-11. *ggplot2 installed*

Notice that there is no `Install` button any longer, and the left pane now shows a circle next to `ggplot2`. This indicates that the package was installed normally.

Next, we need to repeat these same instructions for the `data.table` package. I will leave this as an exercise for you, but you can replicate the instructions I just gave as a repeatable process to install any package you wish.

Summary

Excellent work so far, if you have kept up and gotten everything going as expected. If you haven't, again, please go back and re-run the installations so that you are seeing what I'm showing in the examples. If you aren't, I'm afraid that you really aren't getting the full value of this book. It is going to get a lot more complicated in the next chapters, so you want to be prepared for it. In your spare time, make sure that you are brushing up on R scripting and functions, including the fantastic `ggplot2` package. Big spoiler: We're going to use it a lot later on.

Review time! Let's quickly go over what we did in this chapter.

- Learned about scope creep and the project definition phases
- Learned the importance of a software change process in a completed application
- Became familiar with the initial interface for R Tools for Visual Studio

Next, in Chapter 4, we actually start figuring out what we need and we put it together in a useable report.

PART II

Learning the Basics

CHAPTER 4

Building R Models with RTVS

We left off in Chapter 3 with loading up a solution in RTVS and then testing that we could work successfully within the R environment. In this chapter, we actually build the models in R and then display the information contained within that model in a series of charts. The charts become progressively more advanced until we reach the final culmination of the data visualization.

We also cover most of the syntax for the code that generates the models. We do this by essentially stepping line by line into a file provided by Microsoft in a `.zip` file that we download. This way, we can see what is happening as it happens and hopefully reach a greater understand of not only *what* is happening, but *why* it is happening.

Exploring Samples

Let's go back to the Samples link in the **RTVS Documentation and Samples** menu item in R Tools for Visual Studio (RTVS). That link is `http://microsoft.github.io/RTVS-docs/samples.html`. On that page, there is a download of a `.zip` file containing examples that we will use to get familiar with the new R environment. Unzip that `.zip` file to a location that you can access, navigate to `RTVS-docs-master/examples`, and double-click **README.MD**. This opens the readme doc in RTVS. Figure 4-1 shows this document once opened.

```
1   ### RTVS examples
2
3   Here are two sets of examples to get you started with R Tools
4   for Visual Studio.
5
6   **A First Look at R** shows you how to open and run an R script in
7   the interactive terminal.
8
9   **MRS and Machine Learning** shows you how to use R and Microsoft R Server
10  to create machine learning models and handle large data sets.
11
12  The README files in each directory give more detail about each example.
```

Figure 4-1. *Readme file*

B. Beard, *Beginning SQL Server R Services*, DOI 10.1007/978-1-4842-2298-0_4

We need to get a little bit familiar with R as a language before we attempt to get into any sort of development activity, so let's step through some of the examples given in *A First Look at R*. In the tutorial featured later in this book, we will deal with the R Server aspect a lot more, since we will directly interface with SQL Server R Services to create a report with embedded information.

There is an awful lot of information about R on the internet, so if you already know about it, then you can consider this a refresher course. If not, no worries. I'm not going to get into the complete history of R and I'm not going to make this a comprehensive guide to all of R's functionality. Instead, I highlight the basics—and we can go from there. I think that will be enough to whet the proverbial whistle and get our minds keen on the practicality of using R for serious data analysis.

Navigate to `RTVS-docs-master\examples\A first look at R` and double-click the **README.MD** file in that directory. Figure 4-2 shows what you should see at this point.

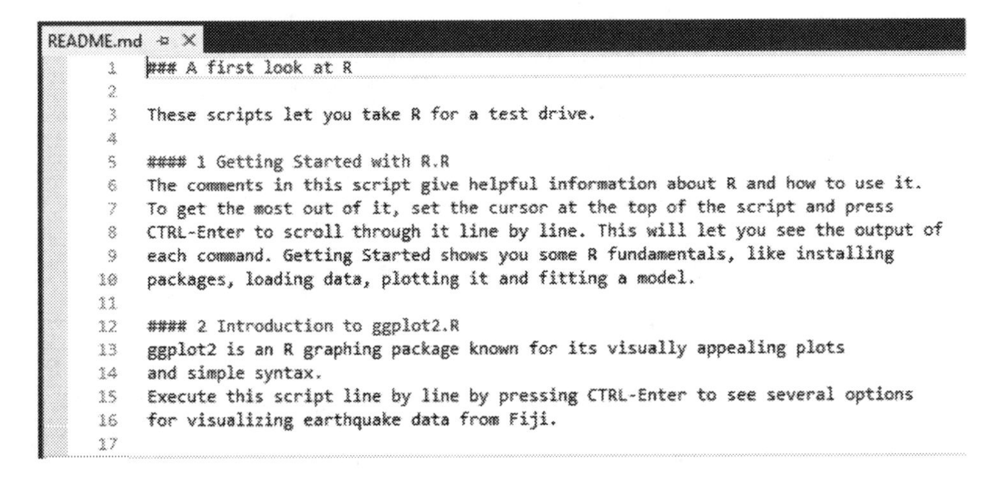

Figure 4-2. *A First Look at R README*

That document tells us that we get to "take R for a test drive" by running R scripts provided by Microsoft in the `.zip` file we downloaded earlier. Navigate back to the `RTVS-docs-master\examples\A first look at R` directory and double-click **1-Getting_Started_with_R.R.** The script is then opened in RTVS, as shown in Figure 4-3.

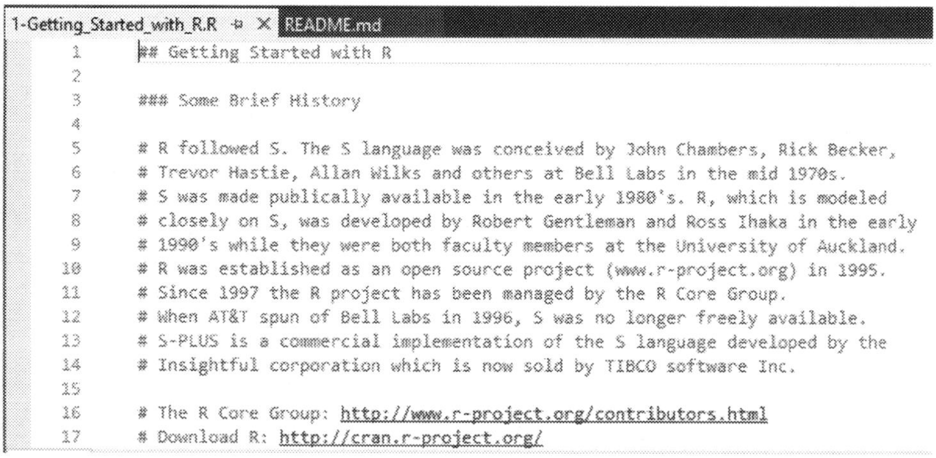

```
1-Getting_Started_with_R.R  X  README.md
1   ## Getting Started with R
2
3   ### Some Brief History
4
5   # R followed S. The S language was conceived by John Chambers, Rick Becker,
6   # Trevor Hastie, Allan Wilks and others at Bell Labs in the mid 1970s.
7   # S was made publically available in the early 1980's. R, which is modeled
8   # closely on S, was developed by Robert Gentleman and Ross Ihaka in the early
9   # 1990's while they were both faculty members at the University of Auckland.
10  # R was established as an open source project (www.r-project.org) in 1995.
11  # Since 1997 the R project has been managed by the R Core Group.
12  # When AT&T spun of Bell Labs in 1996, S was no longer freely available.
13  # S-PLUS is a commercial implementation of the S language developed by the
14  # Insightful corporation which is now sold by TIBCO software Inc.
15
16  # The R Core Group: http://www.r-project.org/contributors.html
17  # Download R: http://cran.r-project.org/
```

Figure 4-3. *Getting Started with R script*

At this point, all that we are going to do is step into the R script and execute some portions in order to get an idea of how R is laid out syntactically and how it might compare to other languages. We will go pretty much line by line through the *Getting Started with R* script so we can really understand what this introduction is getting across.

It is worthwhile to read through the first 75 lines of comments, as this sets up your basis as a new R user, or refreshes your memory if you're a legacy R user. Either way, there is something for everyone here, so be sure you read it thoroughly, particularly the R Resources and R Blogs sections. The Help section is always good, so don't skip over that one either.

Line 76 is the first executable R script. That line is very simply `installed.packages()`. This simple line lets us see what packages are already installed; so highlight line 76 and press **Ctrl+Enter** to execute it. Note that your R Interactive window (which should still be up) starts loading up a lot of information, as shown in Figure 4-4.

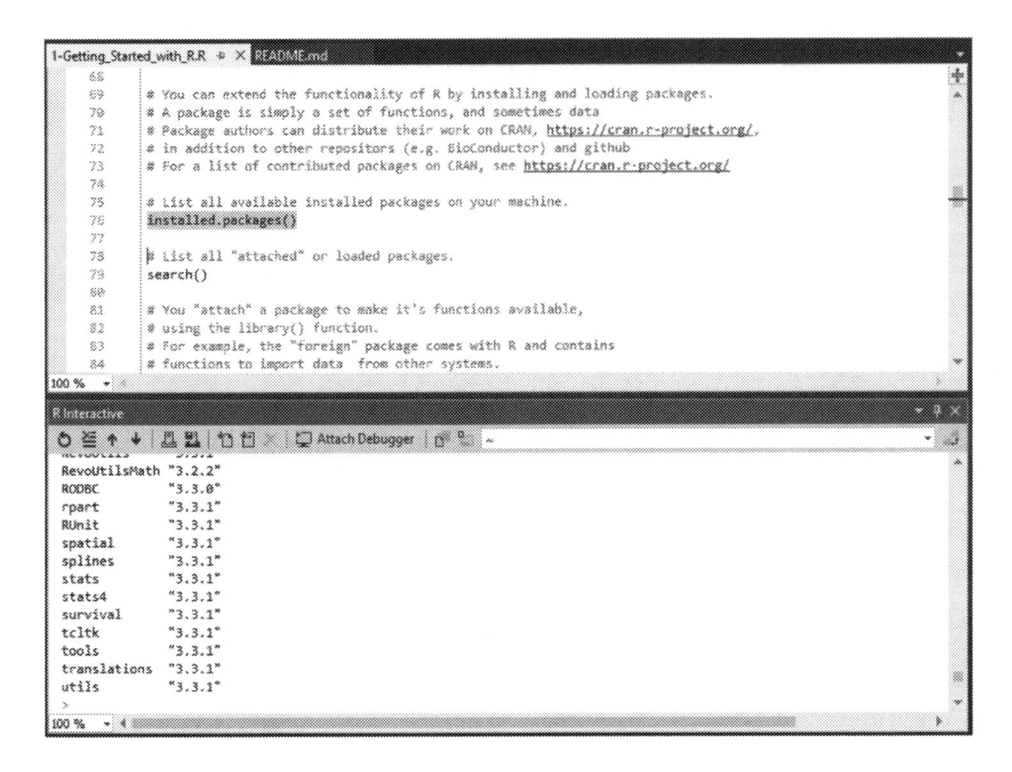

***Figure 4-4.** R Interactive window*

You can scroll up in the R Interactive window to see exactly what happened, but it's strictly informational, at this point. It can be useful to peruse this generated content to ensure that you have the latest version of an installed package, for example, but for general use, it's good to know that it's there.

Note There is now a value in the R History window as well. This is very handy in case we ever need to re-execute a line of code. All you need to do is highlight the code you want to re-execute and press **Enter**. This moves the code from the R History window to the R Interactive window. From here, press **Ctrl+Enter** to execute the line of code as normal.

Highlight line 79, which reads `search()` and execute it. This gives us a listing of the currently loaded packages for this R session. Next, we attach a package using the `library()` function, which is how R makes functionality particular to a specific package available to the session.

Skip down to line 85, which reads `library(foreign)`. This means that we will include the `foreign` library functionality in the current session. Highlight line 85 and press **Ctrl+Enter** to execute it. As soon as you see the caret at the bottom of the R Interactive window turn back to the greater than sign (>), then you know that the code has completed executing. Figure 4-5 shows what you should see at this point.

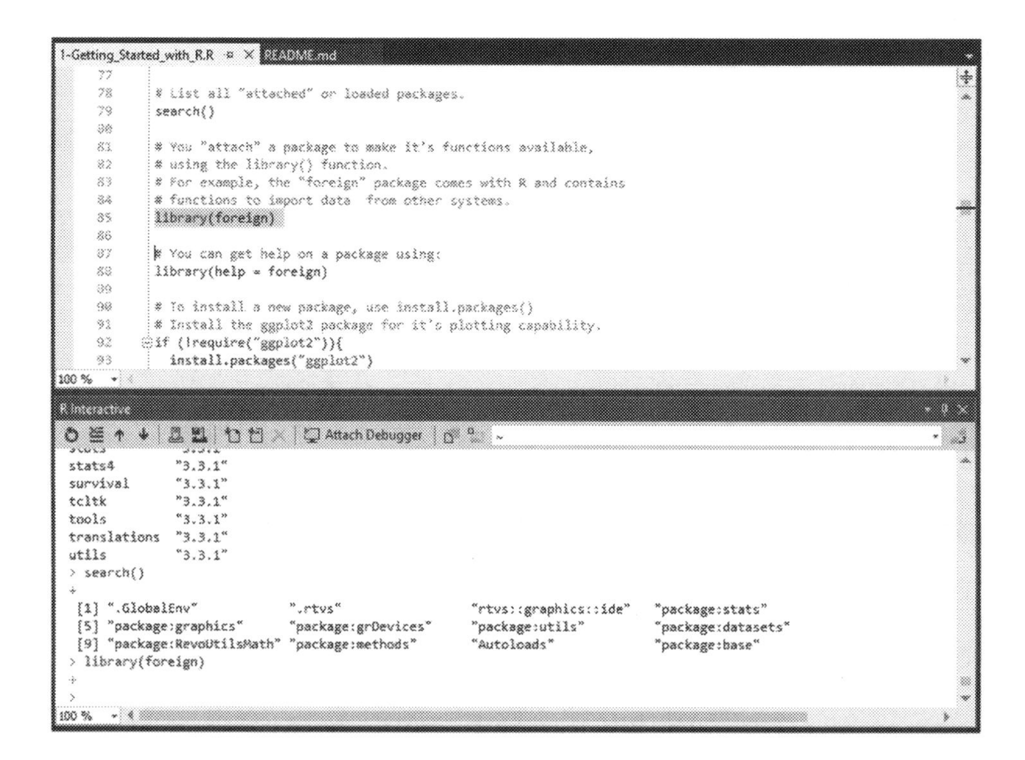

Figure 4-5. *Line 85 execution*

This isn't in the R script we are working with, but if you go back now and execute line 79 again, you should see what is shown in Figure 4-6.

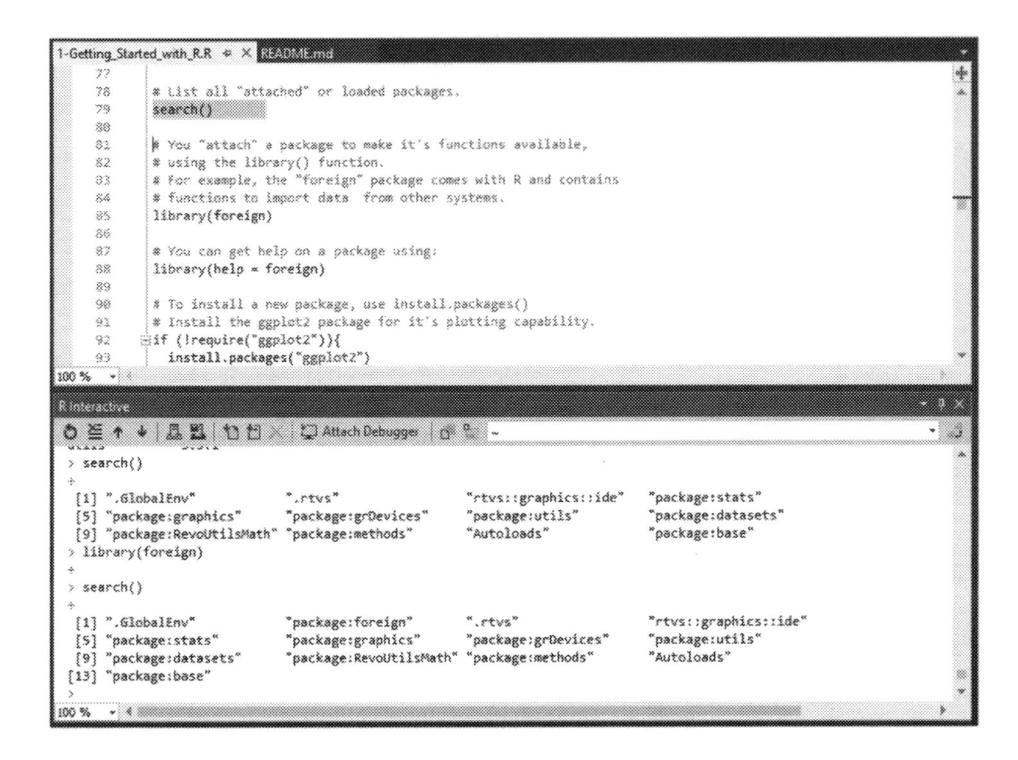

Figure 4-6. *Line 79 execution showing foreign library*

Note how R Interactive shows these listed items? It is done in groups of four, with the starting n-index number being shown as the leftmost column, followed by four packages. The next line begins with the n+4 index, and then lists another four packages, and so on. Now see how the first execution of `search()` showed 12 packages, but we can now see 13 packages returned in the newly returned `search()` command. We can see the addition of the `foreign` package as the reason for this, so that is our proof that the package was successfully added to our current R session.

When packages are added to the R mirrors, they always include a Help section. You can reference this Help section by highlighting line 88 and pressing **Ctrl+Enter**. This opens the help documentation as another page within the top frame in RTVS. Figure 4-7 shows this result.

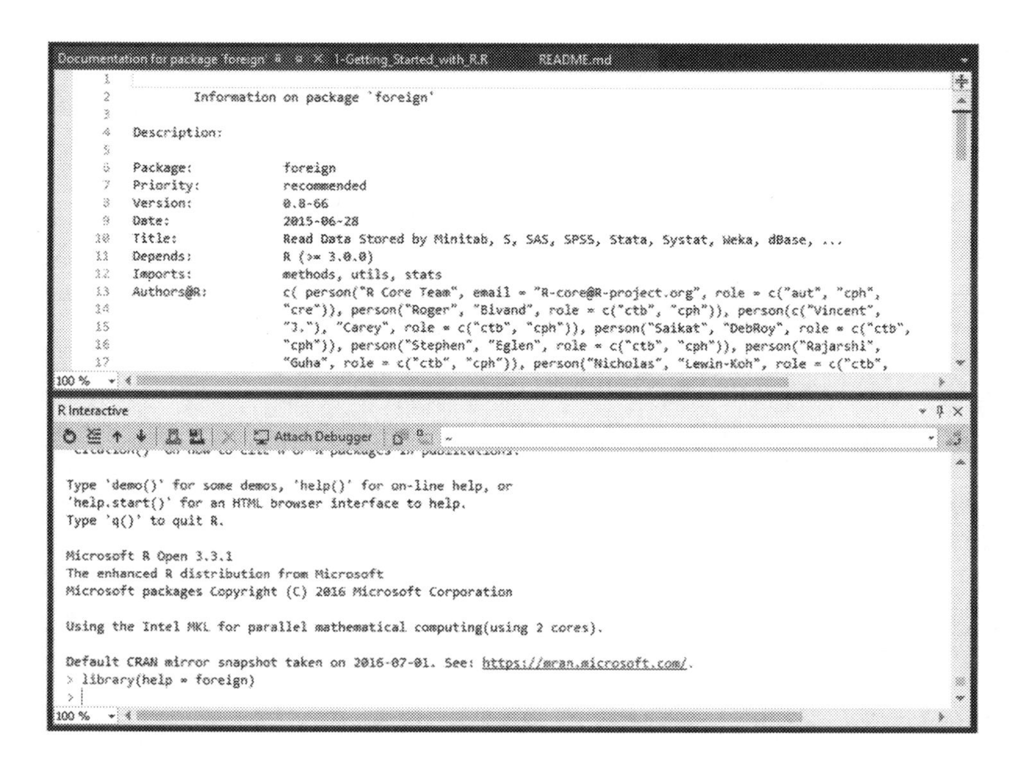

Figure 4-7. *Help documentation for foreign library*

You can close that documentation. Next, skip down to line 90. We are going to install `ggplot2`, which is probably the most popular and robust charting package available for R. Highlight line 92 through 94 and press **Ctrl+Enter** to execute. Figure 4-8 shows the result of this.

```
1-Getting_Started_with_R.R  README.md
83  # For example, the "foreign" package comes with R and contains
84  # functions to import data  from other systems.
85  library(foreign)
86
87  # You can get help on a package using:
88  library(help = foreign)
89
90  # To install a new package, use install.packages()
91  # Install the ggplot2 package for it's plotting capability.
92  if (!require("ggplot2")){
93    install.packages("ggplot2")
94  }
95
96  # Then load the package.
97  library("ggplot2")
98  search()
99  # Notice that package:ggplot2 is now added to the search list.
```

```
R Interactive
Type 'q()' to quit R.

Microsoft R Open 3.3.1
The enhanced R distribution from Microsoft
Microsoft packages Copyright (C) 2016 Microsoft Corporation

Using the Intel MKL for parallel mathematical computing(using 2 cores).

Default CRAN mirror snapshot taken on 2016-07-01. See: https://mran.microsoft.com/.
> if (!require("ggplot2")){
+   install.packages("ggplot2")
+ }
Loading required package: ggplot2
>
```

Figure 4-8. *Loading ggplot2 package*

Now that ggplot2 is loaded, we need to run the library() function in order to load it into the current R session. This is shown in line 97; highlight this line and execute it, and then execute line 98 as well. Line 98 says search(), which if you recall, shows the currently loaded packages. Notice that ggplot2 is now added to the list of currently installed packages for this session.

R Package Manager

The installation of R Tools for Visual Studio includes an R Package Manager. If you navigate to R **Tools ➤ Windows** and choose **Packages**, shown in Figure 4-9, a page opens, showing you a lot of really cool information about the installed R packages, as shown in Figure 4-10.

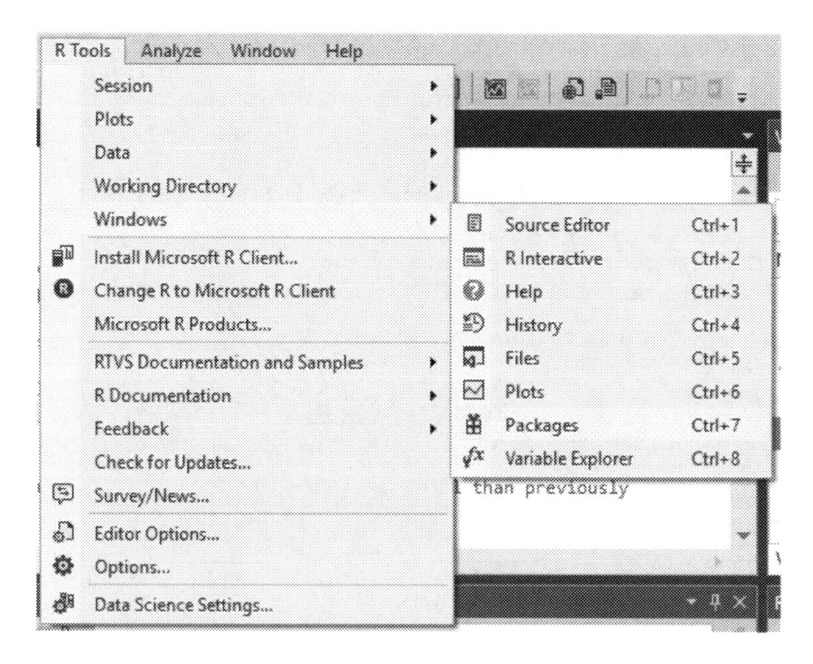

Figure 4-9. *Location of Packages selection*

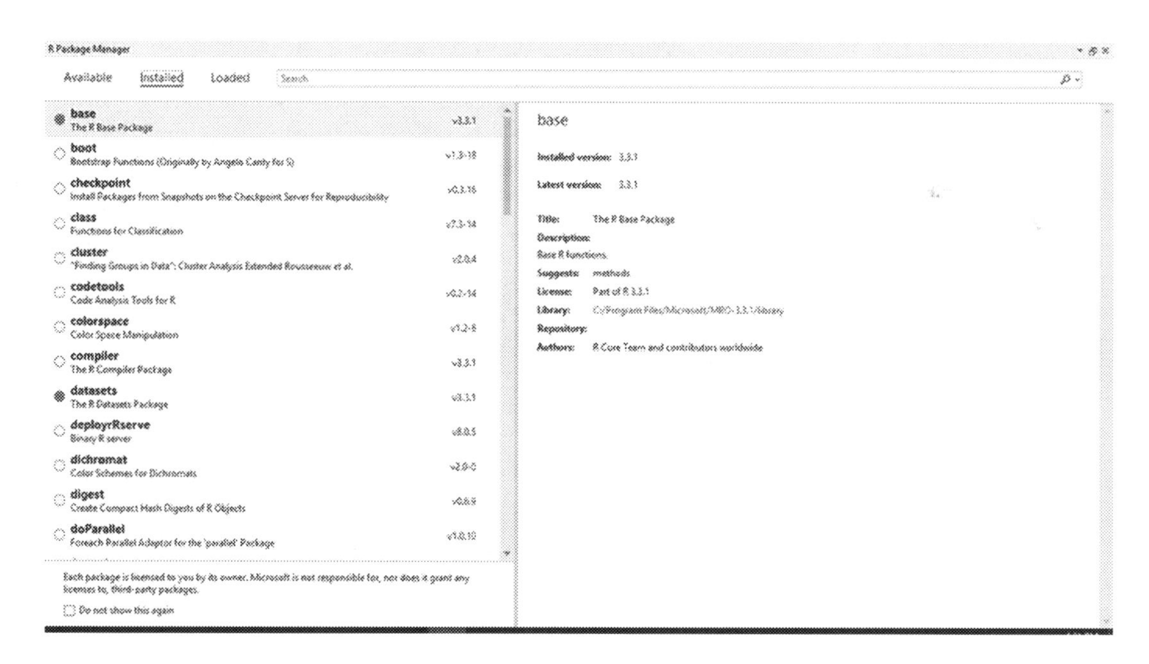

Figure 4-10. *R Package Manager*

The addition of a package manager is very useful, because otherwise, users would be forced to manually manage packages through the command line. While some users may be more comfortable with this approach, it is sort of defeating the purpose of *rapid* application development. With this feature, a user can now view their installed packages and alternatively choose to update them from this interface instead of using `installed.packages()`.

Feel free to explore around in this area a bit. For example, the top-left corner contains three options:

- **Available**: This menu option shows all of the packages available from the current CRAN mirror you are currently pointing to. Packages can be downloaded and installed by clicking the **Install** button located on the right of the interface.
- **Installed**: This option shows all the packages currently installed on your system. If an update is available to an installed package, a blue icon appears in the left pane and an Update button appears in the right pane.
- **Loaded**: This option shows all the packages currently loaded in your project.

To expound on this information, the Available menu option is shown in Figure 4-11.

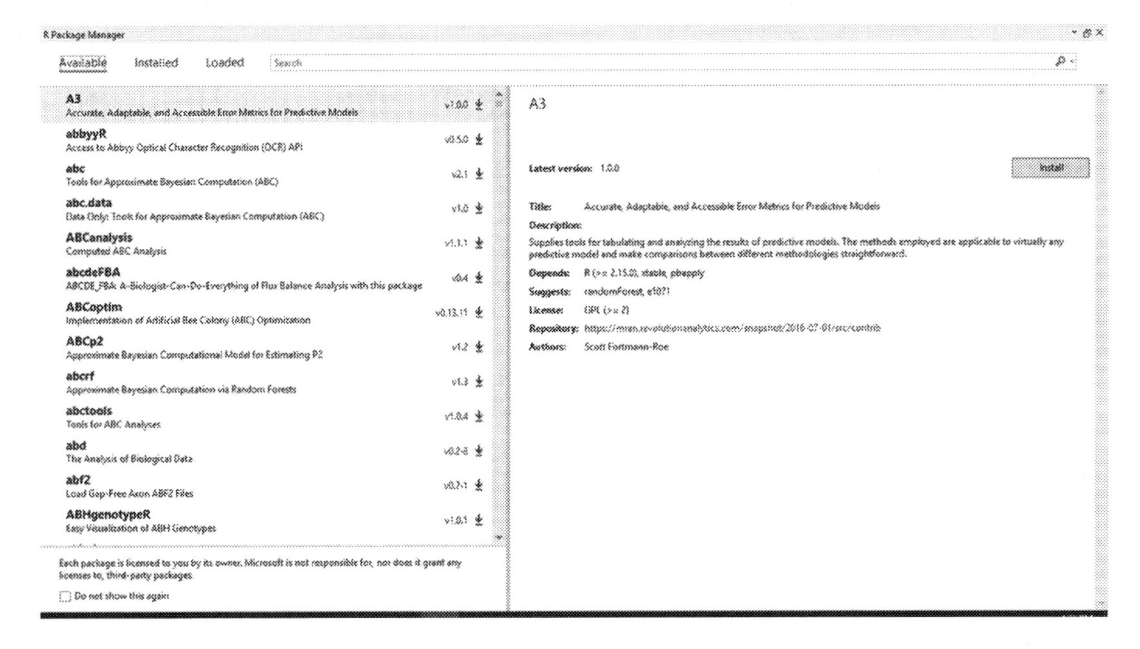

Figure 4-11. *Available menu option*

Notice the Install button on the right. Clicking this button installs the currently selected package into your R installation.

The Installed menu option appears as shown in Figure 4-12. Note that I scrolled to the bottom of the left pane in order to show the updatable packages.

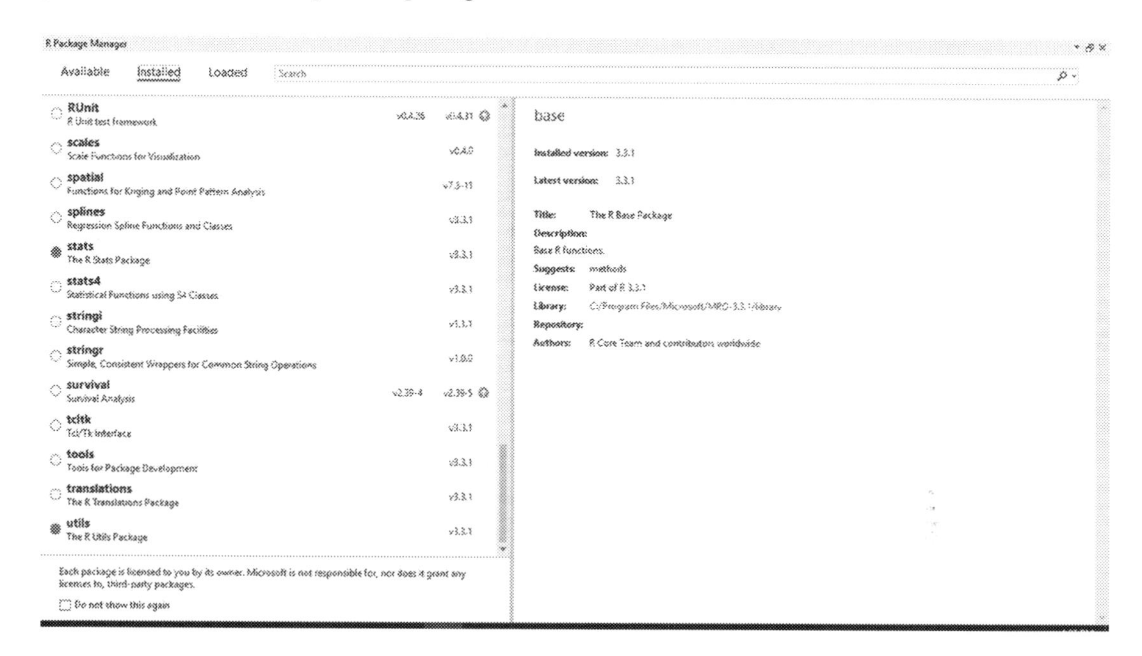

Figure 4-12. *Installed menu option*

Notice that RUnit and survival both need to be updated. Clicking the **RUnit** package name in the left pane changes the interface to what is shown in Figure 4-13.

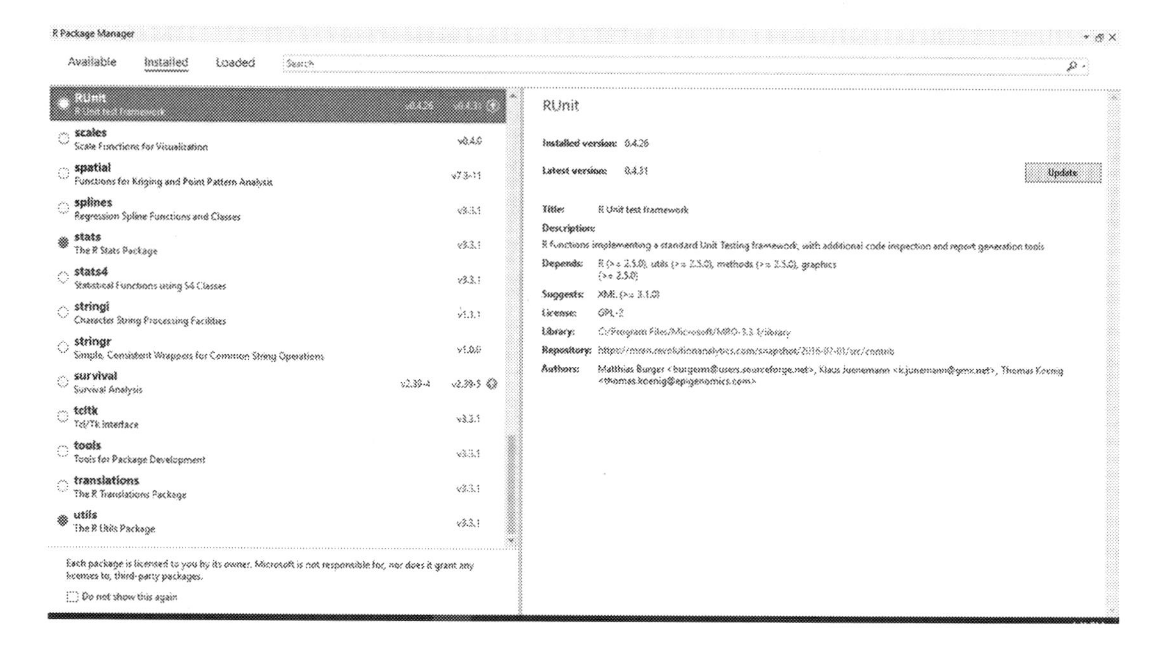

Figure 4-13. *RUnit needs updating*

Notice that the Update button now appears in the right pane.

Finally, the Loaded menu option appears, as shown in Figure 4-14.

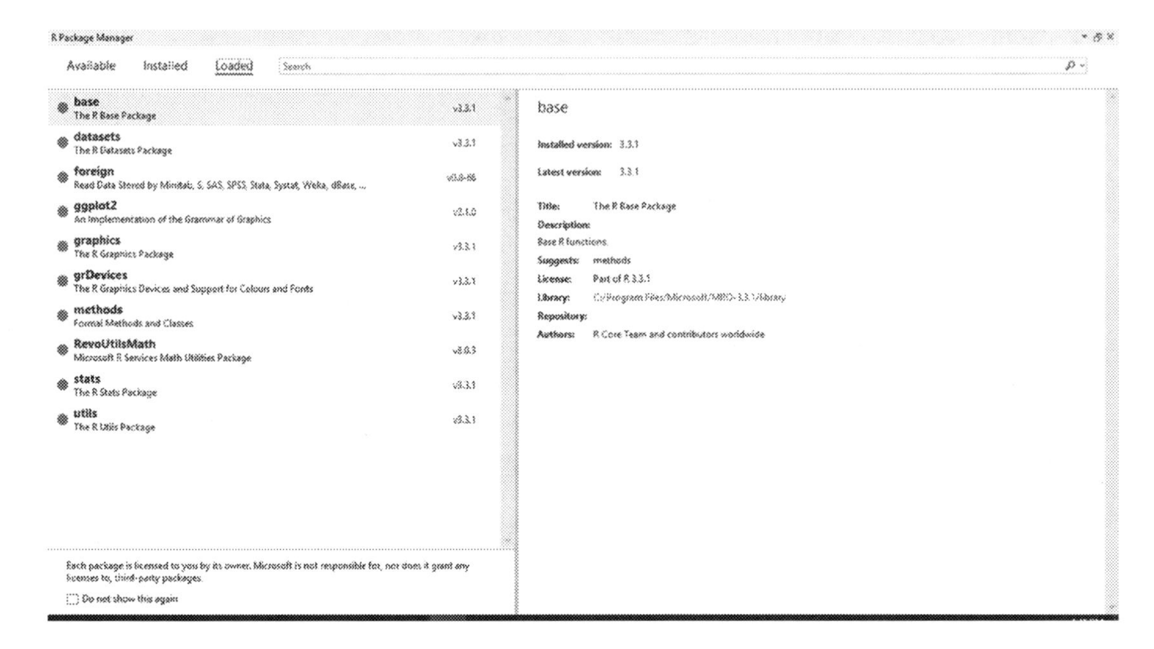

Figure 4-14. Loaded menu option

There isn't much you can do with this area except see what is already loaded. Go ahead and close the R Package Manager window. Let's continue on.

Plotting in R

Next, we look at a simple regression example, as shown in the script. First, I need to point out that the `ggplot2` package comes preloaded with quite a few sets of data that is to be used to test the functionality of the package. This data is accessed as shown on line 105, using the syntax `data(package = "ggplot2")$results`. This syntax says that we want to run the `data()` function against the `ggplot2` package and return the subset of the output referenced as `results` to the screen, as shown in Figure 4-15.

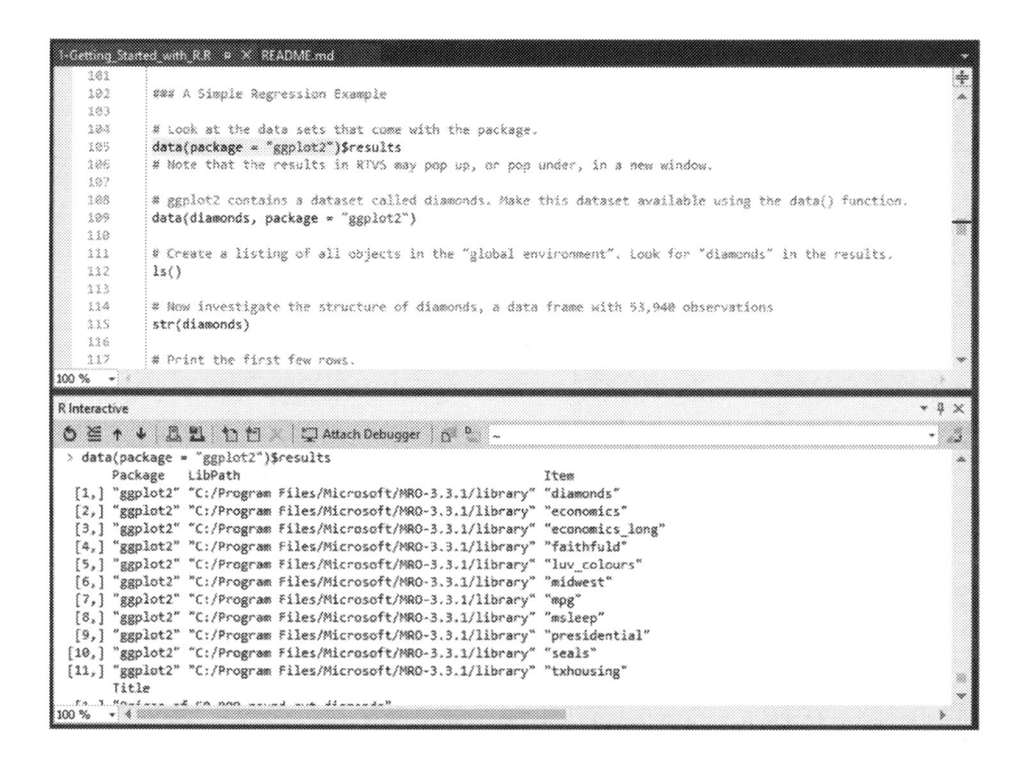

Figure 4-15. *ggplot2 Results*

Consequently, we could also execute the command `data(package = "ggplot2")` to see a listing of the datasets included in the package. Figure 4-16 shows this result.

```
R data sets    1-Getting_Started_with_R.R    README.md
 1  Data sets in package 'ggplot2':
 2
 3  diamonds                 Prices of 50,000 round cut diamonds
 4  economics                US economic time series.
 5  economics_long           US economic time series.
 6  faithfuld                2d density estimate of Old Faithful data
 7  luv_colours              'colors()' in Luv space.
 8  midwest                  Midwest demographics.
 9  mpg                      Fuel economy data from 1999 and 2008 for 38 popular models of car
10  msleep                   An updated and expanded version of the mammals sleep dataset.
11  presidential             Terms of 11 presidents from Eisenhower to Obama.
12  seals                    Vector field of seal movements.
13  txhousing                Housing sales in TX.
14
100 %
```

Figure 4-16. *ggplot2 datasets*

Go ahead and close that window, but keep your R script open. We've got the executed result of `data(package = "ggplot2")$results` shown in the R Interactive window, so skip down to line 109 next. This line says `data(diamonds, package = "ggplot2")`. The syntax of this says that we want to run the `data()` command against the `diamonds` dataset in the `ggplot2` package. Highlight that line and press **Ctrl+Enter** to execute it. There isn't any huge change or anything here; all that happened was that the `diamonds` dataset was just made available for analysis. The way you know that it was just loaded is to check your Variable Explorer window. Figure 4-17 shows what the Variable Explorer window should look like at this point.

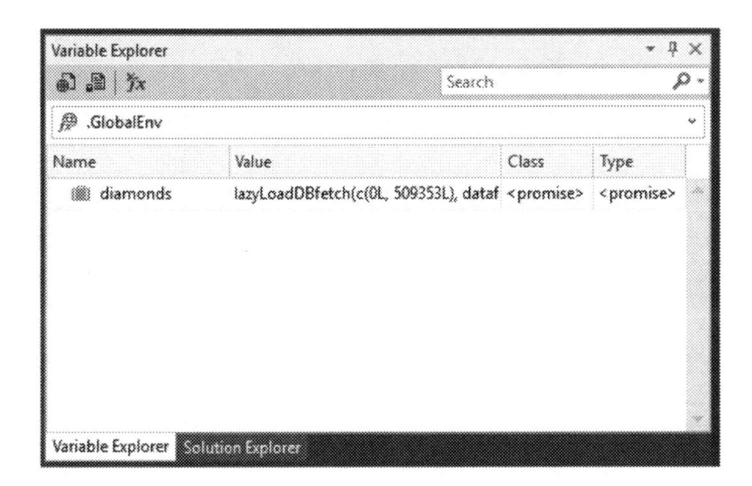

Figure 4-17. *diamonds dataset loaded*

So there is our `diamonds` dataset, loaded and ready to go. Go down to line 112 next, which says `ls()`. That line by itself, without any arguments, only returns the datasets or functions defined by the user in the current session. In this instance, running a simple line of code only results in the output of a single word: *diamonds*. The reason for this is that this is the only dataset loaded for this session. If you were to execute this line inside of a function with no arguments, you would be able to see the local variables for that particular function. As you can tell, this can be a useful debugging tool.

Now go down and execute line 115, which says `str(diamonds)`. This command allows us to examine the structure of the dataset passed in as an argument. Figure 4-18 shows the structure displayed in RTVS.

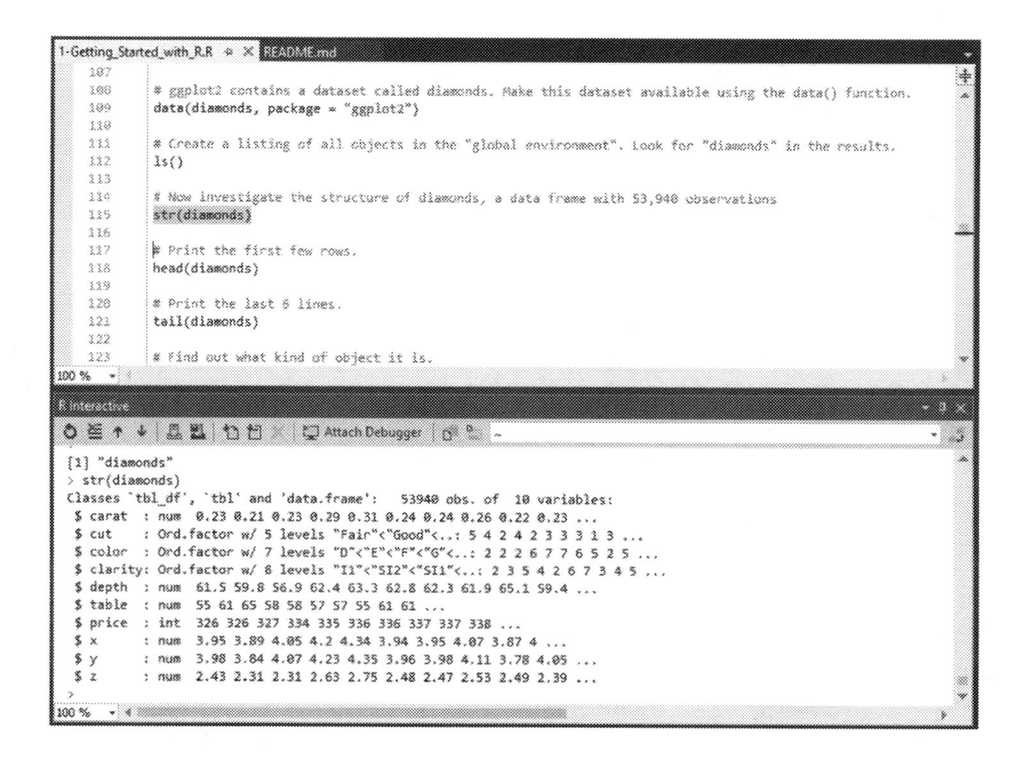

Figure 4-18. *str(diamonds) output*

Briefly look at the returned data there. You can see that there is a lot of information but it appears to be truncated. Have no fear; take a look at your Variable Explorer again. You see that there is a very slight change, in that there is now an indicator next to diamonds, as opposed to just a table graphic. Click this indicator. You should see what is shown in Figure 4-19.

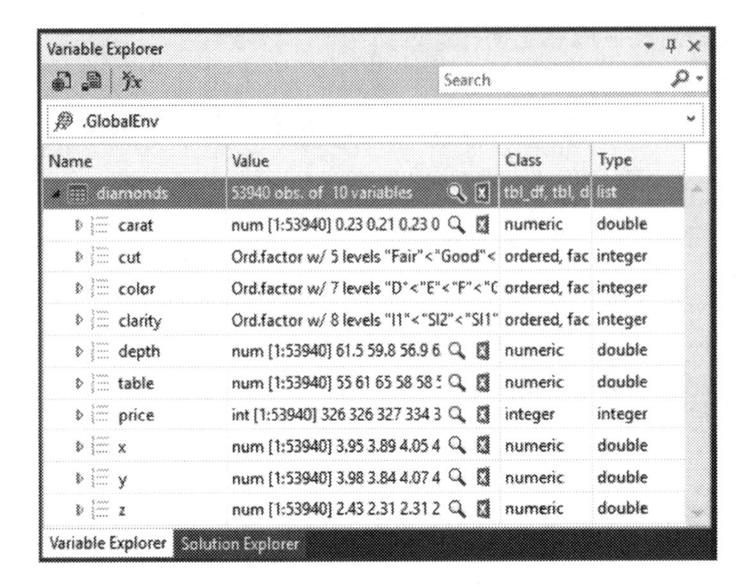

Figure 4-19. *diamonds detail information*

This is much easier for me to read and decipher. Across the top, just above the blue bar shown in Figure 4-19, the columns are defined as Name, Value, Class, and Type. Browse around in there for a minute to get familiar with how it looks. This is a very cool feature that allows the introspection of data before it is really analyzed.

Go to line 118 next. This line says `head(diamonds)`, which simply instructs R to output the first six rows of data in the dataset. Conversely, line 121 says `tail(diamonds)`, which, as you've probably guessed, outputs the last six rows of data in the dataset. The results of these two executions are shown in Figure 4-20.

```
R Interactive
Attach Debugger
> head(diamonds)
+
  carat       cut color clarity depth table price    x    y    z
1  0.23     Ideal     E     SI2  61.5    55   326 3.95 3.98 2.43
2  0.21   Premium     E     SI1  59.8    61   326 3.89 3.84 2.31
3  0.23      Good     E     VS1  56.9    65   327 4.05 4.07 2.31
4  0.29   Premium     I     VS2  62.4    58   334 4.20 4.23 2.63
5  0.31      Good     J     SI2  63.3    58   335 4.34 4.35 2.75
6  0.24 Very Good     J    VVS2  62.8    57   336 3.94 3.96 2.48
> tail(diamonds)
+
      carat       cut color clarity depth table price    x    y    z
53935  0.72   Premium     D     SI1  62.7    59  2757 5.69 5.73 3.58
53936  0.72     Ideal     D     SI1  60.8    57  2757 5.75 5.76 3.50
53937  0.72      Good     D     SI1  63.1    55  2757 5.69 5.75 3.61
53938  0.70 Very Good     D     SI1  62.8    60  2757 5.66 5.68 3.56
53939  0.86   Premium     H     SI2  61.0    58  2757 6.15 6.12 3.74
53940  0.75     Ideal     D     SI2  62.2    55  2757 5.83 5.87 3.64
> |
100 %
```

Figure 4-20. *head() and tail() demonstrated*

Line 127 is next, which says `dim(diamonds)`. The `dim()` function allows us to see the dimension of the object passed as an argument, which would be `diamonds` in this case. Executing this line simply shows what we have already gathered by examining the Variable Explorer earlier: there is one dataset returned with 53940 objects of 10 variables.

Next, we move on to plotting in R. There are three main packages used for charting in R: `base`, `lattice`, and `ggplot2`. This is also referenced on line 136.

Lines 133 and 134 say the following:

```
diamondSample <- diamonds[sample(nrow(diamonds), 5000),]
dim(diamondSample)
```

Let's break the syntax of this apart before anything else.

- **diamondSample**: Declares an object for storage of the results of the command.
- <- : An assignment operator that declares that the result of the command(s) on the right are to be assigned to the object on the left.
- **diamonds**: References the `diamonds` dataset.
- **[sample(nrow(diamonds), 5000),]**: Draws a random sample of 5000 data points from the `diamonds` dataset.
- **dim(diamondSample)**: Shows the dimension of the `diamondSample` object.

As expected, the result of the `dim(diamondSample)` command was the output `[1] 5000    10`, which means that one dataset and 5000 objects in 10 variables were returned.

Next, we go to line 140, which reads: `theme_set(theme_gray(base_size = 18))`. This says that we set the font size to 18pt and use the gray theme. Highlight and execute that line, then skip down to lines 143 and 144. Those lines show the following code:

```
ggplot(diamondSample, aes(x = carat, y = price)) +
  geom_point(colour = "blue")
```

First off, note that this is on two different lines, but it is actually one long command. The presence of the plus sign indicates a line continuation character, outside of usage inside a function or argument.

Let's take a look at the syntax of this command before we move on.

- **ggplot**: Defines the function to be executed.
- **diamondSample**: References the `diamondSample` object as the active dataset.
- **aes**: Defines the aesthetic qualities of the chart; in this example, the x axis is the `carat` set and the y axis is the `price` set.
- **geom_point**: Customizes the points once they are charted.

So this means that we will generate a chart. Highlight these two rows and press **Ctrl+Enter** to execute them. Figure 4-21 shows the result of this action.

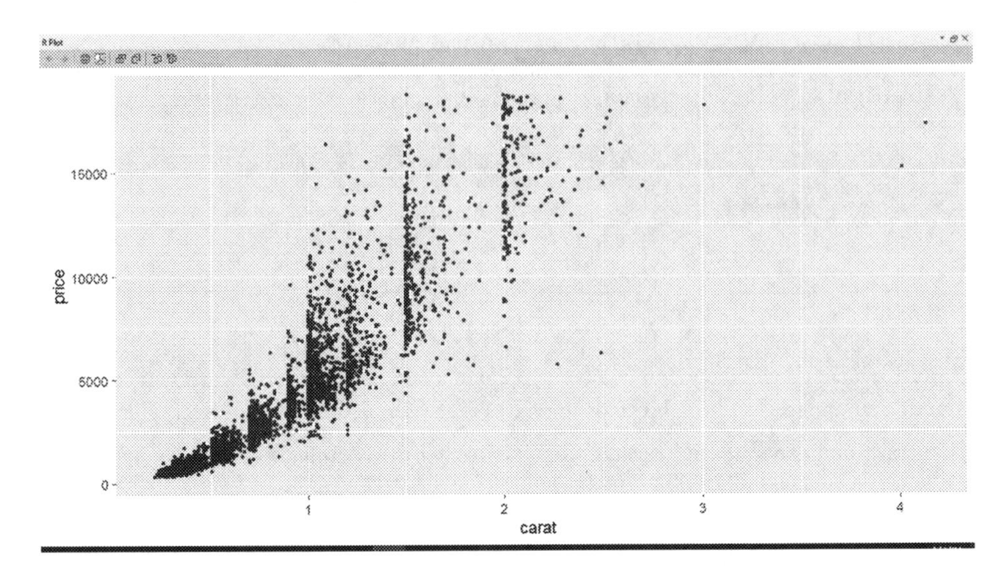

Figure 4-21. *Our first chart!*

Notice that the R Plot window came into focus here? That's pretty handy. Notice also that the x axis is `carat` and the y axis is `price`, as we defined in the `aes()` command.

Go down to lines 147 through 149 next. You'll see there is a very slight addition of `scale_x_log10()`. This gives us a nice little logarithmic scale on the x axis (which is also stated in line 146) so go ahead and highlight and execute lines 146 through 149 now. Figure 4-22 shows what you should see now.

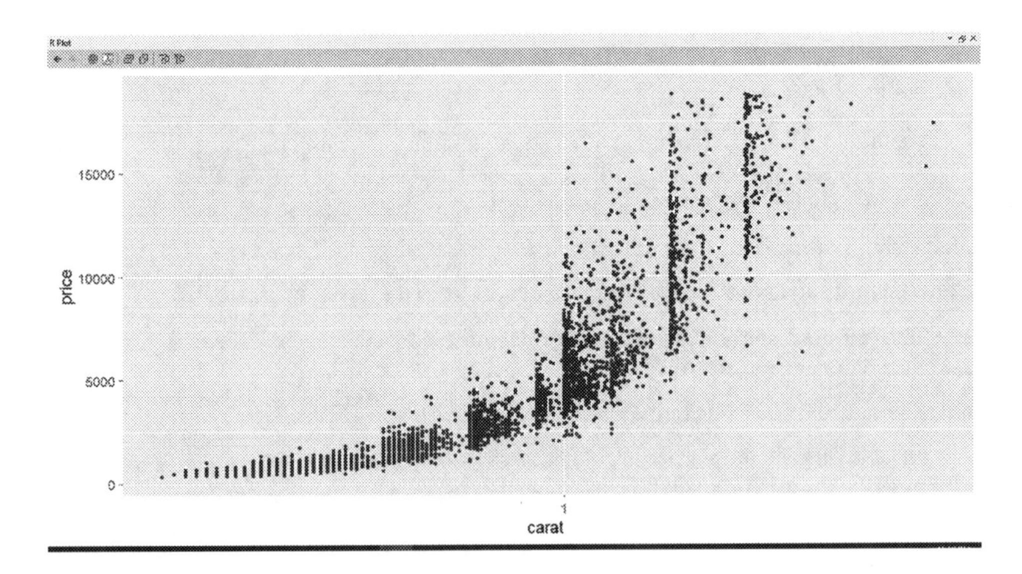

Figure 4-22. *Logarithmic scale addition*

That's starting to make a bit more sense now. Next, another line is added to let us add another logarithmic scale on the y axis. This is on lines 152 through 155; so highlight these lines and press **Ctrl+Enter** to execute them. Figure 4-23 shows what you should see now.

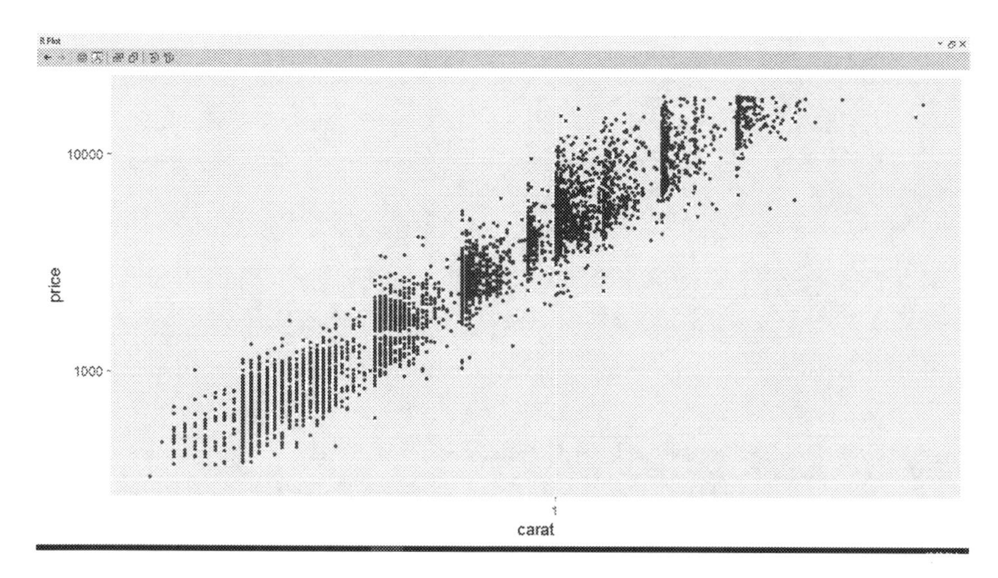

Figure 4-23. *Logarithmic scale on x and y axes*

The chart is even better now, I think.

Linear Regression in R

Next, we look at linear regression in R within the confines of the loaded dataset. Linear regression is basically the way that statisticians define the relationship between a scalar variable and one or many explanatory variables. *Simple linear regression* is when there is only one explanatory variable, which is essentially what we are dealing with in this example.

In R, we must first build a model of the data before we can access that model. To do that, take a look at line 163, which says `model <- lm(log(price) ~ log(carat) , data = diamondSample)`. Again, let's look at the syntax of this before we move on.

- **model**: Defines the object that inherits what is defined in the command.
- <- : An assignment operator that declares that the result of the command(s) on the right are to be assigned to the object on the left.
- **lm**: Defines a linear model.
- **log(price)**: A logarithmic scale of the `price` data (also the scalar variable).
- ~ : Separates the scalar variable and the explanatory variable(s). Scalar variables are shown to the left of this symbol and explanatory variables are shown to the right.
- **log(carat)**: A logarithmic scale of the `carat` data (also the explanatory variable).
- **,data = diamondSample**: Defines the data object to be used (`diamondSample`, in this case).

Highlight line 163 and execute it. Notice that there is no discernible action taken in the IDE, so let's check the Variable Explorer again. There is now another data object in there referenced as `model`. Figure 4-24 shows what you should be looking at now.

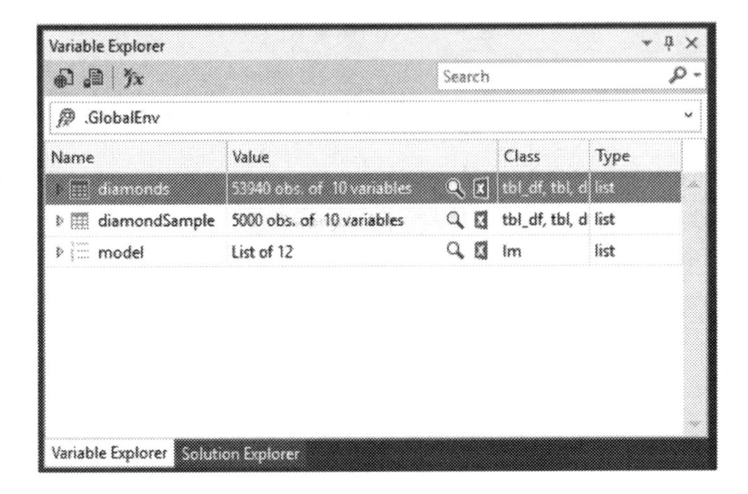

Figure 4-24. *model dataset*

Line 166 simply says `summary(model)`. Go ahead and execute that line. A lot of information is shown in the R Interactive window now. Take a look at Figure 4-25.

```
1-Getting_Started_with_R.R    README.md
160
161   # Build the model. log of price explained by log of carat. This illustrates how linear regression works.
        Later we fit a model that includes the remaining variables
162
163   model <- lm(log(price) ~ log(carat) , data = diamondSample)
164
165   # Look at the results.
166   summary(model)
167   # R-squared = 0.9334, i.e. model explains 93.3% of variance
100 %

R Interactive
Attach Debugger
> summary(model)
+

Call:
lm(formula = log(price) ~ log(carat), data = diamondSample)

Residuals:
     Min       1Q   Median       3Q      Max
-1.23362 -0.17217 -0.00395  0.17259  1.14454

Coefficients:
            Estimate Std. Error t value Pr(>|t|)
(Intercept) 8.453407   0.004448  1900.6   <2e-16 ***
log(carat)  1.673569   0.006321   264.8   <2e-16 ***
---
Signif. codes:  0 '***' 0.001 '**' 0.01 '*' 0.05 '.' 0.1 ' ' 1

Residual standard error: 0.2613 on 4998 degrees of freedom
Multiple R-squared:  0.9334,    Adjusted R-squared:  0.9334
F-statistic: 7.01e+04 on 1 and 4998 DF,  p-value: < 2.2e-16

>
100 %
```

Figure 4-25. *summary(model) information*

Take a look at what is returned by this action. You have a lot of really useful information returned about the model here, so understand that this is a great place to get meaningful statistical information about a model.

```
Head on down to lines 170 through 172, which show the following code:coef(model)
coef(model)[1]
exp(coef(model)[1])
```

Line 169 says that we extract the model coefficients. Let's highlight those lines and execute them now. Figure 4-26 shows what you should see in your R Interactive window now.

```
> coef(model)
+ coef(model)[1]
+ exp(coef(model)[1])
(Intercept)  log(carat)
   8.453407    1.673569
(Intercept)
   8.453407
(Intercept)
   4691.029
```

Figure 4-26. *model coefficients*

Now that we've got that, let's take a look at the next bit of code on lines 175 through 179. This code is defined as follows:

```
ggplot(diamondSample, aes(x = carat, y = price)) +
  geom_point(colour = "blue") +
  geom_smooth(method = "lm", colour = "red", size = 2) +
  scale_x_log10() +
  scale_y_log10()
```

We've already stepped through the syntax of this command, but I see that there is a line that says `geom_smooth(method = "lm", colour = "red", size = 2) +` that we didn't define before. The addition of this line of code adds a thin red trend line to the chart. Highlight these lines and execute them now. Figure 4-27 shows what you should see as a result.

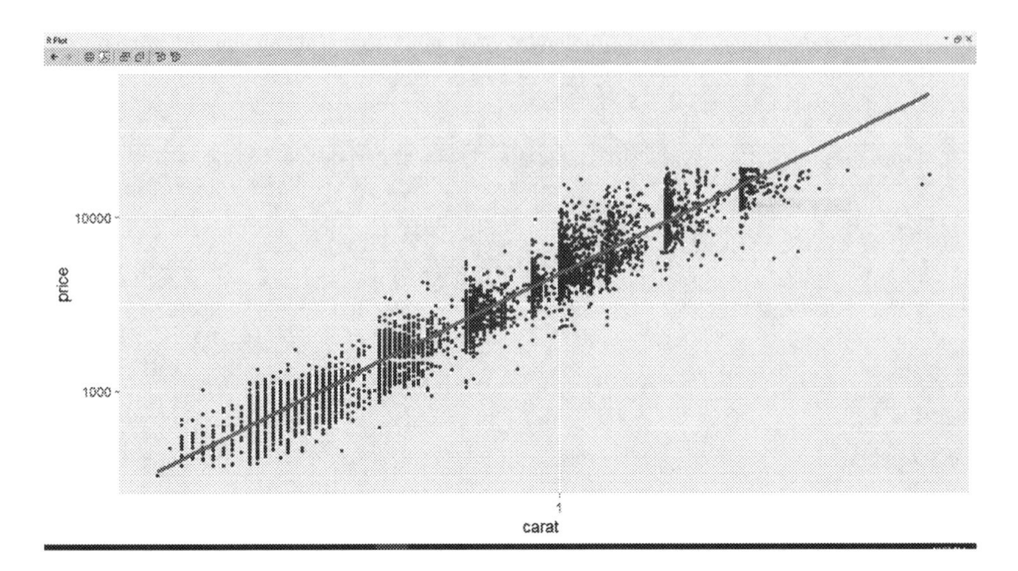

Figure 4-27. *Charted result*

That is one meaningful chart! The charts are getting progressively better and more interesting, as you can see.

Regression Diagnostics

Next, go down and take a look at lines 190 through 192. They are defined as follows:

```
par(mfrow = c(2, 2))
plot(model, col = "blue")
par(mfrow = c(1, 1))
```

So in this example, we set the plot layout to generate a 2×2 grid of charts, show the plotted information within the model dataset, and then reset the plot layout back to a 1×1 grid again. Let's take a further look at the syntax of these statements.

- `par`: This function combines multiple different charts into a single chart as specified by the `mfrow()` argument.
- `mfrow = c(2, 2)`: This argument passes in the values for (`rows` × `columns`), which determine how the grid is laid out on the stage.
- `plot(model, col = "blue")`: This function determines what is to be plotted (`model`) and the color (`col`) of the plotted data points.

Feel free to play around with this or any other function a bit, just to get a feel for what the syntax does and how the subtle changes you make to the functions affect the generated output. I typically find this to be a great way to learn about the functions, so maybe you will like this method as well. For example, change the `col` value to blue or green, and see what happens to your charts. Change the `c()` value to **3, 3** instead of 2, 2. Notice how the charts are realigned.

■ **Tip** If you ever get stuck with an error that says `Error in plot.new(): figure margins too large`, run the commands `dev.off()` and `par(mar=c(1,1,1,1))` to clear the error. They may need to be run multiple times to reset the graphics component. Huge thanks to Stack Overflow (`http://stackoverflow.com`) for this tip!

Highlight lines 190 through 192 and execute them. Figure 4-28 shows what you should now see.

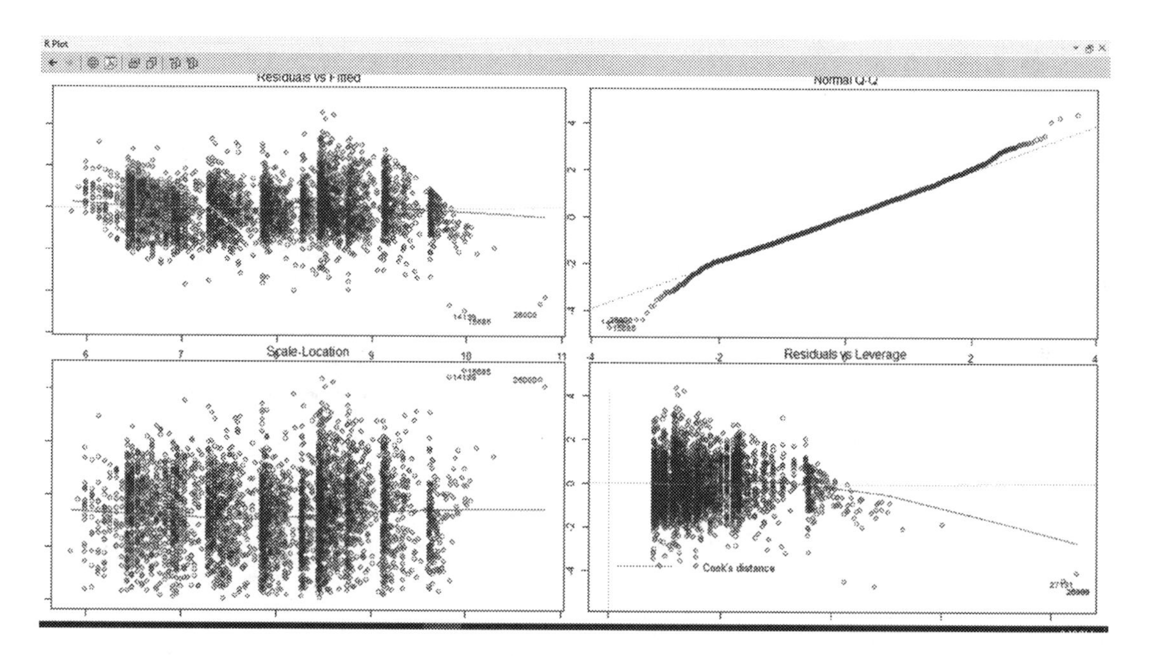

Figure 4-28. *Multiple chart results*

That looks fantastic. Notice how most of the data points lay fairly close the red line? That indicates that this dataset is pretty reliable and does not contain an overwhelming number of outliers.

The Model Object

Line 198 has us look at the structure of the `model` object. Highlight and execute this line. You see a flurry of activity in the R Interactive window. Figure 4-29 shows this returned data. This is to be expected, so go back through the R Interactive window and take a look at how the data was returned.

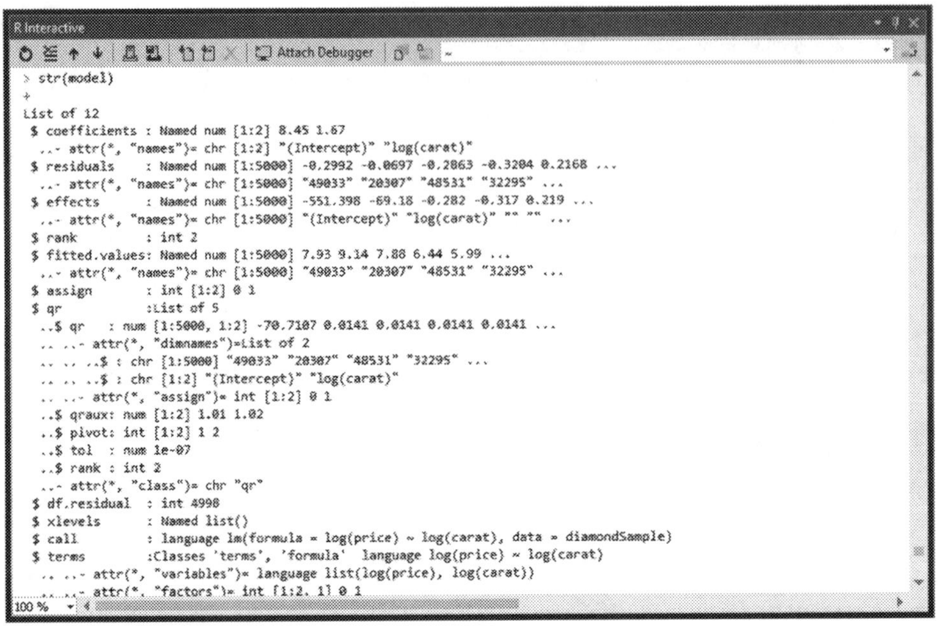

```
R Interactive
> str(model)
+
List of 12
 $ coefficients : Named num [1:2] 8.45 1.67
  ..- attr(*, "names")= chr [1:2] "(Intercept)" "log(carat)"
 $ residuals    : Named num [1:5000] -0.2992 -0.0697 -0.2863 -0.3204 0.2168 ...
  ..- attr(*, "names")= chr [1:5000] "49033" "20307" "48531" "32295" ...
 $ effects      : Named num [1:5000] -551.398 -69.18 -0.282 -0.317 0.219 ...
  ..- attr(*, "names")= chr [1:5000] "(Intercept)" "log(carat)" "" "" ...
 $ rank         : int 2
 $ fitted.values: Named num [1:5000] 7.93 9.14 7.88 6.44 5.99 ...
  ..- attr(*, "names")= chr [1:5000] "49033" "20307" "48531" "32295" ...
 $ assign       : int [1:2] 0 1
 $ qr           :List of 5
  ..$ qr   : num [1:5000, 1:2] -70.7107 0.0141 0.0141 0.0141 0.0141 ...
  .. ..- attr(*, "dimnames")=List of 2
  .. .. ..$ : chr [1:5000] "49033" "20307" "48531" "32295" ...
  .. .. ..$ : chr [1:2] "(Intercept)" "log(carat)"
  .. ..- attr(*, "assign")= int [1:2] 0 1
  ..$ qraux: num [1:2] 1.01 1.02
  ..$ pivot: int [1:2] 1 2
  ..$ tol  : num 1e-07
  ..$ rank : int 2
  ..- attr(*, "class")= chr "qr"
 $ df.residual  : int 4998
 $ xlevels      : Named list()
 $ call         : language lm(formula = log(price) ~ log(carat), data = diamondSample)
 $ terms        :Classes 'terms', 'formula'  language log(price) ~ log(carat)
  .. ..- attr(*, "variables")= language list(log(price), log(carat))
  .. ..- attr(*, "factors")= int [1:2, 1] 0 1
100 %
```

Figure 4-29. *Returned object data*

Notice that the returned data is actually pretty unreadable and might not make a lot of sense right now. This is just one way that we can view the `model` object, fortunately. Look at your Variable Explorer window and expand the `model` entry. Figure 4-30 shows what you should see now.

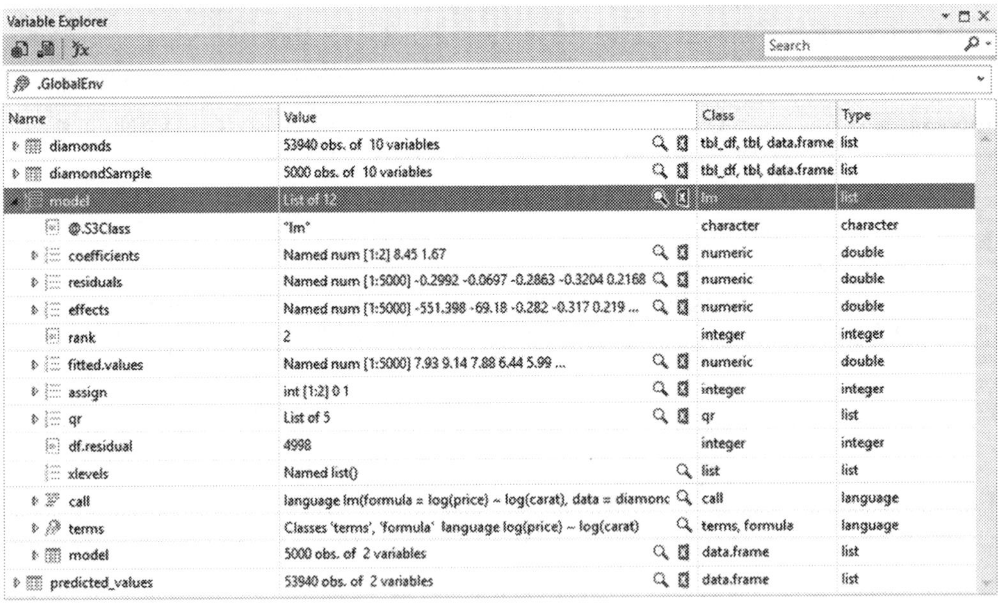

Variable Explorer

.GlobalEnv

Name	Value	Class	Type
diamonds	53940 obs. of 10 variables	tbl_df, tbl, data.frame	list
diamondSample	5000 obs. of 10 variables	tbl_df, tbl, data.frame	list
model	List of 12	lm	list
@.S3Class	"lm"	character	character
coefficients	Named num [1:2] 8.45 1.67	numeric	double
residuals	Named num [1:5000] -0.2992 -0.0697 -0.2863 -0.3204 0.2168	numeric	double
effects	Named num [1:5000] -551.398 -69.18 -0.282 -0.317 0.219 ...	numeric	double
rank	2	integer	integer
fitted.values	Named num [1:5000] 7.93 9.14 7.88 6.44 5.99 ...	numeric	double
assign	int [1:2] 0 1	integer	integer
qr	List of 5	qr	list
df.residual	4998	integer	integer
xlevels	Named list()	list	list
call	language lm(formula = log(price) ~ log(carat), data = diamonc	call	language
terms	Classes 'terms', 'formula' language log(price) ~ log(carat)	terms, formula	language
model	5000 obs. of 2 variables	data.frame	list
predicted_values	53940 obs. of 2 variables	data.frame	list

Figure 4-30. *model entry*

Notice that the entries in the Name column follow those columns that were aligned next to the dollar signs in the R Interactive window. If you were to double-click the **residuals** column in the Variable Explorer window, another window opens on the top-left pane that is titled **R Data: $residuals**. Figure 4-31 shows this window.

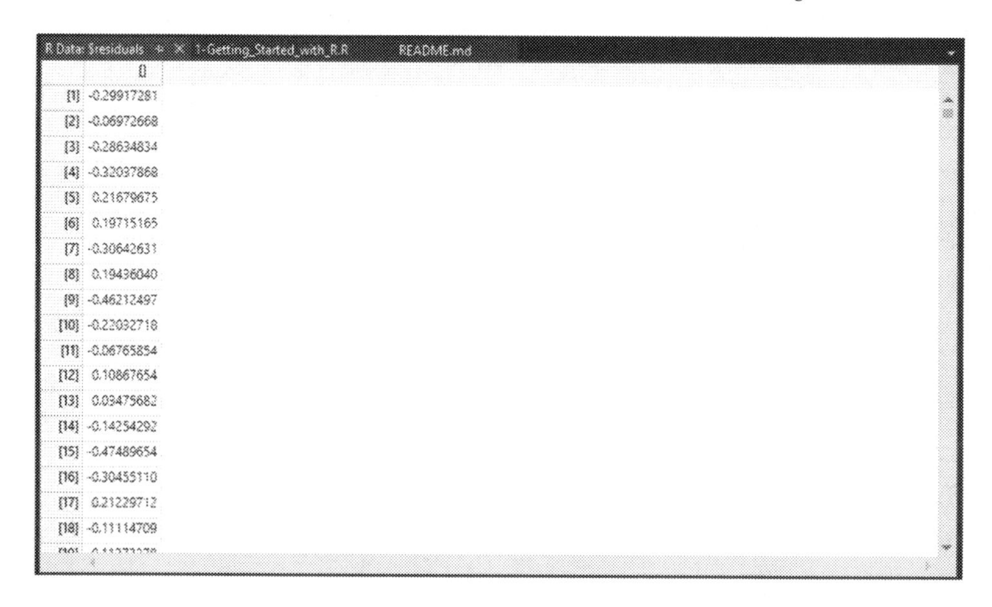

Figure 4-31. *R Data: $residuals window*

This window contains the data that makes up the dataset. This procedure can be repeated for any column that can be expanded, as seen in the Name column of the Variable Explorer.

Next is line 199, defined simply as `model$coefficients`. This line basically means that we want to view the `coefficients` column in the `model` object. Execute line 199. Figure 4-32 shows you what you should see now.

```
> model$coefficients
(Intercept)  log(carat)
   8.453407    1.673569
```

Figure 4-32. *coefficients displayed in text format*

Consequently, we could also see this same data by double-clicking the **coefficients** column in Variable Explorer, as shown in Figure 4-33.

R Data: $coefficients

	[]
[1]	8.453407
[2]	1.673569

Figure 4-33. *coefficients displayed in tabular format*

So again, there are two ways to do the same thing and we have verified that the data is correct.

Skip down to line 202 now. This line says `model <- lm(log(price) ~ log(carat) + ., data = diamondSample)` and it is closely related to the model we built before, but this time, we declared `log(price) ~ log(carat) + .`, which means that we want to model the log of the price column against all the other columns in the dataset. In other words, this is true linear regression and not simple linear regression.

Highlight line 202 and execute it. Notice that Variable Explorer now shows **List of 13** where is once showed **List of 12**, which means that we have added the `model` to the list and it is now available in the dataset. Figure 4-34 shows this updated value. You can always double-click **model** to view the data contained within, if you would like.

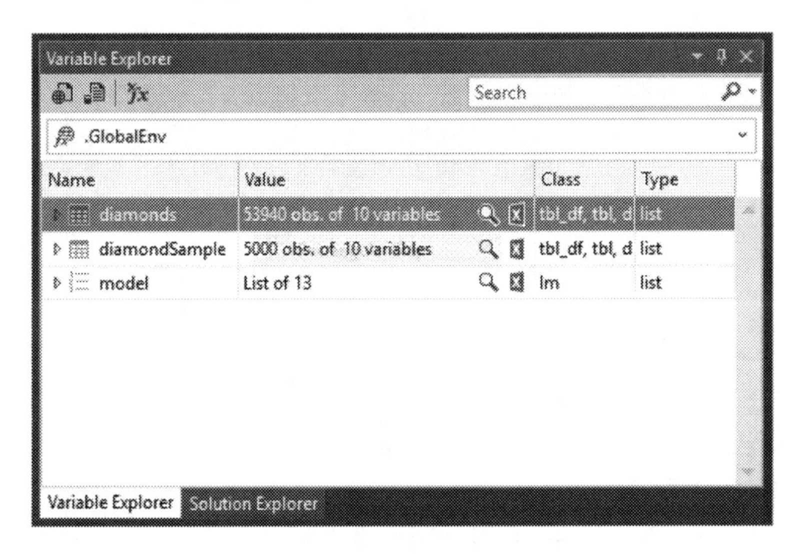

Figure 4-34. *model value increased*

Line 204 shows the summary of the model; so highlight this line and execute it. Figure 4-35 shows the result of this action.

```
Call:
lm(formula = log(price) ~ log(carat) + ., data = diamondSample)

Residuals:
     Min       1Q   Median       3Q      Max
-0.59504 -0.08640  0.00063  0.08060  1.32134

Coefficients:
              Estimate Std. Error t value Pr(>|t|)
(Intercept)  7.867350   0.224753  35.004  < 2e-16 ***
log(carat)   1.737683   0.026715  65.046  < 2e-16 ***
carat       -0.067950   0.019644  -3.459 0.000547 ***
cut.L        0.115715   0.008827  13.109  < 2e-16 ***
cut.Q       -0.034995   0.007016  -4.988 6.32e-07 ***
cut.C        0.011090   0.006078   1.824 0.068141 .
cut^4       -0.005822   0.004834  -1.205 0.228412
color.L     -0.437917   0.006594 -66.409  < 2e-16 ***
color.Q     -0.104306   0.006029 -17.300  < 2e-16 ***
color.C     -0.008947   0.005581  -1.603 0.108988
color^4      0.009815   0.005115   1.919 0.055041 .
color^5     -0.003020   0.004867  -0.620 0.534972
color^6     -0.002552   0.004471  -0.571 0.568176
clarity.L    0.886421   0.011428  77.564  < 2e-16 ***
clarity.Q   -0.250352   0.010607 -23.602  < 2e-16 ***
clarity.C    0.121507   0.009075  13.390  < 2e-16 ***
clarity^4   -0.075253   0.007287 -10.327  < 2e-16 ***
clarity^5    0.025475   0.005997   4.248 2.19e-05 ***
clarity^6   -0.011436   0.005232  -2.186 0.028883 *
clarity^7    0.030808   0.004602   6.694 2.40e-11 ***
depth        0.001640   0.002087   0.786 0.431986
table       -0.001954   0.001148  -1.703 0.088674 .
x            0.093449   0.020742   4.505 6.78e-06 ***
y           -0.003096   0.002834  -1.092 0.274785
z            0.022161   0.017820   1.244 0.213690
---
Signif. codes:  0 `***' 0.001 `**' 0.01 `*' 0.05 `.' 0.1 ` ' 1

Residual standard error: 0.1314 on 4975 degrees of freedom
Multiple R-squared:  0.9833,    Adjusted R-squared:  0.9832
F-statistic: 1.217e+04 on 24 and 4975 DF,  p-value: < 2.2e-16
```

Figure 4-35. *summary(model) information*

This is significant (bad pun) because the R-squared value is 98%, so that means that 98% of the data fits closely to the regression. Not bad!

Lines 211 through 214 show that we create a data frame, which is how R structures data within its models. These lines are defined as follows:

```
predicted_values <- data.frame(
  actual = diamonds$price,
  predicted = exp(predict(model, diamonds))
)
```

So, again, let's step through this to get a feel for what the syntax is doing here.

- **predicted_values**: The object that holds the result of the command.
- **data.frame**: The way that R handles structured data; similar to a table.
- **actual = diamonds$price**: Sets a column variable named `actual` equal to the value represented by the `price` values in the `diamonds` dataset.
- **predicted = exp(predict(model, diamonds))**: Sets a column variable named `predicted` equal to the exponential value of the predicted values in the `model` linear model in the `diamonds` dataset.

Execute lines 211 through 214. You'll see that nothing happens again, which is expected. Look at your Variable Explorer again. You'll see that there is now another dataset available called `predicted_values`, which we just added. Figure 4-36 shows what you should now see.

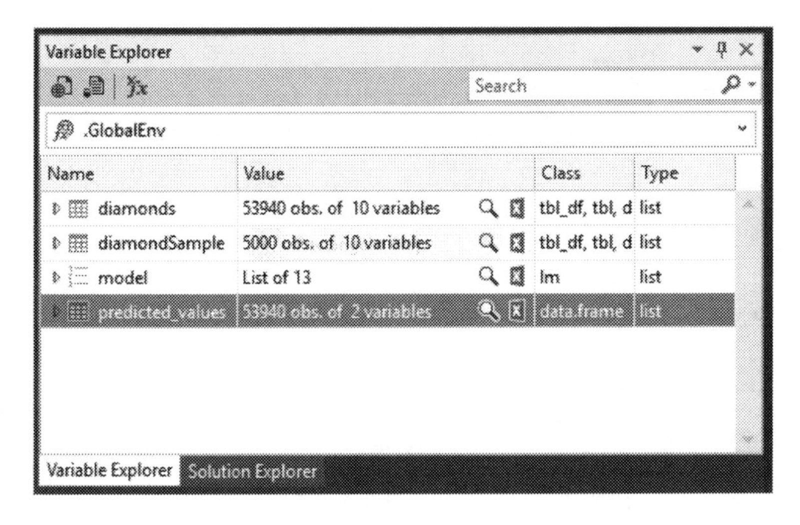

Figure 4-36. *predicted_values dataset*

Line 217 is next, which says `head(predicted_values)`. We saw this before, so recall that the `head()` command allows us to see the first six rows of data. Execute this. You should see what is shown in Figure 4-37.

```
> head(predicted_values)
+
  actual predicted
1    326  284.2928
2    326  272.5036
3    327  375.3411
4    334  418.4202
5    335  288.7658
6    336  303.4273
```

Figure 4-37. *head(predicted_values)*

Next is the big show. Lines 220 through 224 contain the `ggplot` command that charts all of this out for us. So far, we've laid out the `data.frame` object with the `actual` and `predicted` columns, Now we see what that data looks like visualized. The R code for this command is defined as follows):

```
ggplot(predicted_values, aes(x = actual, y = predicted)) +
  geom_point(colour = "blue", alpha = 0.01) +
  geom_smooth(colour = "red") +
  coord_equal(ylim = c(0, 20000)) +
  ggtitle("Linear model of diamonds data")
```

I'm pretty sure that we can read that and decipher what the syntax says, but basically, we are running the `ggplot` command against the `predicted_values` dataset. We are using the aesthetic value for the x axis to be the `actual` data and the y axis to be the `predicted` data. The data points are blue with a varying alpha (transparency), depending on the data value, with a red trend line, a y-limit of 20000 (which forces scale), and a title.

Figure 4-38 shows what you should see after you execute lines 220 through 224.

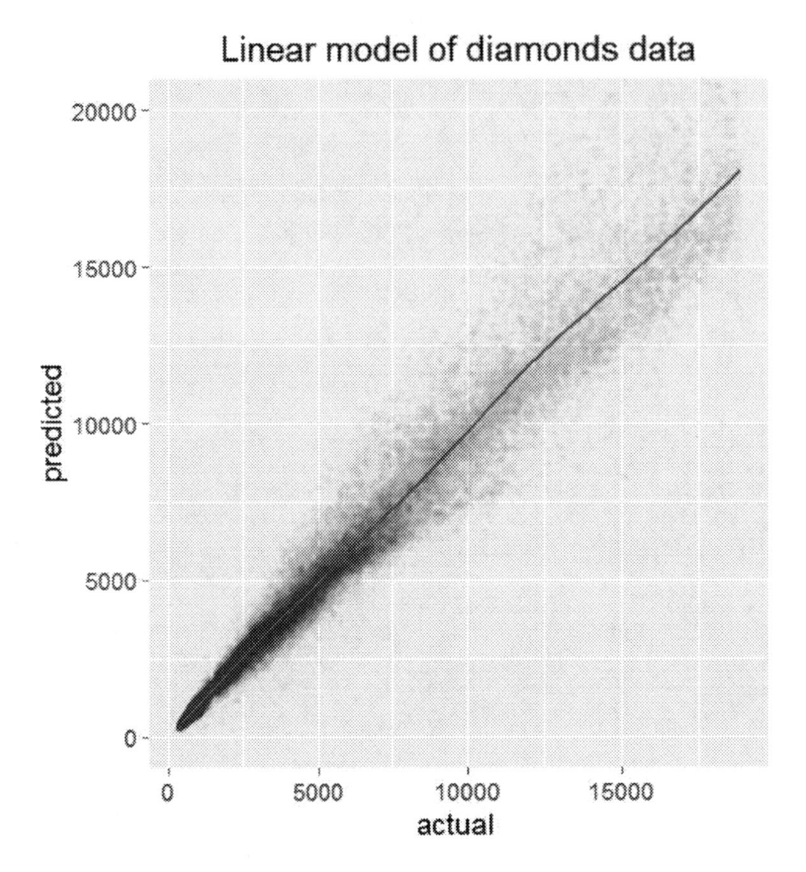

Figure 4-38. *ggplot of predicted_values*

Really, really useful chart there, don't you think? If you got this far with no problems, excellent work! You've actually learned quite a bit about the basics of R syntax now and you have seen real-world application of the most commonly used R plotting package: `ggplot2`.

Summary

We have done quite a bit in this chapter, including getting the data models built and verified that we are returning data into our object variable. We also looked at charting these values and the different ways that we can look at the data via different kinds of charts. Next, we get into more advanced plotting in R with R Tools for Visual Studio.

CHAPTER 5

Plotting in RTVS

Picking up right where we left off in Chapter 4, we are probably comfortable enough with R to begin getting into the task of creating the reports specified in the software requirements document created in Chapter 3. Understand that I am in no way intimating that we are experts in this language. Quite the opposite, actually. We are far enough along to understand some basics and we are now going to expound on that knowledge by pressing forward, while still recognizing that we are crawling and not yet walking.

Let's review the reports that were promised as outlined in our software requirements document. These are the reports:

- Average Wind Speed per AirportID
- Average Temperature per AirportID (°F)

With this in mind, let's start planning out the solution in R Tools for Visual Studio and seeing if we can get them working correctly.

Report 1: Average Wind Speed by Airport ID

In this first part, we calculate the Average Wind Speed per Airport. This is done by simply adding all the wind speed values for each Airport ID, and then dividing by the number of records. R provides a quick way for us to do this by using core functionality called `data.table`. What is `data.table`? It's a package freely available with a lot of really cool ways of slicing and dicing large amounts of numerical data. You need the `data.table` package if you want to calculate mean, sum, or just about any R built-in function easily.

Open R Tools for Visual Studio. You should see what is shown in Figure 5-1. Notice that we are working with a fresh instance of RTVS and not using the old interface from the previous chapters.

B. Beard, *Beginning SQL Server R Services*, DOI 10.1007/978-1-4842-2298-0_5

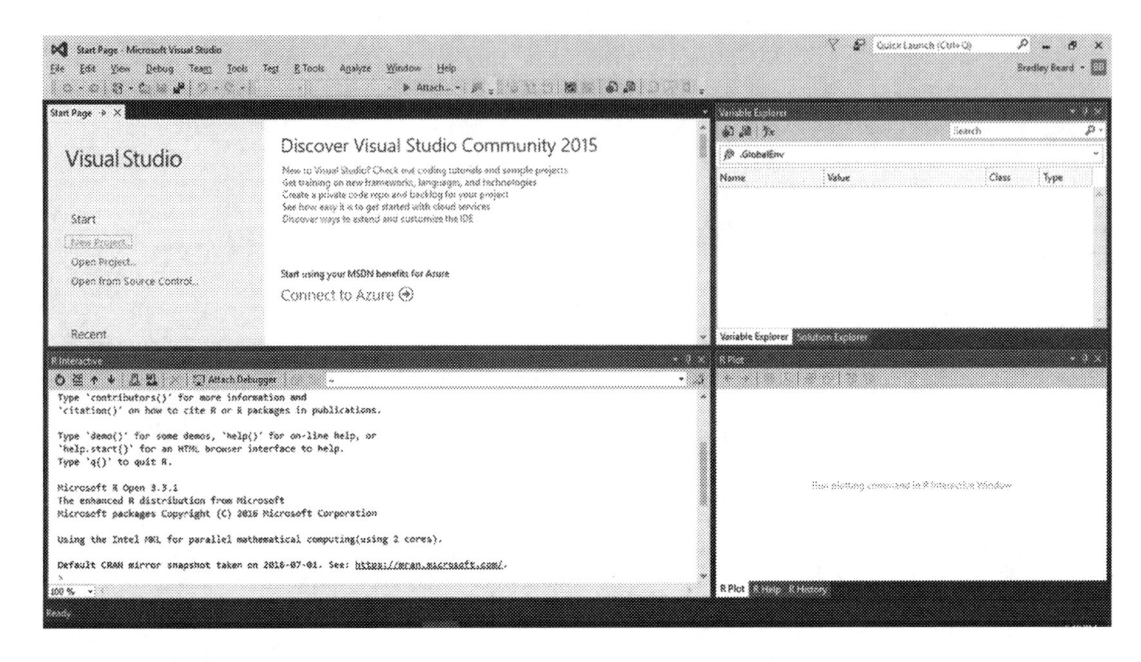

Figure 5-1. *R Tools for Visual Studio*

The reason that we are working in a fresh environment is because we are importing datasets and working with that data. I don't want there to be any chance of working with the wrong data and possibly getting incorrect results.

Importing the Dataset

The first thing that we have to do, as data scientists, is to gather our data. This sounds easy when you have a file location or URL to get the data from. When you don't have the data available and you have to define it, it is consequently *much* more difficult to ascertain. Luckily, that is not the case, since we have a complete set of data ready to be imported.

First, make sure that you have unzipped the file we downloaded from Chapter 4, RTVS-docs-master. Within this `.zip` file, navigate to `RTVS-docs-master\examples\MRS_and_Machine_Learning\Datasets`. In this directory is a file named `Weather_Sample.csv`. This file is our data source for this example, so you can either leave it there or copy it to another location on your PC that is easier to remember.

Now, with RTVS up and running, we go to R **Tools ➤ Data ➤ Import Dataset into R Session from Text File...** to get that data into our environment. Once you click the **Import Dataset into R Session from Text File** menu option, an interface appears that prompts you to navigate to the file that you want to import. We will use `Weather_Sample.csv`, so navigate to the location where you saved that file and click the **Open** button.

You should see what is shown in Figure 5-2.

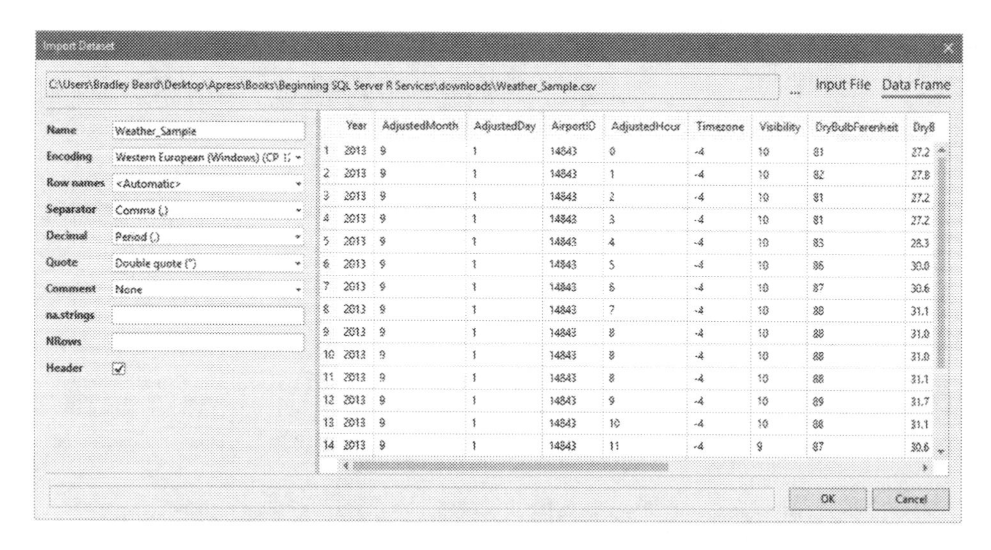

Figure 5-2. *Initial data view*

Pretty impressive! We can see that the first 19 rows of data were brought in, which is probably meant to give us a good sample size of the data types and lengths.

Note that you can click the top-right corner between Input File and Data Frame to see the raw data vs. the formatted data.

When you are done looking at this, click **OK**. You then see what is shown in Figure 5-3.

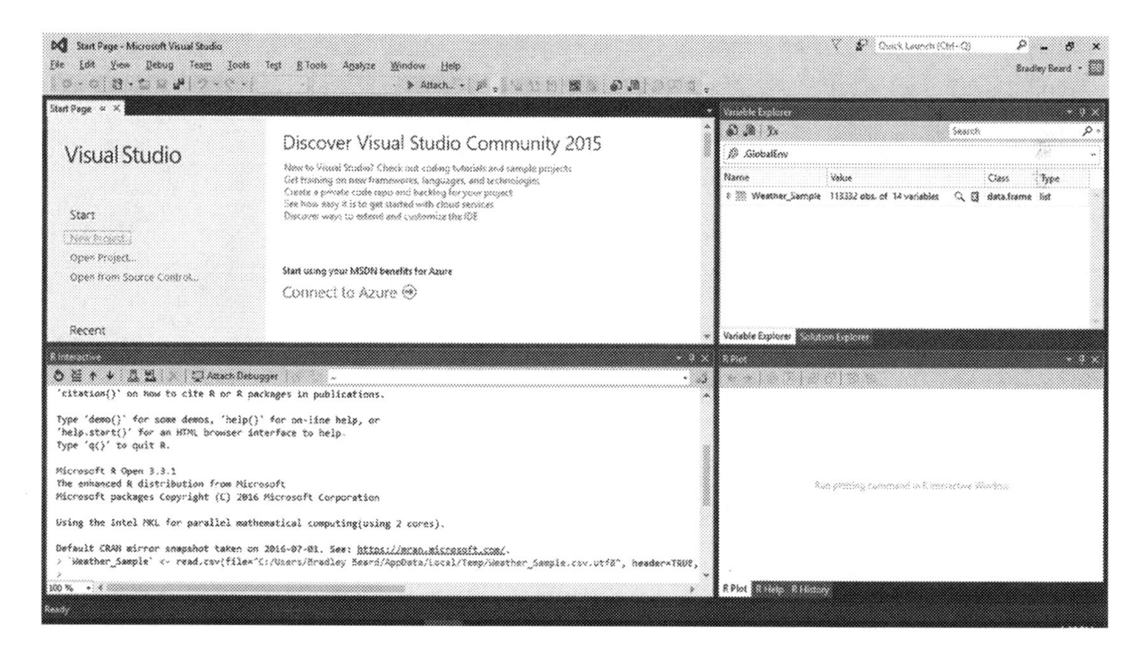

Figure 5-3. *Variable Explorer*

Note that the R Interactive window also generated the following code:

```
`Weather_Sample` <- read.csv(file="C:/Users/Bradley Beard/AppData/Local/Temp/Weather_Sample.
csv.utf8", header=TRUE, row.names=NULL, encoding="UTF-8", sep=",", dec=".", quote="\"",
comment.char="")
```

This is interesting, because this shows us from R's point of view how data gets imported into the current working project.

Opening a Script Pane

Now that we've loaded the dataset, we can open a Visual Studio pane for executing R scripts to prepare and analyze our data. Close the **Start Page** shown in the top-left pane of Figure 5-3. Then open a fresh R script by pressing **Ctrl+N**, clicking **R** in the left pane, and choosing **R Script**, as shown in Figure 5-4.

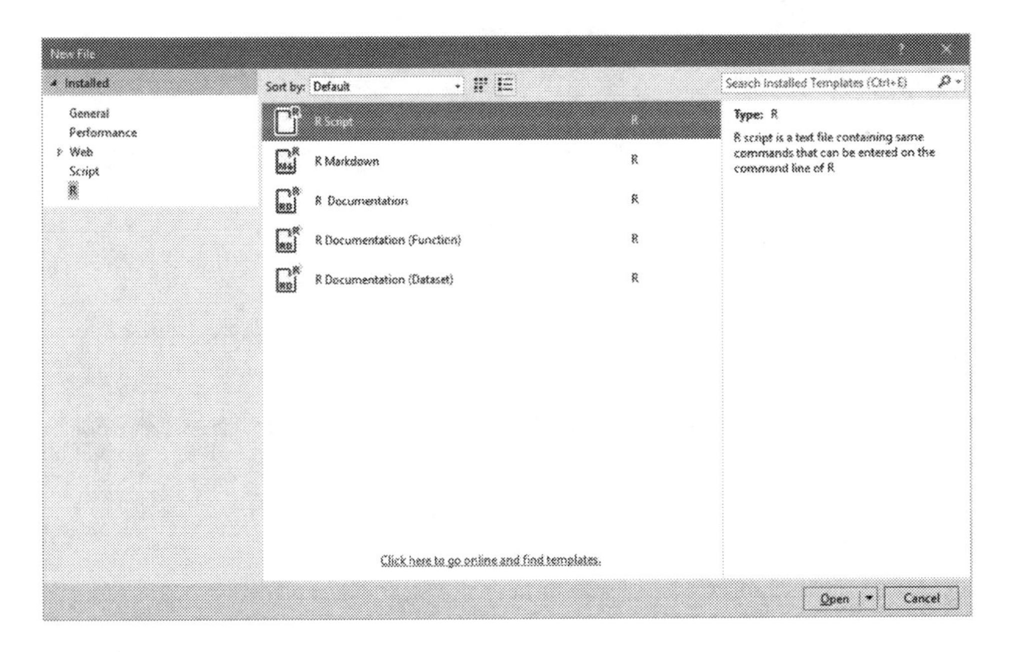

Figure 5-4. *New R Script*

Click the **Open** button to open a blank script window. You should now have an interface similar to what is shown in Figure 5-5.

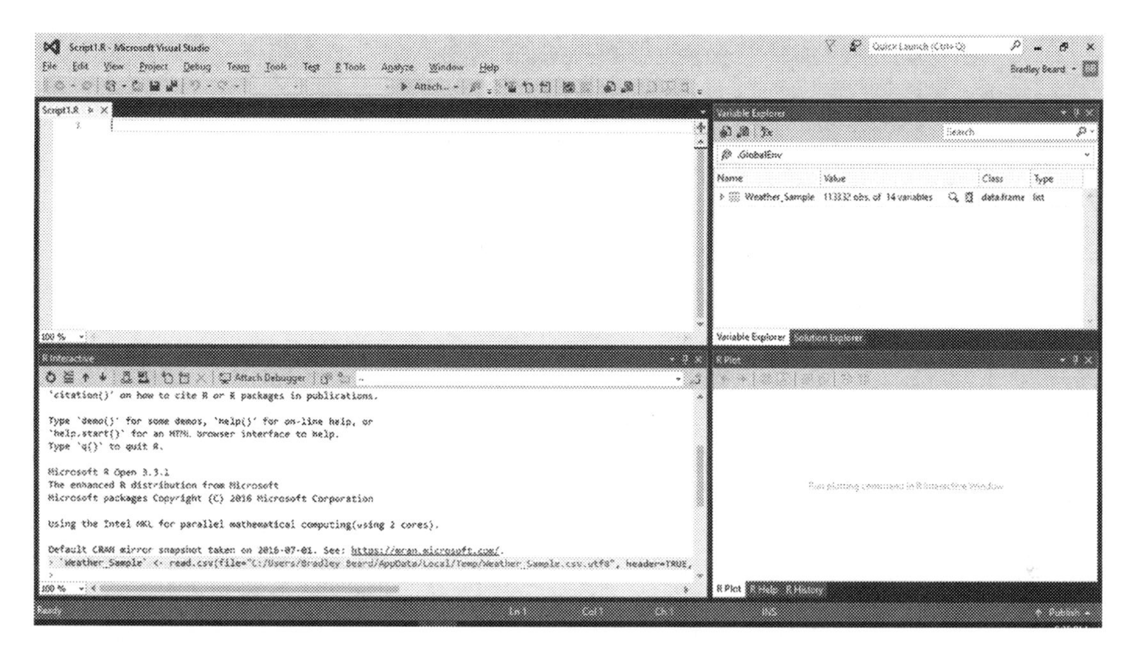

Figure 5-5. *Ready to work!*

Now that our data is loaded correctly and we have a fresh interface to work in, we can move on to the next section.

Preparing the Dataset

How on earth do you prepare a dataset? Isn't it already prepared? Well, yes and no: yes, because the data is already well-formed and meaningful, so now we get to see what it is really trying to say; no, because it isn't ready to be analyzed by R yet because it's still raw. We need to do some auxiliary things to the data first, to get it ready for analysis.

To get the data ready for analysis, type the following code into your Script1.R window:

```
install.packages("data.table")

library(data.table)

Weather_Sample <- data.table(Weather_Sample)

setkey(Weather_Sample, AirportID)
```

Figure 5-6 shows how your Visual Studio window should look after entering the code that I've just given.

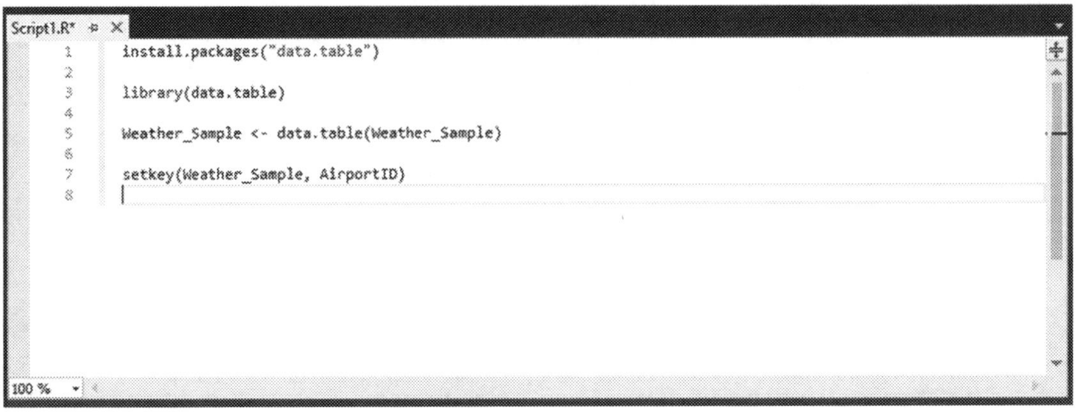

```
Script1.R*
1  install.packages("data.table")
2
3  library(data.table)
4
5  Weather_Sample <- data.table(Weather_Sample)
6
7  setkey(Weather_Sample, AirportID)
8
100 %
```

Figure 5-6. *Preparation*

Now, let's go over what this code is actually doing.

- `install.packages("data.table")`: This code simply installs the `data.table` package
- `library(data.table)`: This code makes `data.table` package available to the current R session
- `Weather_Sample <- data.table(Weather_Sample)`: This code sets an empty object named `Weather_Sample` equal to the `data.table` representation of the `Weather_Sample` dataset.
- `setkey(Weather_Sample, AirportID)`: This code allows for grouping on a specific column.

Now that we understand what is happening here, we see that there is no code that actually provides analysis. You are correct in that generalization, because we haven't actually begun the analysis, we have only begun to prepare the dataset for analysis. This is an important first step, because it is sort of like the foundation for what we do next, which is create the reports we need to generate as part of the software requirements document.

Executing the lines of code that I provided in Figure 5-6 shows that the `chron` package was also downloaded along with `data.table`. It is important to point out that any dependent packages to a requested package for download is always downloaded as well. This eliminates the possibility of errors occurring because of package dependencies. In case you haven't already executed those lines of code, go ahead and do it now.

Average Wind Speed by Airport ID (Tabular)

The first report that we are going to work on is Average Wind Speed by Airport ID. This report takes the data available and calculates the average wind speed per airport, and then displays the values in a chart. The columns we need to concern ourselves with are named `WindSpeed` and `AirportID`. We know that we have to have an object instantiated to hold the plot data and we know that we need to return the data to the object as a `data.frame` so that it can be interpreted into a chart correctly. We can start writing our code as follows:

```
avg_windspeed_by_ID <- as.data.frame(Weather_Sample)
```

▒ **Tip** Using the as keyword in front of the `data.frame` declaration indicates that we want to force the referenced dataset to be returned as a data frame.

That code doesn't quite do what we need yet, because we haven't yet figured out the `mean()` of the `WindSpeed` column yet. To do this, we need to update our code as follows:

```
avg_windspeed_by_ID <- as.data.frame(Weather_Sample[, mean(WindSpeed)])
```

That code looks closer to what we need, but it's not complete yet. What if there are blank rows in the data? This will surely affect our output, so let's get rid of those values by updating our code to the following:

```
avg_windspeed_by_ID <- as.data.frame(Weather_Sample[, mean(WindSpeed, na.rm = TRUE)])
```

We can now see that `na.rm = TRUE` will remove the "missing" values in a dataset.

Finally, we add a sorting column on the end. Keep in mind that this is different than the `setkey()` function we called earlier, because the `setkey()` function grouped the data. To sort the data by a column, update your code to the following:

```
avg_windspeed_by_ID <- as.data.frame(Weather_Sample[, mean(WindSpeed, na.rm = TRUE),
by = AirportID])
```

Now, highlight the command we just finished writing and press **Ctrl+Enter**. After that, type **avg_windspeed_by_ID**, highlight what you just wrote, and press **Ctrl+Enter** to execute it. You should see the following output in the R Interactive window at this time:

```
> avg_windspeed_by_ID
   AirportID        V1
1      10140  7.594268
2      10299  6.708183
3      10397  6.554401
4      10423  5-991983
5      10529  5-377292
6      10693  4.686593
7      10713  6.993095
8      10721  8.713368
9      10800  4.193095
10     10821  5-095719
```

Note that I scrolled up to the top of the returned data in order to show the top 10 results returned out of 66 rows. The following is the entire script to generate this result:

```
install.packages("data.table")

library(data.table)

Weather_Sample <- data.table(Weather_Sample)

setkey(Weather_Sample, AirportID)
```

```
avg_windspeed_by_ID <- as.data.frame(Weather_Sample[, mean(WindSpeed, na.rm = TRUE),
by = AirportID])

avg_windspeed_by_ID
```

Save your script as avgWindspeedByAirportID.tabular.R. This is the first part of the report that we are going to generate.

Average Wind Speed by Airport ID (Plot)

Next, open a new R script window. In that new window, you want to type in the following code:

```
library(data.table)

Weather_Sample <- data.table(Weather_Sample)

chart_by_ID <- as.data.frame(Weather_Sample[, mean(WindSpeed, na.rm = TRUE), by =
AirportID])

library(ggplot2)

ggplot(chart_by_ID, aes(x = AirportID, y = V1)) + geom_point(stat = "identity") + geom_
smooth(method = "lm", formula = y ~ splines::bs(x, 3)) + scale_x_continuous(name = "Airport
ID") + scale_y_continuous(name = "Average Wind Speed") +
geom_text(aes(label = AirportID), size = 3, vjust = 1.0) +
geom_text(aes(label = round(V1, digits = 2)), size = 3, vjust = 2.0)
```

Let's take a look at this code before we move on.

- `Weather_Sample <- data.table(Weather_Sample)`: We are using `Weather_Sample` as our object variable name, and then populating that object with the `data.table` representation of the data contained in the dataset `Weather_Sample` (which is shown by reference). There is a reference to the data, in other words, to create the analysis.
- `chart_by_ID <- as.data.frame(Weather_Sample[, mean(WindSpeed, na.rm = TRUE), by = AirportID])`: First, we're setting the variable name as before to `chart_by_ID`. Then, we're setting this variable equal to the data frame equivalent of the mean of the `WindSpeed` column in the dataset (`na.rm=TRUE` means to remove the N/A values) grouped by the `AirportID`. Does this make sense? It is imperative that you understand how this statement is structured, so that you can go back in and create custom reports from this data on your own. If you haven't gotten it yet, keep looking at it until it makes sense.
- `library(ggplot2)`: This command makes the newly installed ggplot2 package available to this project.
- `ggplot(chart_by_ID, aes(x = AirportID, y = V1)) + geom_point(stat = "identity") + geom_smooth(method = "lm", formula = y ~ splines::bs(x, 3)) + scale_x_continuous(name = "Airport ID") + scale_y_continuous(name = "Wind Speed") + geom_text(aes(label = AirportID), size = 3, vjust = 1.0) + geom_text(aes(label = round(V1, digits = 2)), size = 3, vjust = 2.0)`: This is a big one. This is just one huge command that controls how the chart looks and renders. The individual parts of this statement are:

- `ggplot(`: This is the plotting function.
- `chart_by_ID,`: Feeds the data.frame of Weather_Sample into this variable
- `aes(x = AirportID, y = V1)) +`: aes is the aesthetic definition for ggplot2. It is saying that the x axis (horizontal) shows AirportID and that the y axis (vertical) shows V1, which is going to the average wind speed, in this instance.
- `geom_point(stat = "identity") +`: The `geom_point()` function lets you define how you want the points on the plot defined. In this case, we want them shown as an identity.
- `geom_smooth(method = "lm", formula = y ~ splines::bs(x, 3)) +`: This lets us add a conditional average to the plot.
- `scale_x_continuous(name = "Airport ID") +`: This lets us continuously scale the x axis.
- `scale_y_continuous(name = "Wind Speed") +`: This lets us continuously scale the y axis.
- `geom_text(aes(label = AirportID), size = 3 , vjust = 1.0) +`: This lets us label and offset the x-axis label.
- `geom_text(aes(label = round(V1, digits = 2)), size = 3 , vjust = 1.0)`: This lets us label and offset the y-axis label.

Note that there is a plus sign at the end of some of the lines. This counts as a continuous character; in other words, the statement continues on the next line, which is not a separate statement.

Once you've got that all typed in, press **Ctrl+A** to select all, and then **Ctrl+Enter** to run it. You then see what is shown in Figure 5-7.

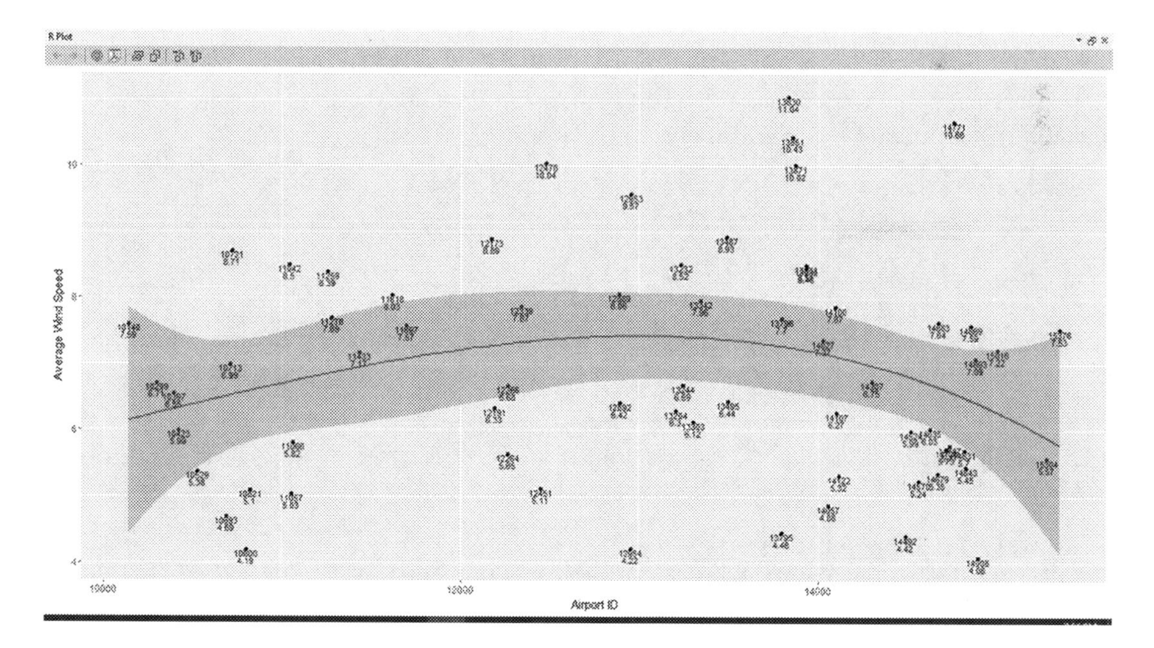

Figure 5-7. *Plotted information*

So we've got Airport ID on the x axis and Average Wind Speed on the y axis. Save that script as `avgWindspeedByAirportID.plot.R`.

Very nice so far! That finishes up the basics of the first report, so let's move on to the second one.

Report 2: Average Temperature by Airport ID (°F)

Our next task is to report the average temperature by airport ID. The code involved is very similar to that for the Average Wind Speed by Airport ID report, so you shouldn't have any trouble in following along.

We'll approach our task in two steps. First we'll compute the average temperature values. Then we'll plot them in the graph. This is exactly how we created the first report as well.

Average Temperature by Airport ID (Tabular)

Notice that the data is still loaded and available to us, so we can skip the step of loading the data and get right into computing our average temperatures. Following is the code to do that:

```
install.packages("data.table")

library(data.table)

Weather_Sample <- data.table(Weather_Sample)

setkey(Weather_Sample, AirportID)

avg_temperature_by_ID <- as.data.frame(Weather_Sample[, mean(DryBulbFarenheit,
na.rm = TRUE), by = AirportID])

avg_temperature_by_ID
```

Type that in, press **Ctrl+A** to select all, and then **press Ctrl+Enter** to execute it. At this point, your R Interactive window should show the following information:

```
   AirportID       V1
1      10140 62.76688
2      10299 45-91877
3      10397 68.61980
4      10423 74.13576
5      10529 58.44648
6      10693 65-93558
7      10713 57.59134
8      10721 60.77466
9      10800 71.38043
10     10821 62.85850
```

That's the first 10 rows of what you should see.

If you scroll down, you see that all 66 rows have been returned. Now save the script as `avgTemperatureByAirportID.tabular.R`.

Average Temperature by Airport ID (Plot)

We have the data that we need, so let's plot it onto a chart now. The code for this plot is based on the previous plot, so let's look at that code now:

```
Weather_Sample <- data.table(Weather_Sample)

chart_by_ID <- as.data.frame(Weather_Sample[, mean(DryBulbFarenheit, na.rm = TRUE),
by = AirportID])

install.packages("ggplot2")

library(ggplot2)

ggplot(chart_by_ID, aes(x = AirportID, y = V1)) + geom_point(stat = "identity") + geom_
smooth(method = "lm", formula = y ~ splines::bs(x, 3)) + scale_x_continuous(name = "Airport
ID") + scale_y_continuous(name = "Average Temperature") +
geom_text(aes(label = AirportID), size = 3, vjust = 1.0) +
geom_text(aes(label = round(V1, digits = 2)), size = 3, vjust = 2.0)
```

This code should look very familiar. I only changed the `WindSpeed` text in the `chart_by_ID` section to `DryBulbFarenheit`, and the `scale_y_continuous` name to `Average Temperature` instead of `Average Wind Speed`.

Once you get this code typed into RTVS, press **Ctrl+A** to select all, and then press **Ctrl+Enter** to execute the code. Figure 5-8 shows what you should see at this point.

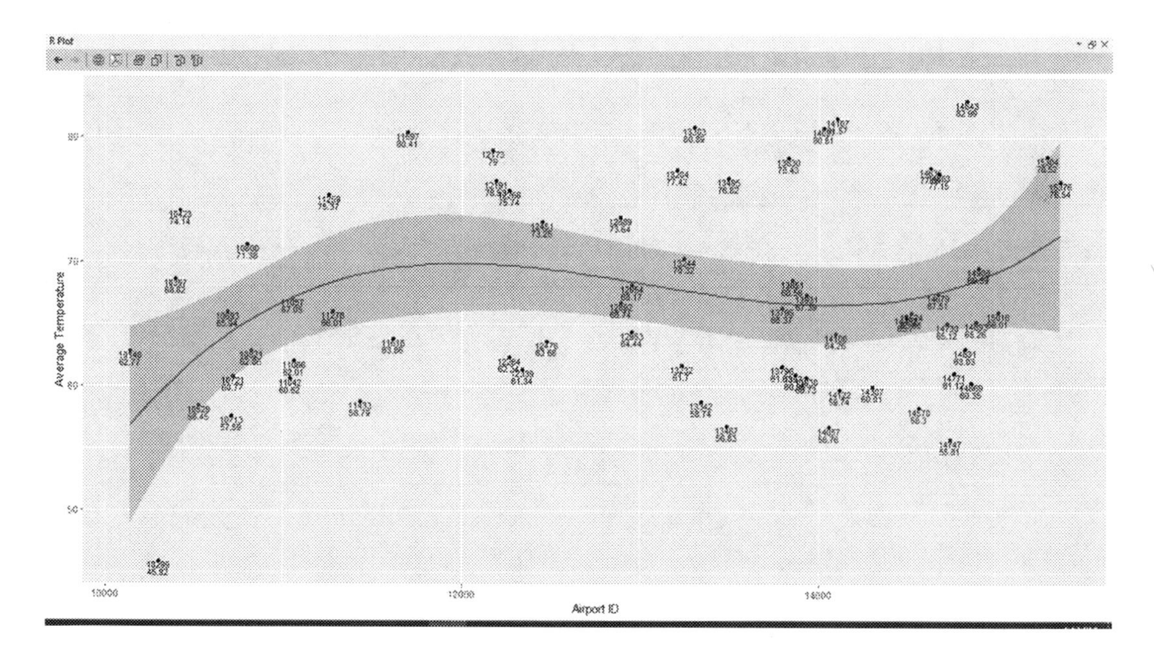

Figure 5-8. *Plotted data*

Excellent! That shows us our data—and even a moving average line thrown in for good measure.

Save that script as `avgTemperatureByAirportID.plot.R`. The following are the four files that you should have saved:

- `avgTemperatureByAirportID.plot.R`
- `avgTemperatureByAirportID.tabular.R`
- `avgWindspeedByAirportID.plot.R`
- `avgWindspeedByAirportID. tabular.R`

Summary

Let's review this chapter really quick. We actually did an awful lot here. You should be feeling pretty good about yourself and your growing knowledge of R. Here's what we've accomplished:

- Gone through the basics of plotting in R
- Loaded an external data source (`Weather_Sample.csv`) into Visual Studio
- Wrote custom R scripts based on this data in accordance with the software requirements document
- Generated both tabular data and plots based on this data

The next chapter is gets into the Reporting Services aspect of the delivery of these reports.

PART III

Creating and Viewing Reports

CHAPTER 6

Configuring SQL Server Reporting Services

Recall back in Chapter 1 that we installed Reporting Services as part of the initial SQL Server installation. When we installed Reporting Services, we selected the **Install and configure** option. Now we have to configure the Reporting Services instance that we installed. Keep in mind that Reporting Services was already installed and configured (and operational) at the point that the **Install and configure** option was selected when SQL Server was installed. The purpose of this chapter is to familiarize you with the options available within Reporting Services Configuration Manager, so that if you need to change a setting in your own installation, you have the knowledge and confidence of how to make the required change and where to look for the solution.

First, we'll run Reporting Services Configuration Manager and connect to our newly installed instance, and then we'll go through a number of configuration options at our disposal.

Connecting to an Instance

Go to the Windows Start menu. Reporting Services Configuration Manager should show in the Recently Added box. If it's not there, open the Windows Start menu, click **All Apps**, and then scroll down to Microsoft SQL Server 2016. It should be inside that folder. If not, just start typing **Reporting Services** and it will pop up. Click it to continue. You then see what is shown in Figure 6-1.

B. Beard, *Beginning SQL Server R Services*, DOI 10.1007/978-1-4842-2298-0_6

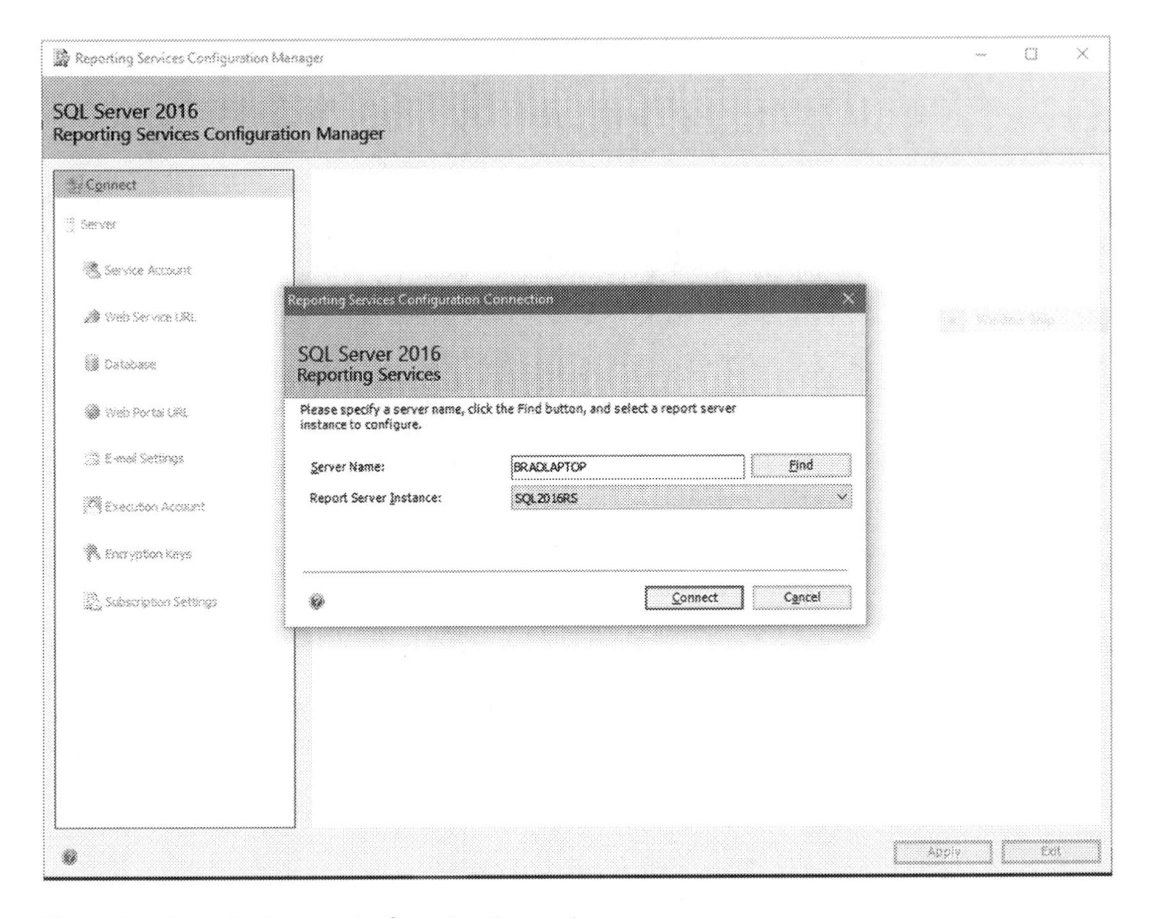

__Figure 6-1.__ Reporting Services Configuration Connection

Click the **Connect** button here to connect to your instance. You'll then see the home configuration page shown in Figure 6-2.

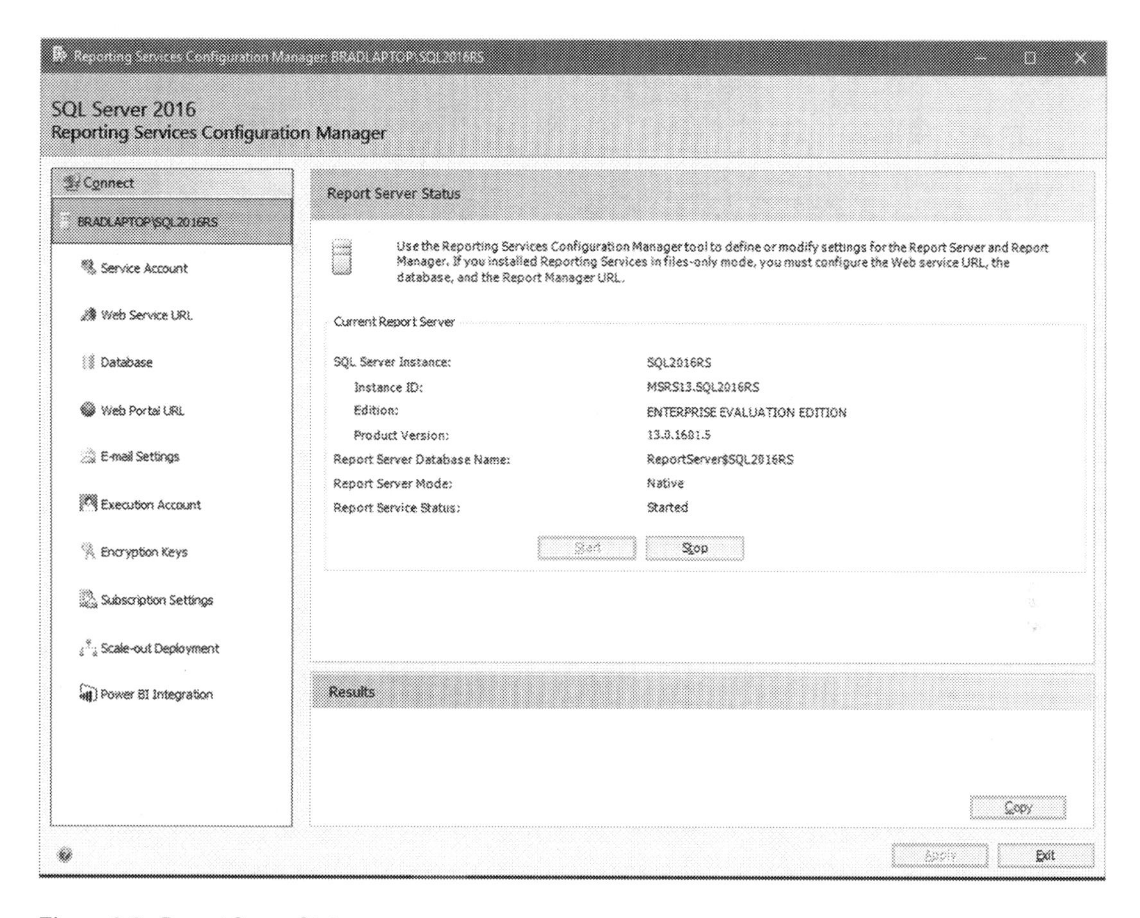

Figure 6-2. *Report Server Status*

The home configuration page, which is my own name for it, tells you general information about your Reporting Services instance. Most importantly, note that the service has started. We won't get very far otherwise.

Let's take a quick look at this interface before we move on. I will introduce the areas in the following subsections, and then we go back and update the fields so that our report server starts and runs successfully.

Service Account

The settings in the Service Account area can actually stay just as they are by default. If you've changed the **Use built-in account** option shown in Figure 6-3, the default setting is Virtual Service Account. This is an account that was created when you installed Reporting Services in order to connect to the report server. You could also choose Local System, Local Service, or Network Service. I would advise against that, because these accounts have heightened privileges that aren't necessary to run the report server.

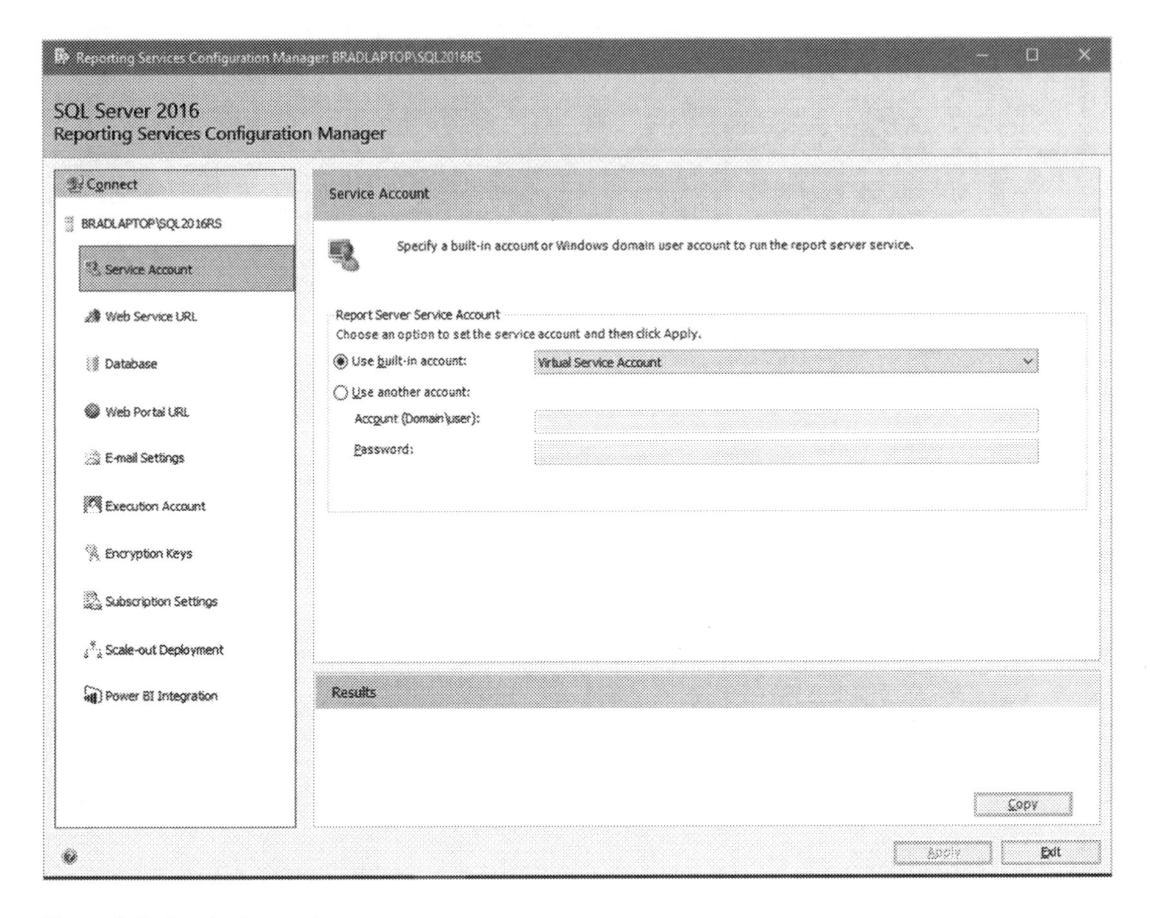

Figure 6-3. *Service Account*

You can also specify another domain account to interact with this service, if you prefer. This is the choice for a lot of professionals, because they can tailor the user account to be specific to this instance of this application, for example. If you choose this method, go right ahead and put the username and password for this account in the appropriate boxes.

Keep in mind that if you choose a custom service account, the account specified more than likely needs to be reviewed by a server or domain administrator in your organization for security purposes and for adherence to corporate policies. In my experience, it is best to have the server or domain administrator provide a list of service accounts that can be used for the various purposes needed within a database or application server.

Web Service URL

Figure 6-4 shows that this area allows you to configure a URL where the reports generated by the Reporting Server are to be generated and provided to authorized users. This area is already set up, unless you wanted to give your virtual directory a different name. In that case, go ahead and do that.

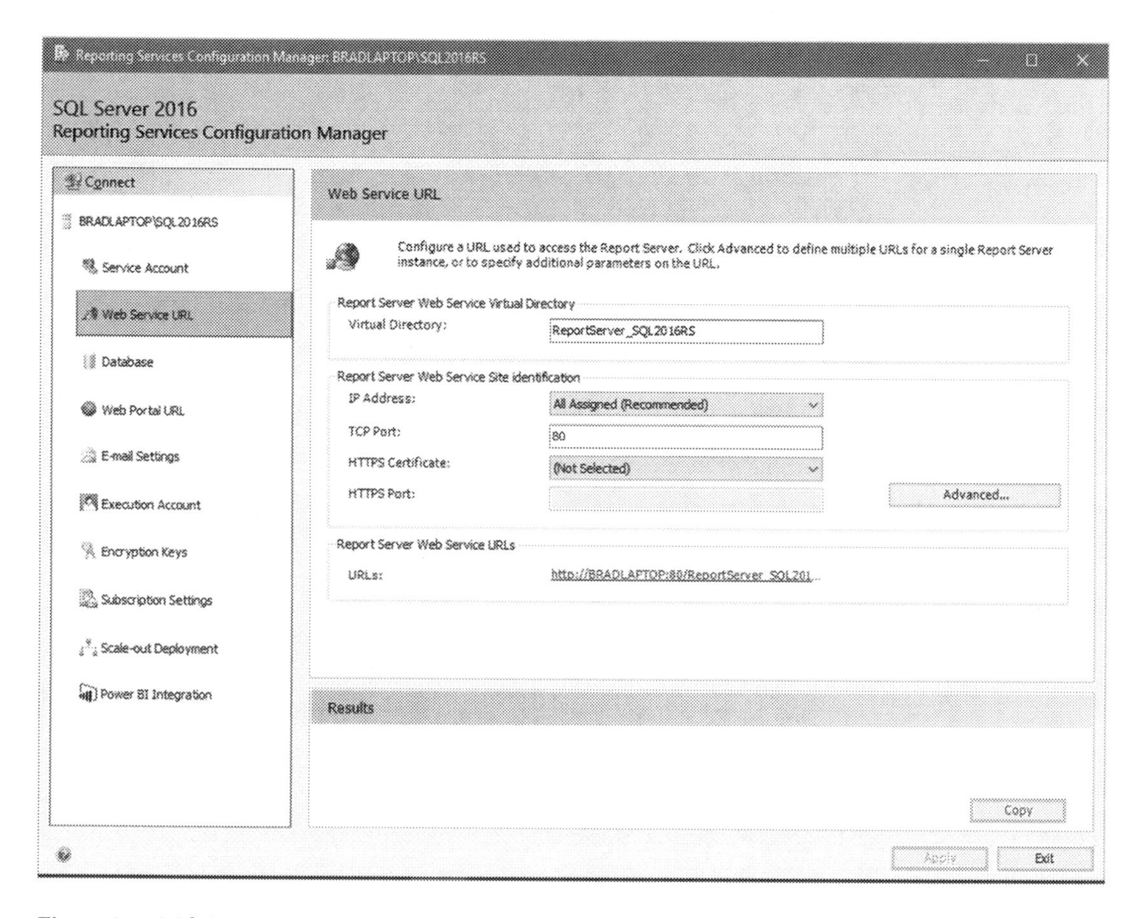

Figure 6-4. *Web Service URL*

Note that there are default values that have been pre-populated for you.

The only setting I would advise on adding is the HTTPS Certificate option. If you happen to have an SSL certificate, make sure that the certificate is provided here, if you want to run over HTTPS port 443.

Database

Figure 6-5 shows that this area allows you to configure the database that the report server is going to connect to for data.

Note that the Current Report Server Database options are also filled in. This indicates the database server that hosts the SQL Server Reporting Services databases. Microsoft did this for us when we selected the **Install and configure** option back in Chapter 1.

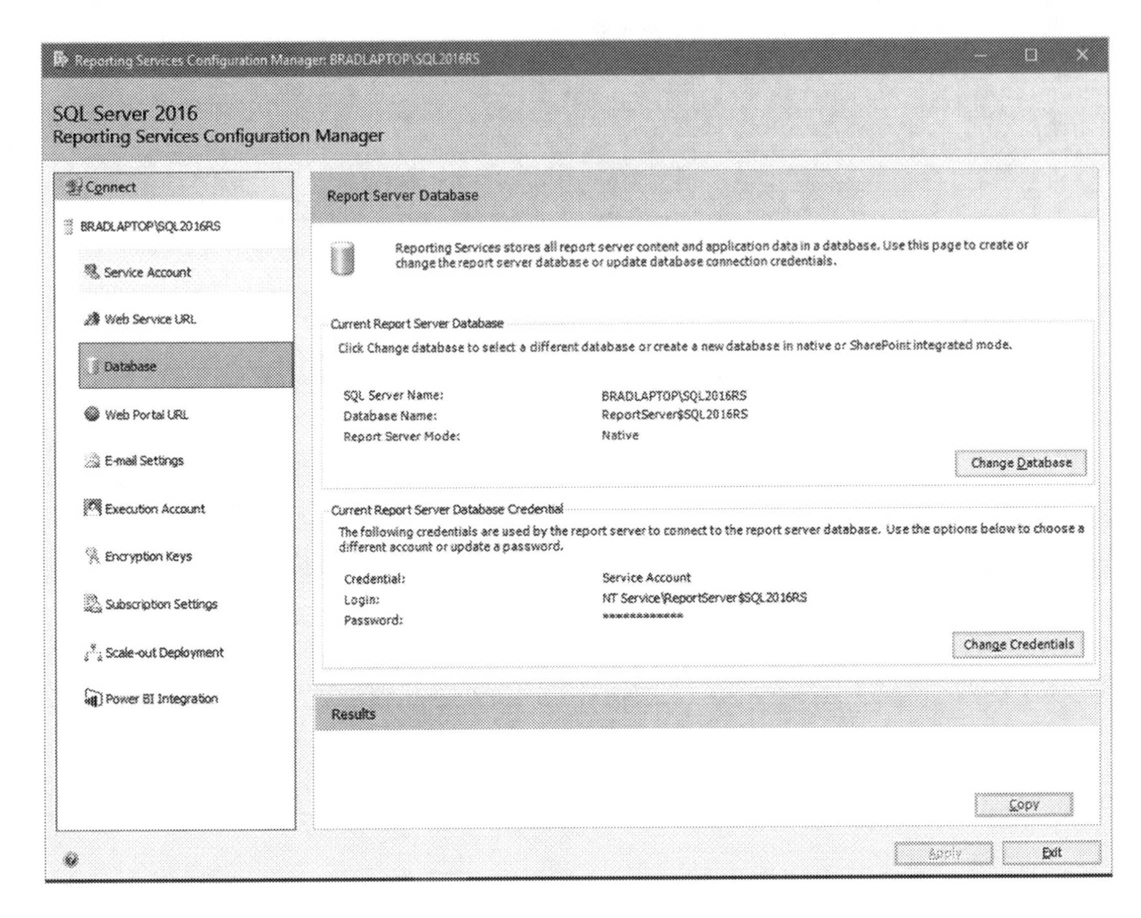

Figure 6-5. Database

Next, we get to set up the database that we want to point the report server to. This means that the report server looks at this specific database, by default. Obviously, this can be changed with separate installations and instances of Report Server, but for now, we'll just stick with this one since that is sort of beyond the scope of this book. This also implies that other databases, apart from the Reporting Services databases, can be used as sources of data.

Web Portal URL

Notice that we now have a hyperlink on the screen that reads `http://BRADLAPTOP:80/Reports_SQL2016RS`? If you are happy with that default URL, then everything is fine for this area. Figure 6-6 shows that this area allows you to define a custom URL for access to the web portal, in case you aren't happy with the default value. I advise you to leave the Virtual Directory value alone, since the default value works just fine and this is just for evaluation purposes.

Note that the Web Service URL (discussed earlier in this chapter) must be defined before this area can be defined.

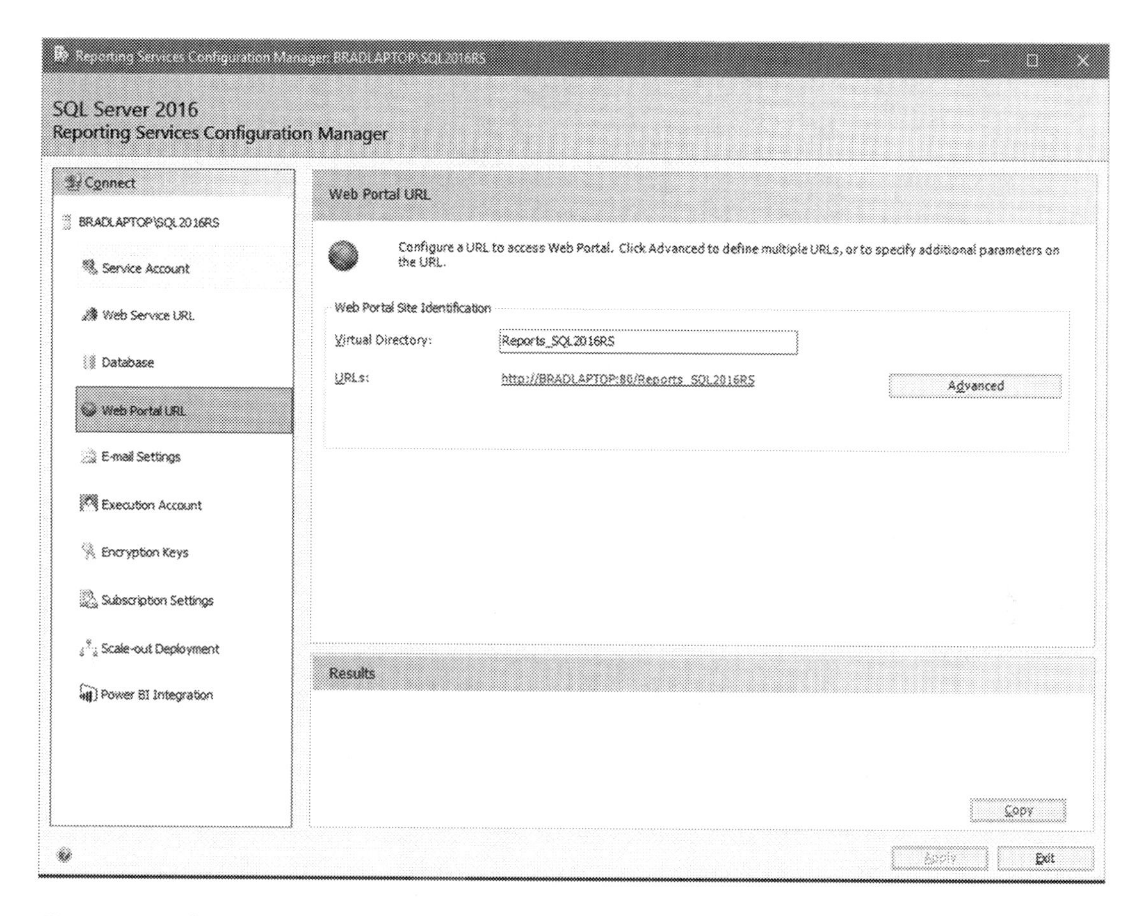

Figure 6-6. *Web Portal URL*

E-mail Settings

Figure 6-7 shows that this area allows you to specify e-mail settings. Obviously, this would be useful if you wanted to send e-mails with the reports attached. The default values are not selected. You must fill these options in yourself.

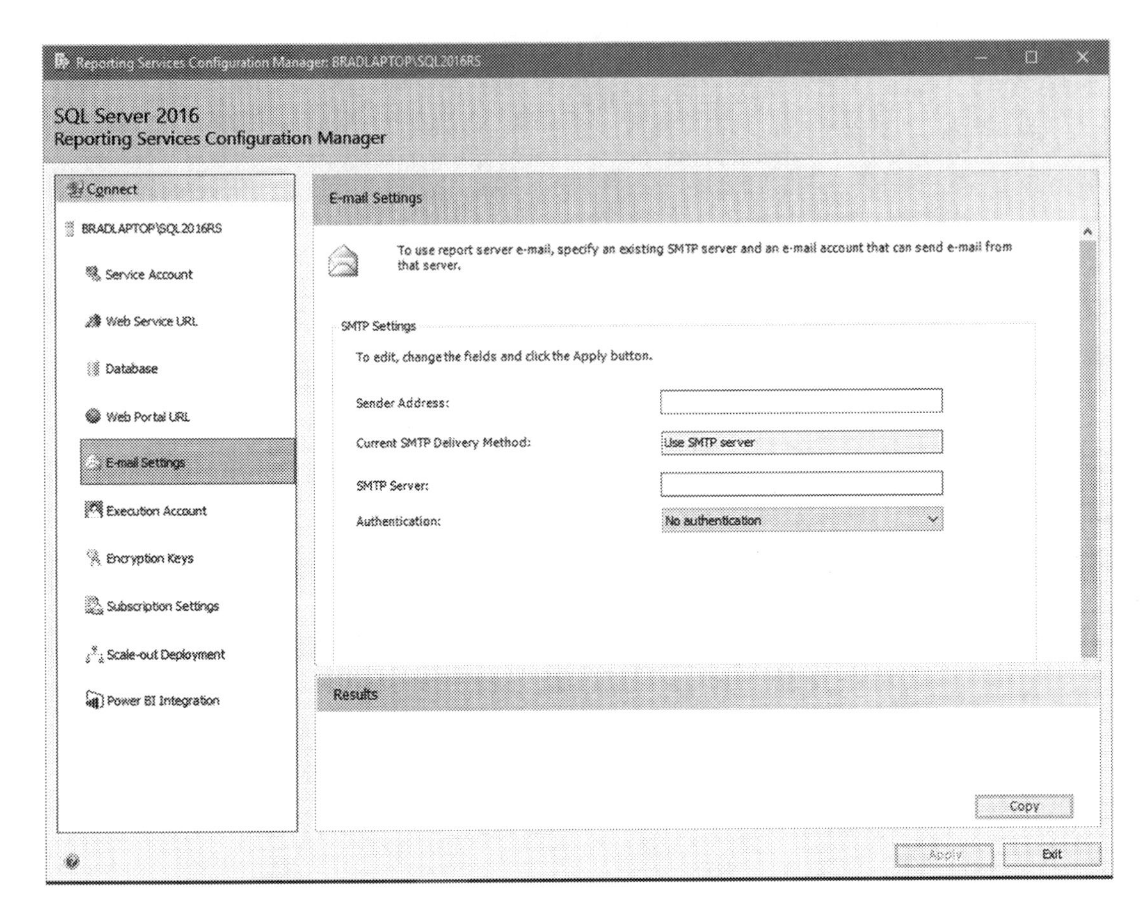

Figure 6-7. *E-mail Settings*

As you can see in Figure 6-8, I have provided the general format for how you want to fill in the fields shown.

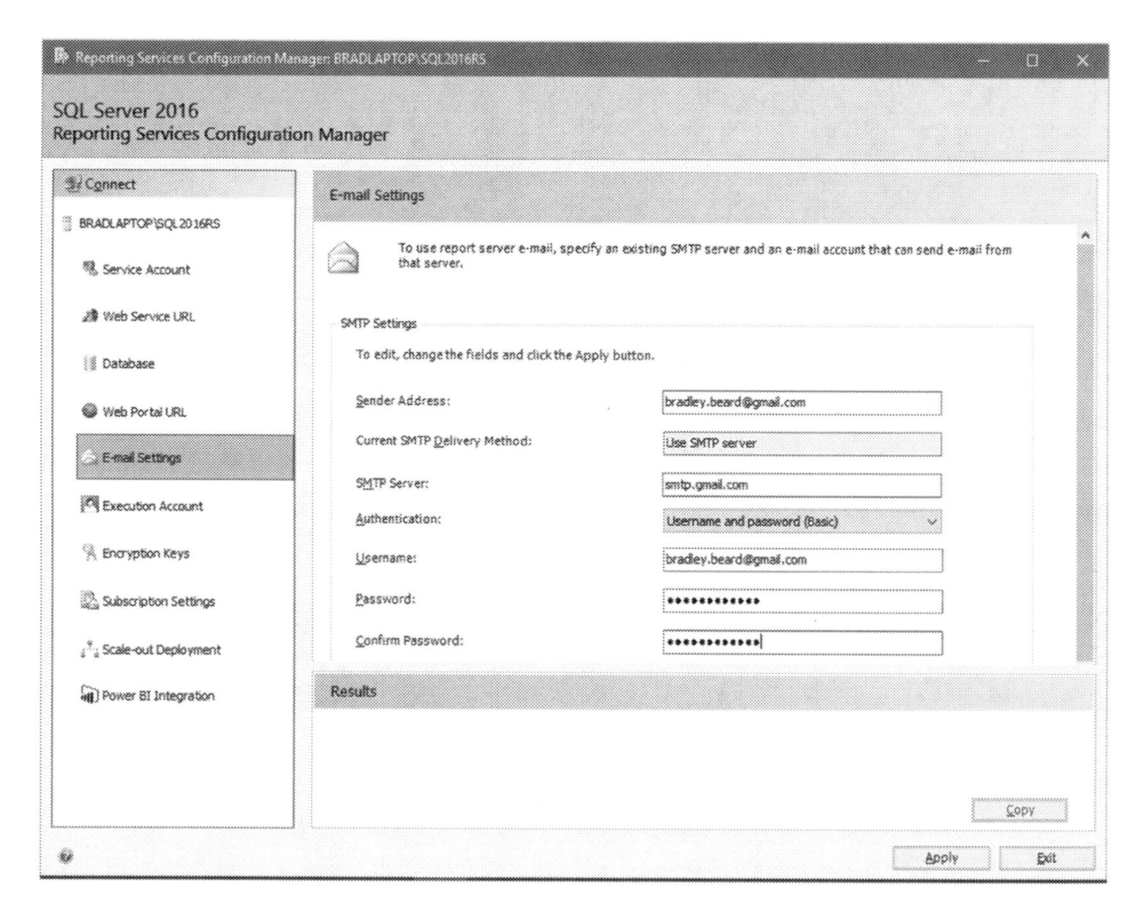

Figure 6-8. E-mail Settings, updated

When you have updated your settings, click **Apply** to save them.

■ **Note** Obviously, your settings will be different than the ones shown here.

Execution Account

Figure 6-9 shows that this area allows you to specify an execution account.

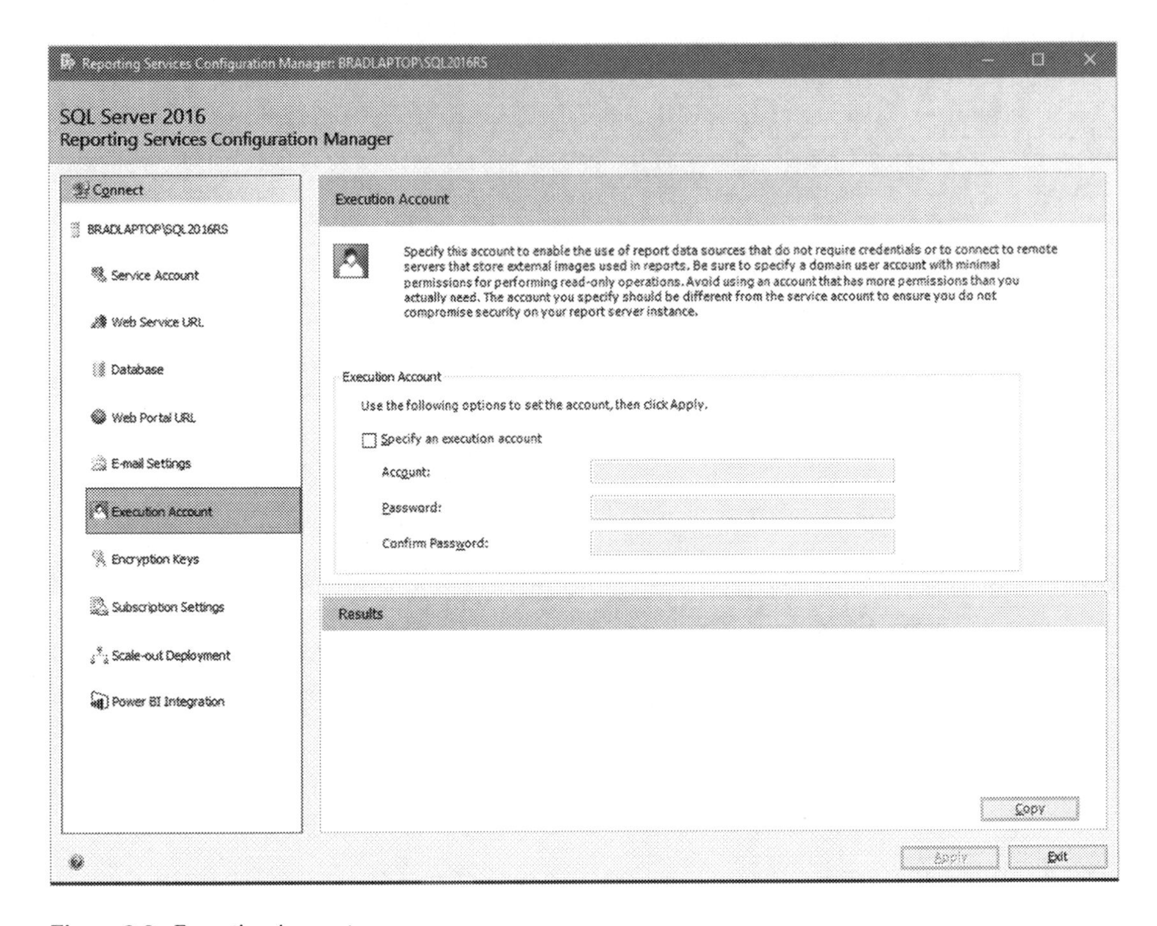

Figure 6-9. *Execution Account*

We don't need to worry about an execution account for this book, so just bypass this for now.

Encryption Keys

Figure 6-10 shows that this area allows you to backup, restore, change, or delete encryption keys.

I highly recommend backing up your encryption keys immediately. Without these keys, you cannot decrypt any encrypted database. It is of vital importance that the keys be backed up regularly and stored securely.

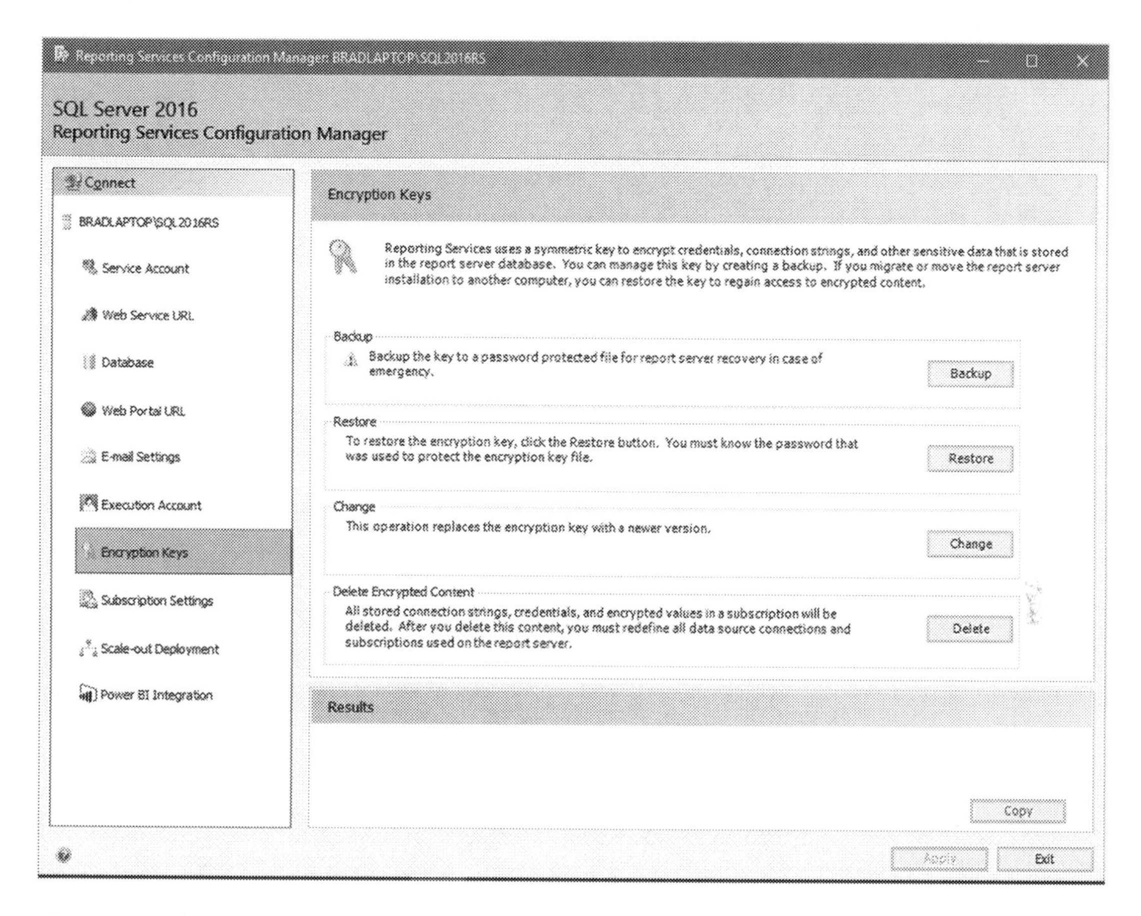

Figure 6-10. *Encryption Keys*

Subscription Settings

Figure 6-11 shows that this area allows you to configure an account that is used by remote users to access your available file share subscriptions. Note that a file share subscription is necessary when e-mail delivery is not allowed.

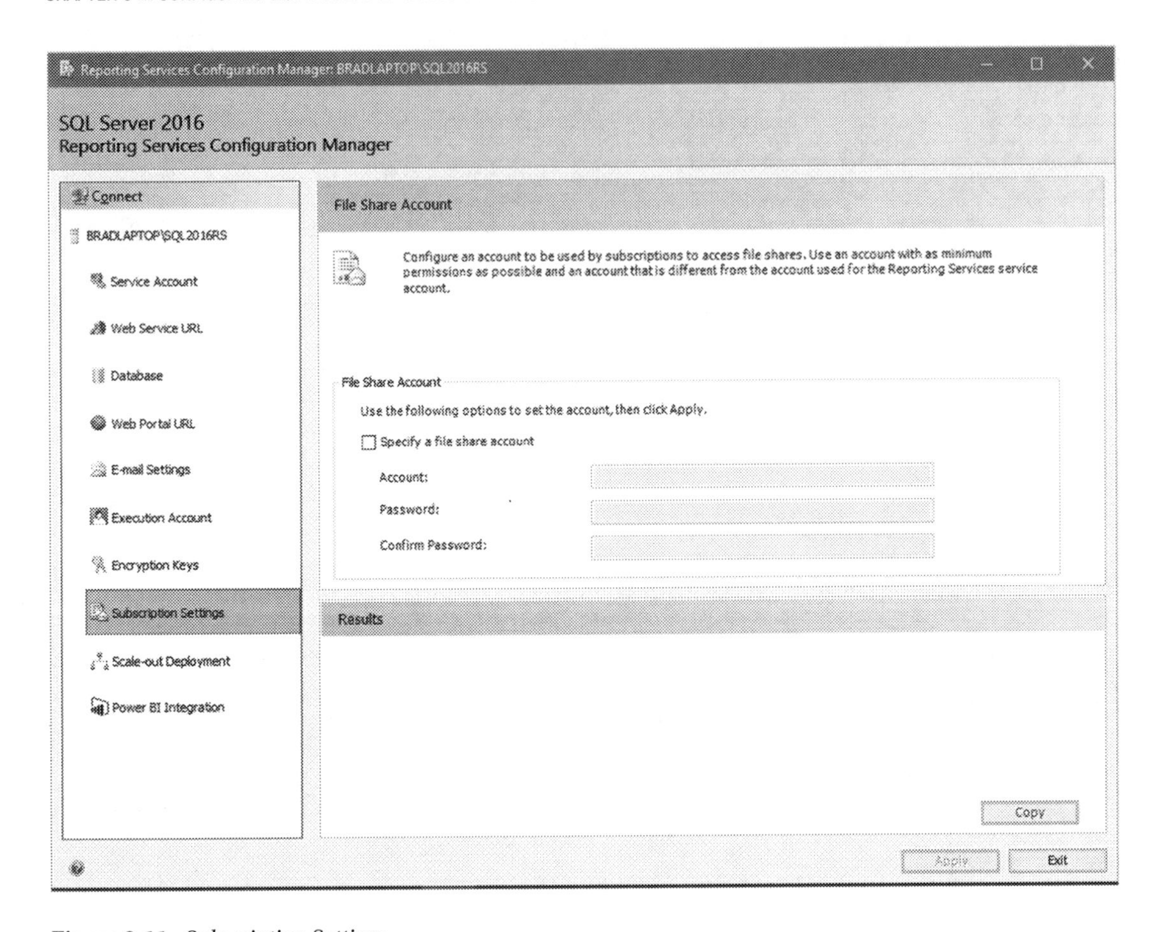

__Figure 6-11.__ Subscription Settings

Since this deployment is strictly for testing, we can leave this area alone.

Scale-out Deployment

Figure 6-12 shows that this area allows you to view information about a scale-out deployment. A scale-out deployment is essentially a load-balancing model, which increases scalability in a server cluster. As more users of the Reporting Services instance consume resources, the load can be shared to another server that shares the same encrypted data as the original server. At this time, the scale-out deployment option is only available in Enterprise edition for production instances.

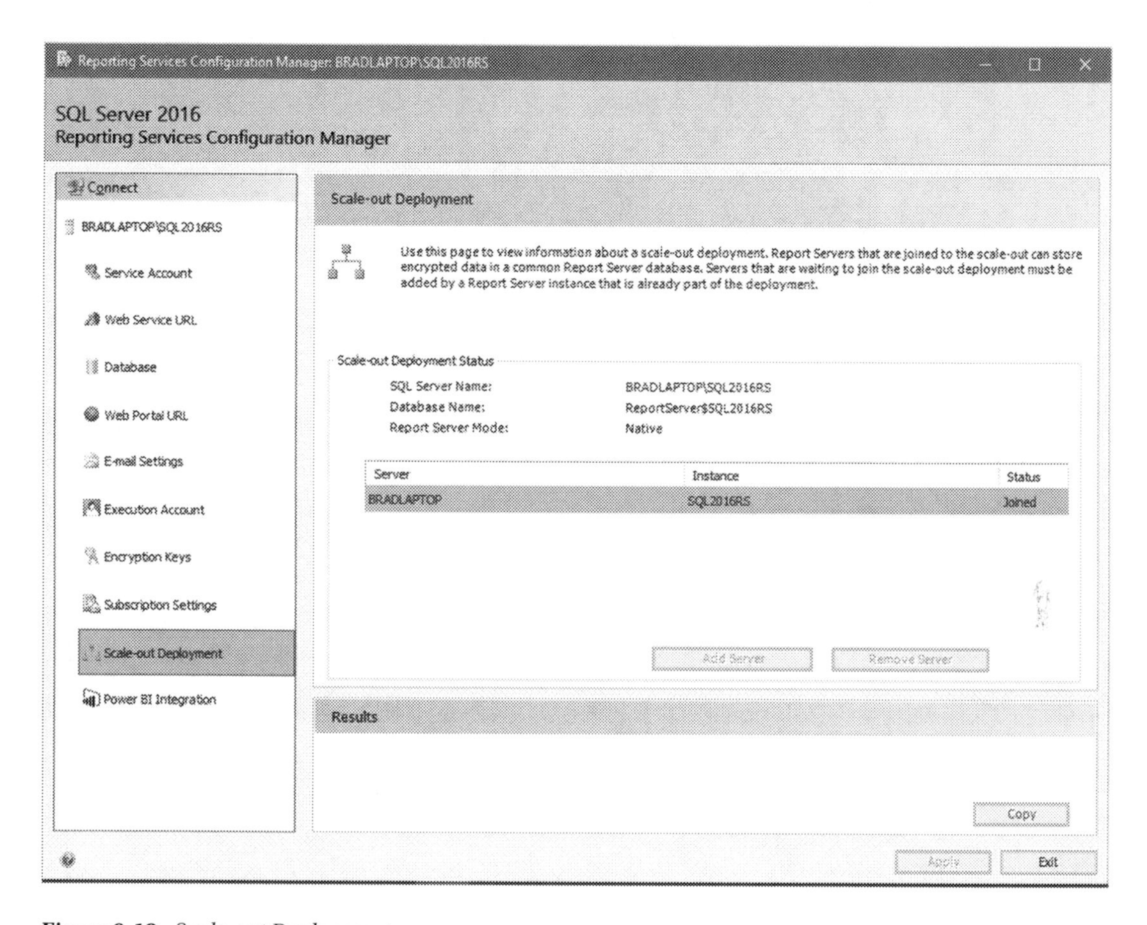

Figure 6-12. *Scale-out Deployment*

Power BI Integration

Figure 6-13 shows that this area allows you to set up integration with Power BI with a single click. Unfortunately, Power BI is beyond the scope of this book, so leave these settings at their defaults.

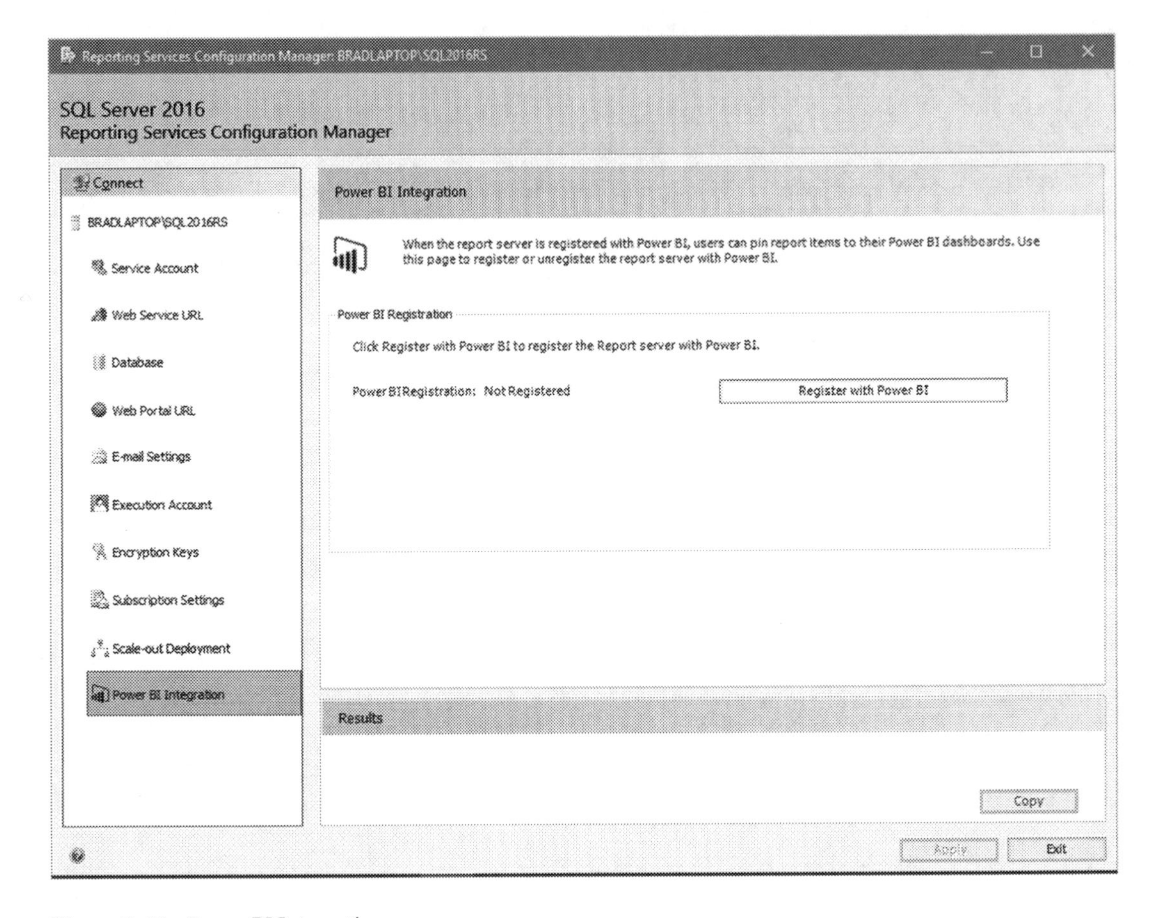

Figure 6-13. *Power BI Integration*

First of all, how cool is it that Power BI can integrate right into Reporting Services—and with a single click?

Now, go back to the Web Portal URL page and click the hyperlink shown on the page. The location of this hyperlink is shown in Figure 6-14.

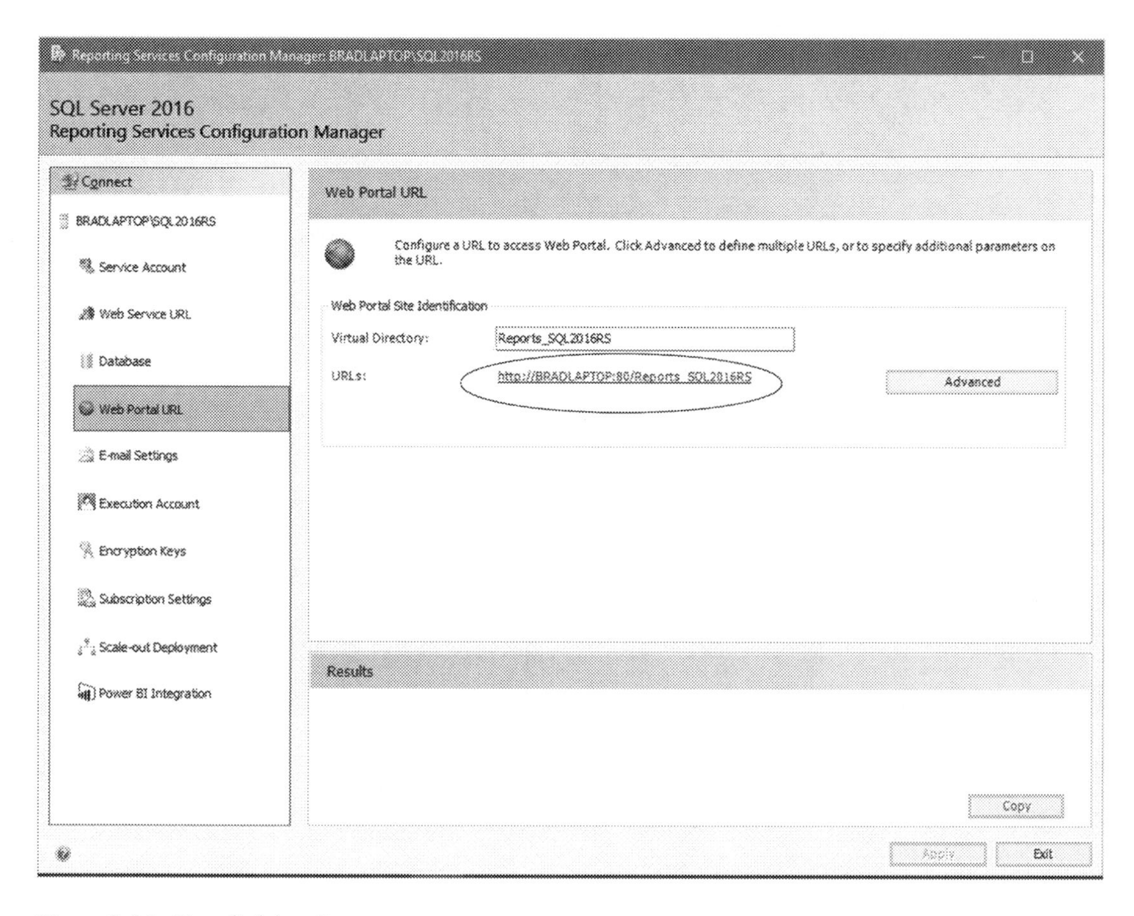

Figure 6-14. *Hyperlink location*

When you click the hyperlink, you should see what is shown in Figure 6-15.

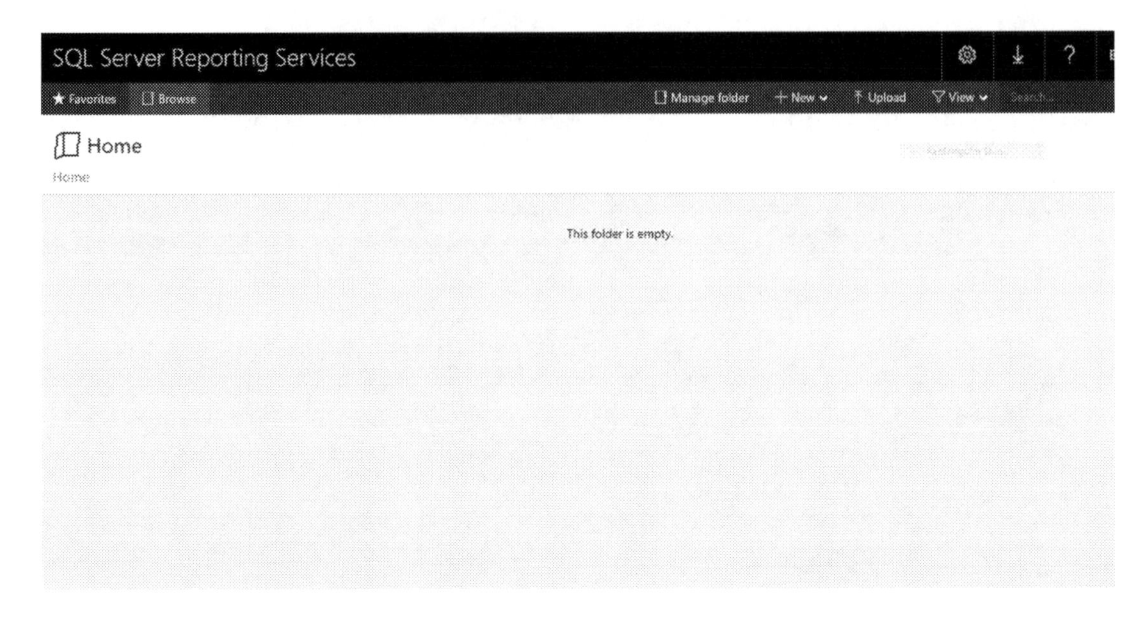

Figure 6-15. *Initial reports interface*

If you see anything except this screen, then something went wrong with your installation or configuration of SQL Server Reporting Services. Sometimes, depending on the operating system and the browser settings that you currently have in place, you might get an error screen. In this instance, you may need to run your browser instance as a local administrator (if you're running on a local computer). It's doubtful that you need to uninstall everything, so that's the good news. In the event of an error at this point, it is recommended to take a look at the Reporting Services error log.

Summary

This is a good time to stop and take a breather for a second. We've actually done quite a bit in these past few pages, so let's recap.

- Verified that Reporting Services installed correctly
- Configured our new installation of Reporting Services
- Verified that the web portal is working as advertised

Note that there is no content yet in the web portal. This is to be expected, as it is a brand-new installation and we haven't begun writing reports yet.

In the next chapter, we install and configure Report Builder, which we use to create the reports that consume the R data for use in our reports. Configure an account to be used by subscriptions to access file shares. Use an account with as minimum permissions as possible and an account that is different from the account used for the Reporting Services service account.

CHAPTER 7

Report Builder Installation and Data Preparation

Now that we have SQL Server Reporting Services installed, we need to figure out how to create and deliver the reports requested by the customer in our software requirements document. Probably the best way to create the reports is using Report Builder, a free download from Microsoft. The best way to deliver the reports, obviously, is to use the built-in functionality from SQL Server Reporting Services, as previously configured in Chapter 6.

Keep in mind that, if the customer only wanted flat-file reports (meaning just the data in CSV or XML format, for example) we could do that with a simple export of the data from SQL Server without interaction from SQL Server Reporting Services in the slightest. But the real point of this exercise is to show the power of R when integrated with SQL Server; so, using Report Builder, we're going to incorporate those charts that we created in Chapter 5 into a slick-looking report. We can then deal with little things like formatting and such, but the most important part is making sure that the report can be built first.

Download Report Builder

To start this chapter off, we need to download Report Builder. The link for this is at `https://www.microsoft.com/en-us/download/details.aspx?id=52674`, but that could change. Google is your friend, remember.

Tip You can also download Report Builder from the SQL Server Reporting Services Web Portal that we saw at the end of Chapter 6 by clicking the down arrow in the top-right corner.

Figure 7-1 shows the page where Report Builder can be downloaded. Note that this interface could change in the future, but the Download button should stay clearly visible.

B. Beard, *Beginning SQL Server R Services*, DOI 10.1007/978-1-4842-2298-0_7

Microsoft® SQL Server® 2016 Report Builder

Select Language: English Download

Report Builder provides a productive report-authoring environment for IT professionals and power users.

⊕ Details

⊕ System Requirements

⊕ Install Instructions

⊕ Additional Information

Figure 7-1. *Report Builder download screen*

Click the button to download. You should see what is shown in Figure 7-2.

Microsoft® SQL Server® 2016 Report Builder

If your download does not start after 30 seconds, Click here

Figure 7-2. *Download in progress*

A file called ReportBuilder3.msi is being downloaded in the background. Once the download gets done, you should be able to click it in the browser bar to begin the installation. If you don't see the downloaded file here, just go to your Downloads folder and sort by date. The file will show up there.

Figure 7-3 shows the first screen that is shown for the installation.

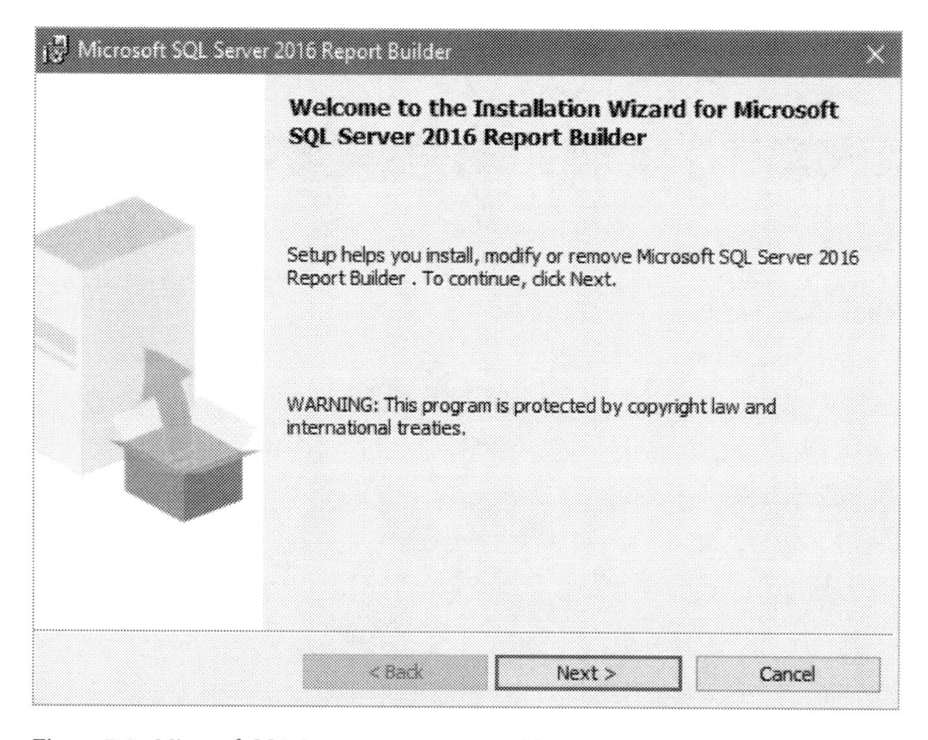

Figure 7-3. *Microsoft SQL Server 2016 Report Builder Installation*

Obviously, we click **Next** to continue to the license terms shown in Figure 7-4.

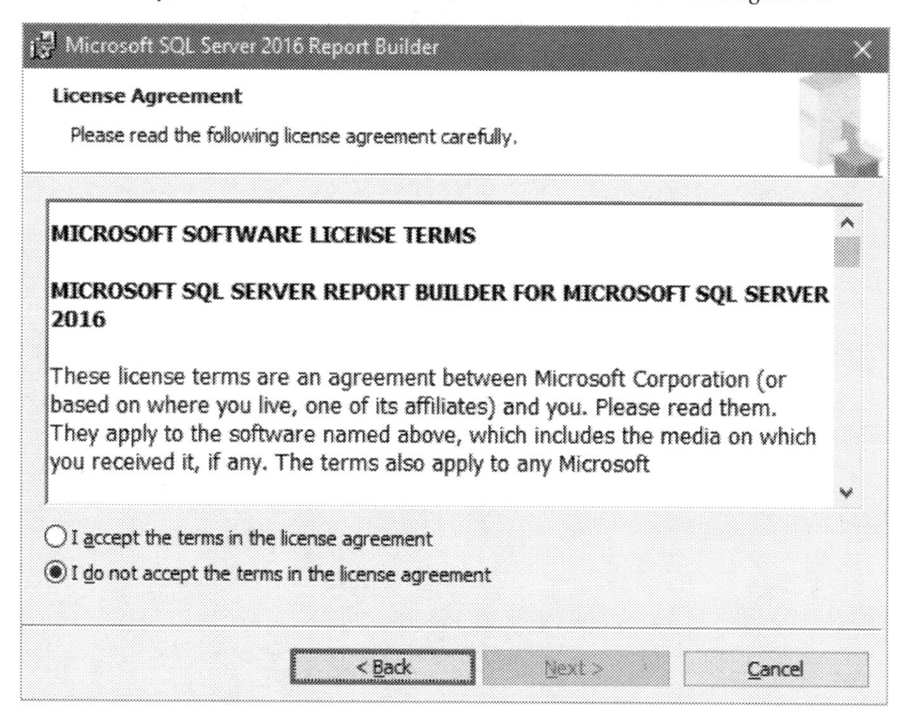

Figure 7-4. *License Agreement*

Click the **I accept the terms in the license agreement** radio button and then click **Next**.

Next, we get to choose our features. Figure 7-5 shows the default feature selection screen.

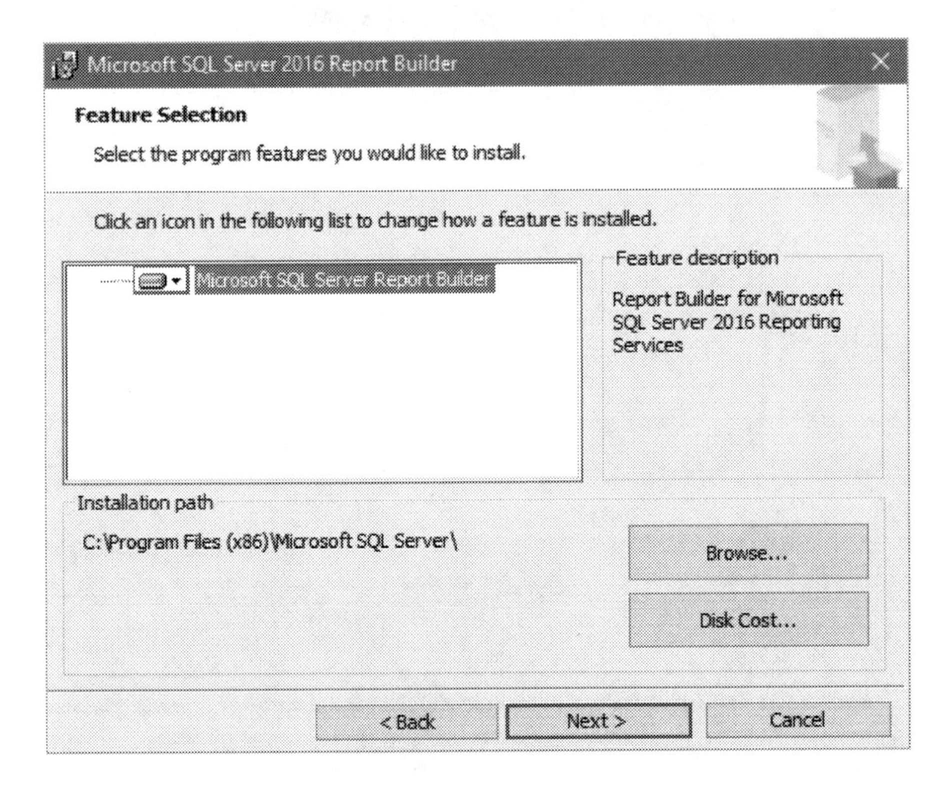

Figure 7-5. *Feature Selection*

We want to install everything for this, so pull down the menu on the disk in the white area. Choose **Entire feature will be installed on local hard drive** to continue.

Figure 7-6 shows the correct selection for this.

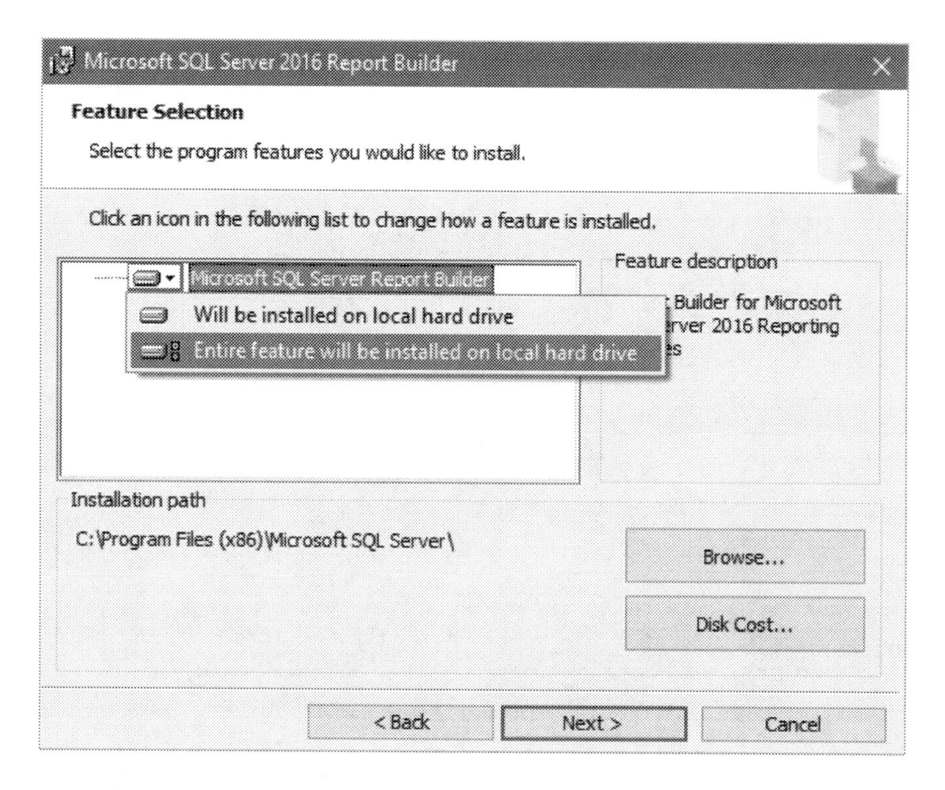

Figure 7-6. *Correct selection*

This makes sure that we have absolutely everything possible for Report Builder. When you're ready, click **Next** to continue.

We are now presented with an interface that asks for the Default Target Server, as shown in Figure 7-7.

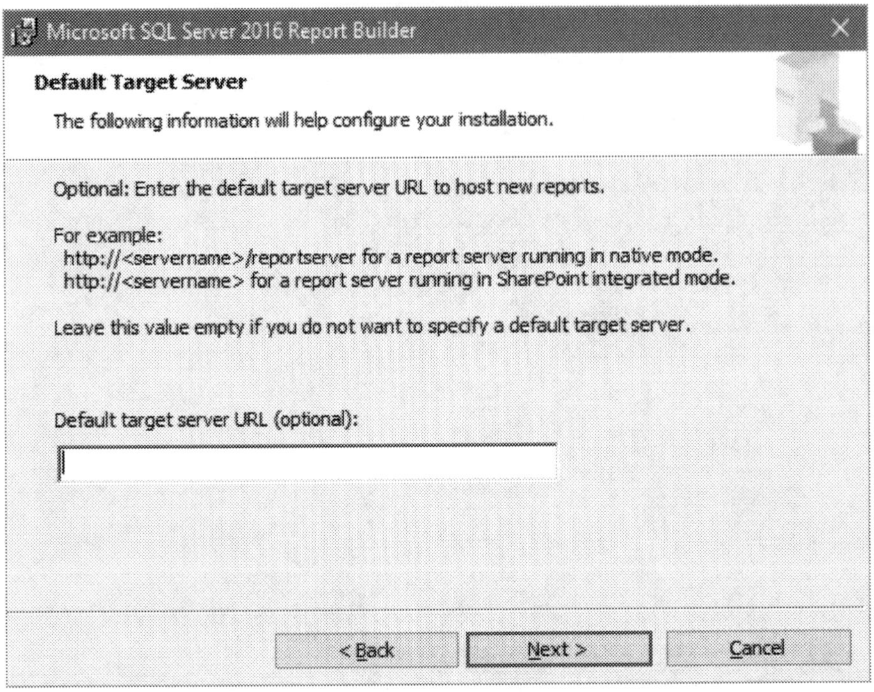

Figure 7-7. *Default Target Server*

The value that needs to be entered into **Default target server URL (optional)** is the value from the Web Service URL section of the Reporting Services Configuration Manager that we did in Chapter 6, which in my case is `http://bradlaptop/ReportServer_SQL2016RS`, so enter your own specific URL in the box. You should have something similar to what is shown in Figure 7-8.

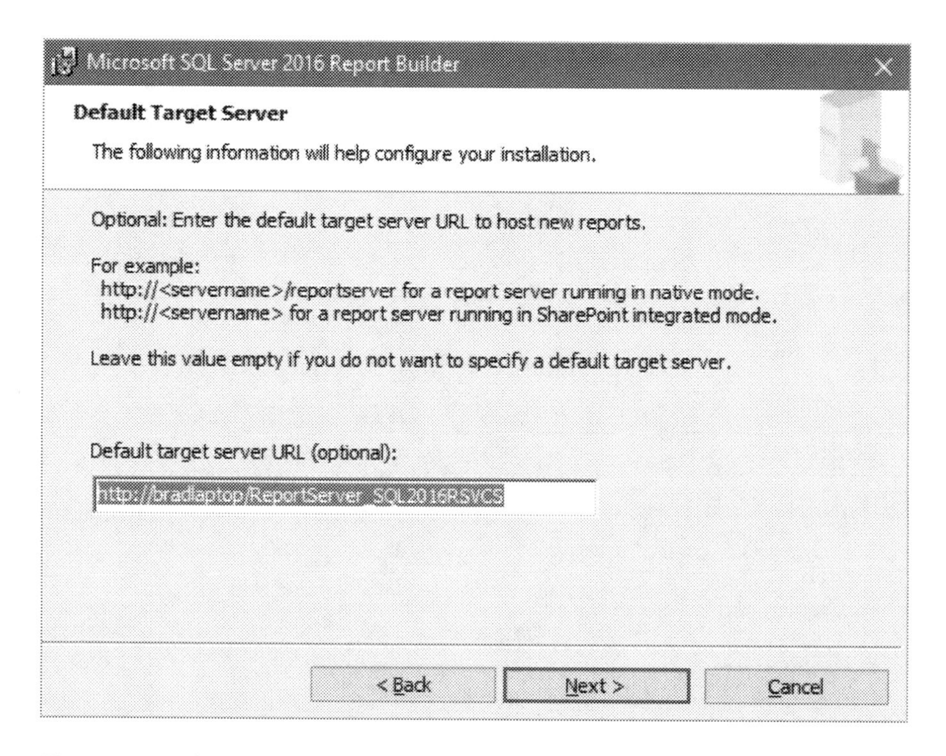

Figure 7-8. *Default Target Server, updated*

Click **Next** when you are ready to move on. You are then shown Figure 7-9, which says that we are ready to install.

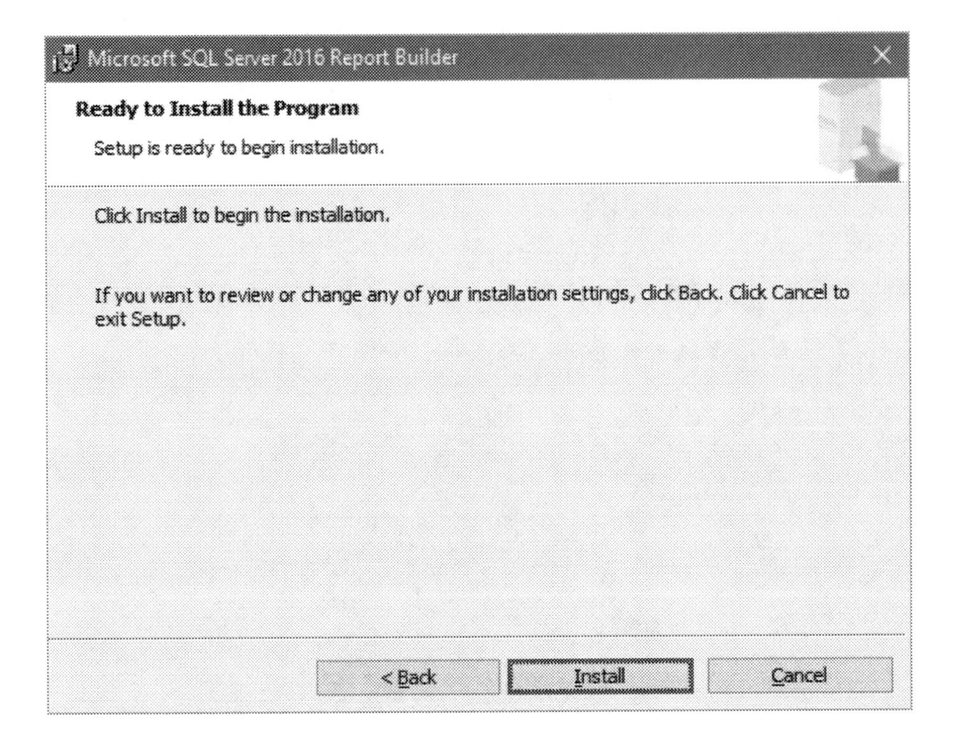

Figure 7-9. *Ready to Install*

If you need to make any changes, click **Back** and take care of them; otherwise, click **Install**.

Report Builder is a fairly small installation, so it loads fairly quickly. Figure 7-10 shows the completed installation screen, which you should see at this point.

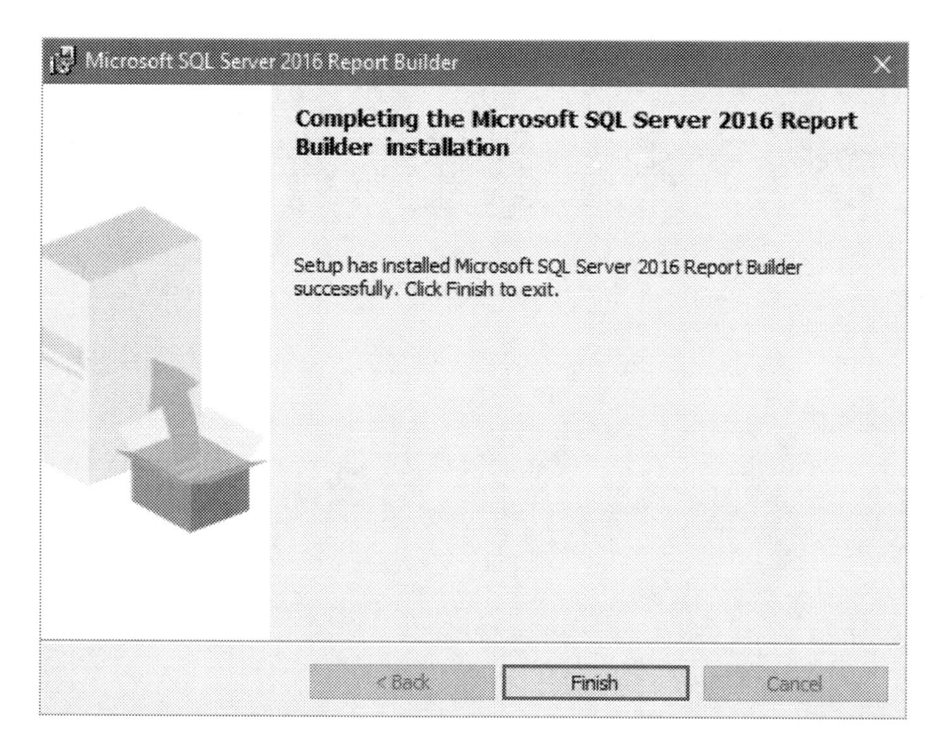

Figure 7-10. *Completed installation*

Click **Finish** and you're done. Now that we've got Report Builder installed, click **Report Builder** in your Start menu. You should initially see what is shown in Figure 7-11.

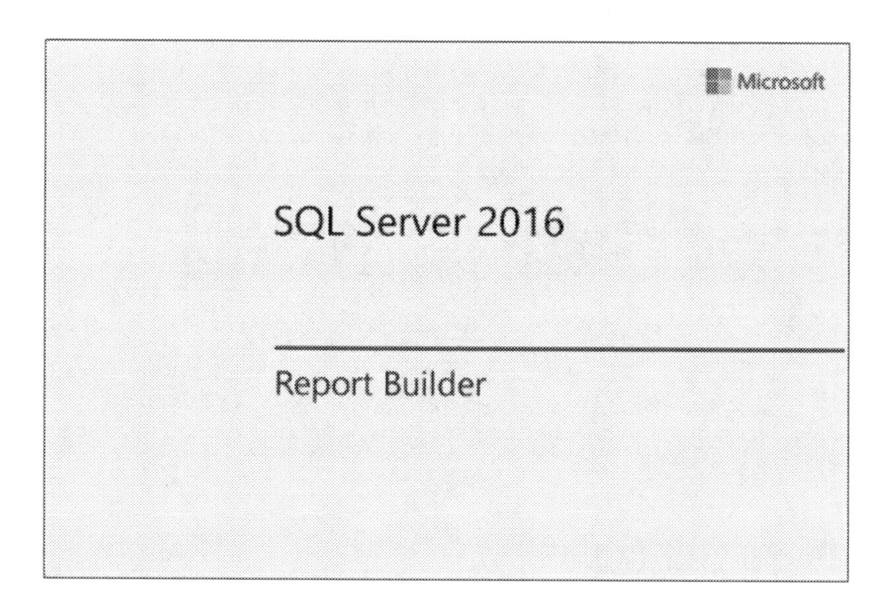

Figure 7-11. *Report Builder splash screen*

Eventually, the interface loads and we see that Report Builder is authenticating to the SQL2016RS instance we specified earlier. This is shown in Figure 7-12.

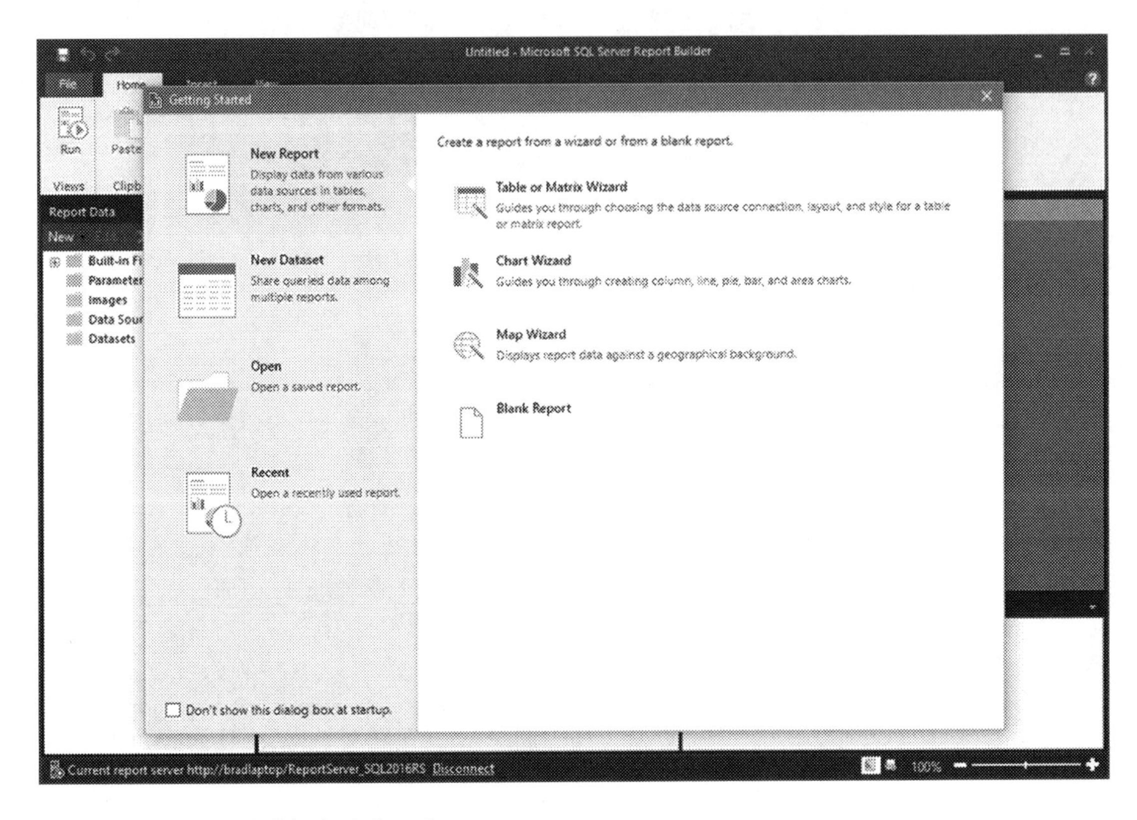

Figure 7-12. *Report Builder login interface*

It takes a minute to load because it has to authenticate to the Report Server specified during setup. Take a quick look at the bottom-left corner of the initial login interface. I've shown this in Figure 7-13 zoomed in so we can see it.

Current report server http://bradlaptop/ReportServer_SQL2016RS Disconnect

Figure 7-13. *Report Server detail*

This quick reference section tells us right away which instance we are connected to. It gives us the opportunity to disconnect from this instance and either reconnect this same instance or connect to another SQL Server Reporting Services instance, if we like.

Setup New Database and Tables

So now that we are all set up and ready to go, let's do a bit of refresher work.

Recall back in Chapter 5 that we created a couple of reports that were to be delivered to the customer. These reports were based on the `Weather_Sample.csv` data. Later on, we will import this data into SQL Server and then manipulate it. First, let's take a cursory look at the data though.

Let's set up some of the database infrastructure before we go any further. We need to set up a new database and then create one table to act as the data source for the data.

In SSMS, right-click **Databases** and choose the **New Database** option. A screen appears, which allows for the creation of a new database. The initial tab, shown in the left pane of the New Database screen shown in Figure 7-14, is named General. Make the changes shown in this screen.

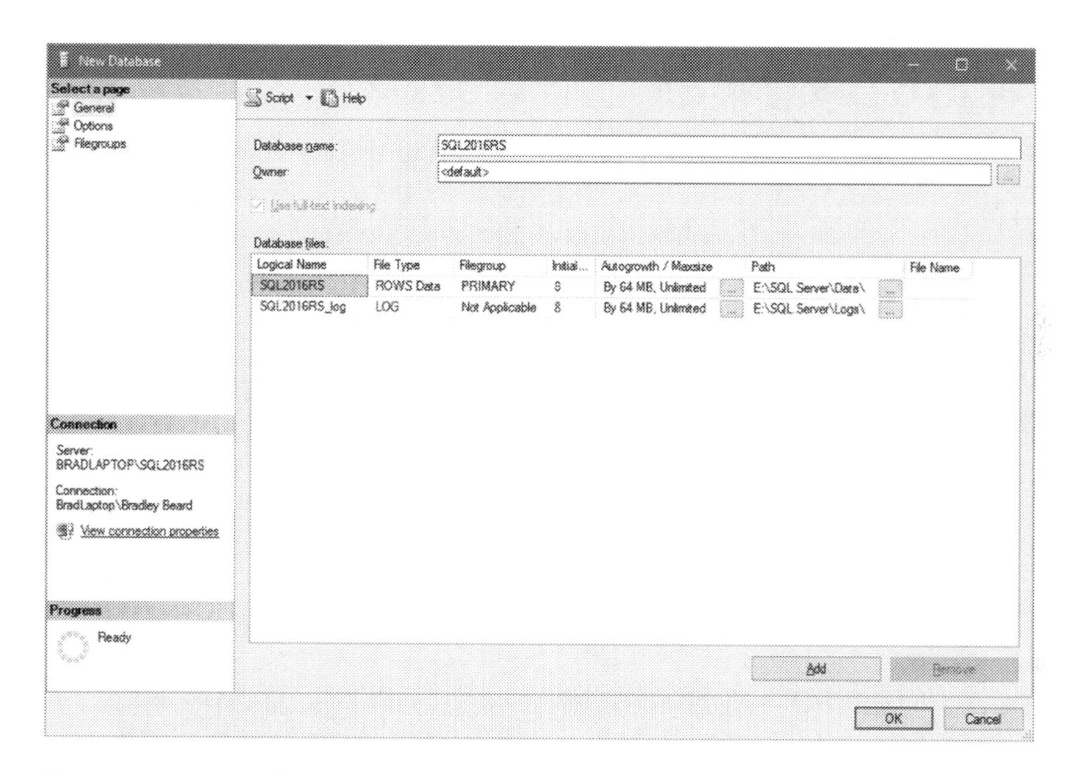

Figure 7-14. *New Database creation*

Click the **OK** button when you are finished. The new database named SQL2016RS is created.

Next, open a new Query window, change the active database to **SQL2016RS**, and then type the following:

```
CREATE TABLE [dbo].[chartBinary] (
[uid] int identity(1,1) PRIMARY KEY,
[title] varchar(100) NULL,
[binData] varbinary(max) NULL
)
```

Press F5 to execute that code—and voilà! There's our table. Figure 7-15 shows this table.

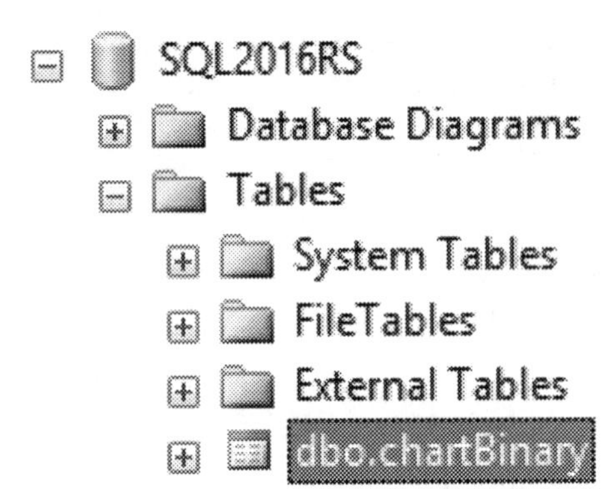

Figure 7-15. *chartBinary table*

Now we've got a table to store the binary data and a title for the chart. Next, we need to generate the actual binary data. I realize that there are ways to pull this binary data directly into Reporting Services, but I think that it is important to understand how the data works with the Report Server to take the generated binary data and create a chart. You will certainly learn more about the chart creation process this way, as opposed to learning the point-and-click method some may prefer.

Recall back in Chapter 5 how we were able to import the `Weather_Sample.csv` file into the interface and then generate the charts. This is the code to do that:

```
Weather_Sample <- read.csv(file="C:/Users/Bradley Beard/AppData/Local/Temp/Weather_Sample.
csv.utf8", header=TRUE, row.names=NULL, encoding="UTF-8", sep=",", dec=".", quote="\"",
comment.char="")
```

Instead of pointing to the `Temp` directory, I am copying the `Weather_Sample.csv` file from the `.zip` file that we downloaded earlier and put that right in the root of C, strictly to make it a shorter file location. This is the updated code:

```
Weather_Sample <- read.csv(file="C:\\Weather_Sample.csv", header=TRUE, row.names=NULL,
encoding="UTF-8", sep=",", dec=".", quote="\"", comment.char="")
```

Note that I changed the single backslash to a double backslash in order to escape the `\W` command. R throws an error when there is an escaped character, so ensure that you have converted any single backslashes to double backslashes.

So, I assume that we will be able to pop that into the stored procedure and have that data available. Is that a correct assumption? As it turns out... yes!

Consider the following code.

```
exec sp_execute_external_script
@language =N'R',
@script=N'Weather_Sample <- read.csv(file="C:\\Weather_Sample.csv",
header=TRUE, row.names=NULL, encoding="UTF-8", sep=",", dec=".", quote="\"", comment.
char="");
print(unique(Weather_Sample$AirportID));'
```

Breaking it apart, we can see that the `@script` attribute contains a declaration of `Weather_Sample` to be the result of a `read()` operation of the CSV located at `C:\Weather_Sample.csv`, with the `header` information, without `row.names`, encoded to `UTF-8`, comma separated, with decimal points if needed, with escaped quotes and comments. We are then simply printing the unique values in the `AirportID` column of the `Weather_Sample` dataset. Type that into SSMS. Figure 7-16 shows what happens once the code is executed.

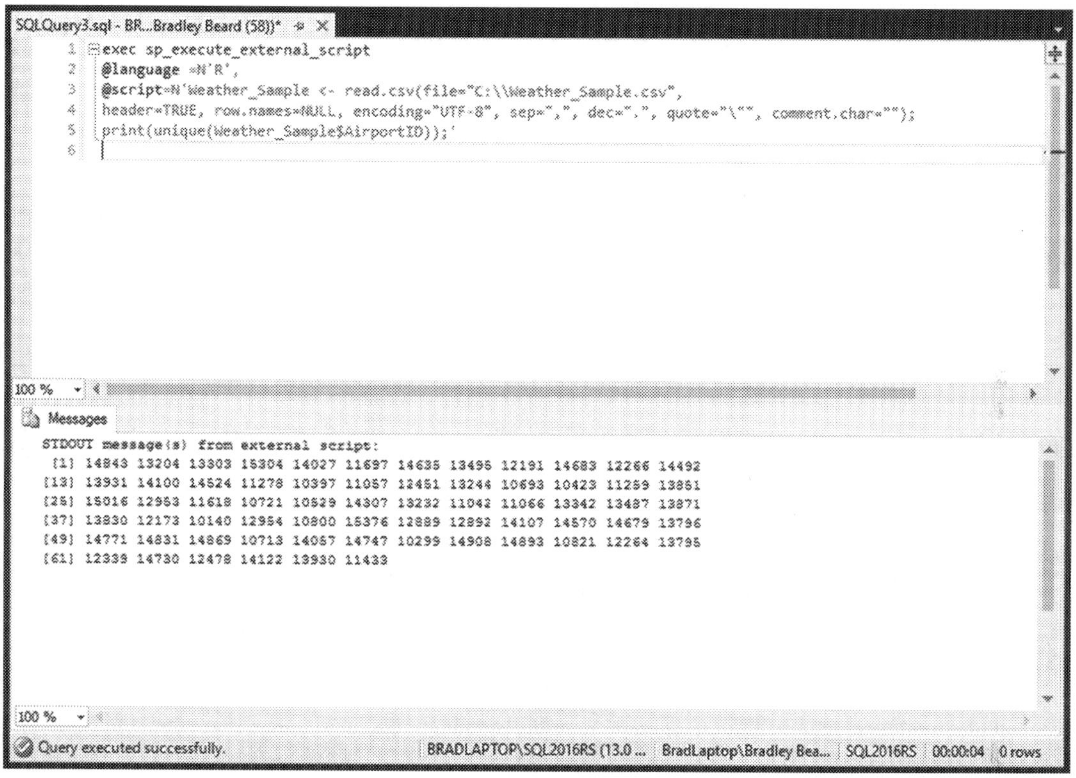

Figure 7-16. *Script execution*

Those are the unique Airport IDs in the data. We can do this with any other column you would like as well. How about AdjustedDay? Just change the last bit to AdjustedDay from AirportID. Your code should look like the following.

```
exec sp_execute_external_script
@language =N'R',
@script=N'Weather_Sample <- read.csv(file="C:\\Weather_Sample.csv",
header=TRUE, row.names=NULL, encoding="UTF-8", sep=",", dec=".", quote="\"", comment.
char="");
print(unique(Weather_Sample$AdjustedDay));'
```

Run that in SSMS. You should see what's shown in Figure 7-17.

```
exec sp_execute_external_script
@language =N'R',
@script=N'Weather_Sample <- read.csv(file="C:\\Weather_Sample.csv",
header=TRUE, row.names=NULL, encoding="UTF-8", sep=",", dec=".", quote="\"", comment.char="");
print(unique(Weather_Sample$AdjustedDay));'
```

```
STDOUT message(s) from external script:
 [1]  1  2  3  4  5  6  7  8  9 10 11 12 13 14 15 16 17 18 19 20 21 22 23 24 25
[26] 26 27 28 29 30 31
```

Figure 7-17. AdjustedDay script

So essentially, this shows us that we could use the existing data in CSV format as part of a local report. Recall that I mentioned earlier that we are going to add this data to SQL Server; so let's do that now.

Importing Weather Data

What we want to do is pretty simple: import the dataset so we can work with it. The trick this time is that we will import it directly into SQL Server as a table named `Weather_Sample`. To do this, we need to expand SQL Server Management Studio until you see the Tables menu in the SQL2016RS database, as shown in Figure 7-18.

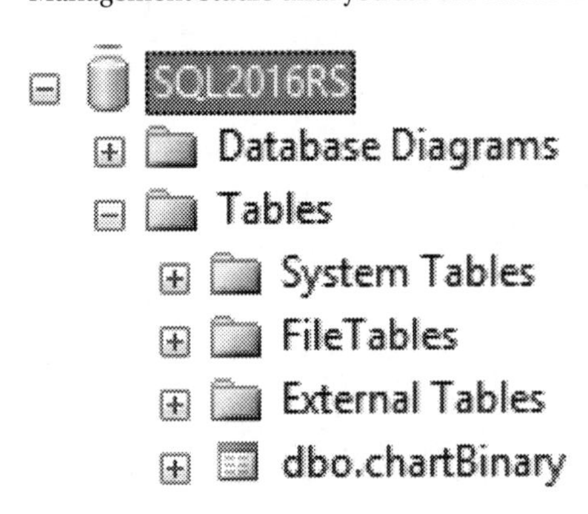

Figure 7-18. Tables location

Right-click **SQL2016RS** and choose **Tasks ➤ Import Data**. That opens up the SQL Server Import and Export Wizard window. You are probably very familiar with this, if you regularly deal with manual data manipulation. This is shown in Figure 7-19.

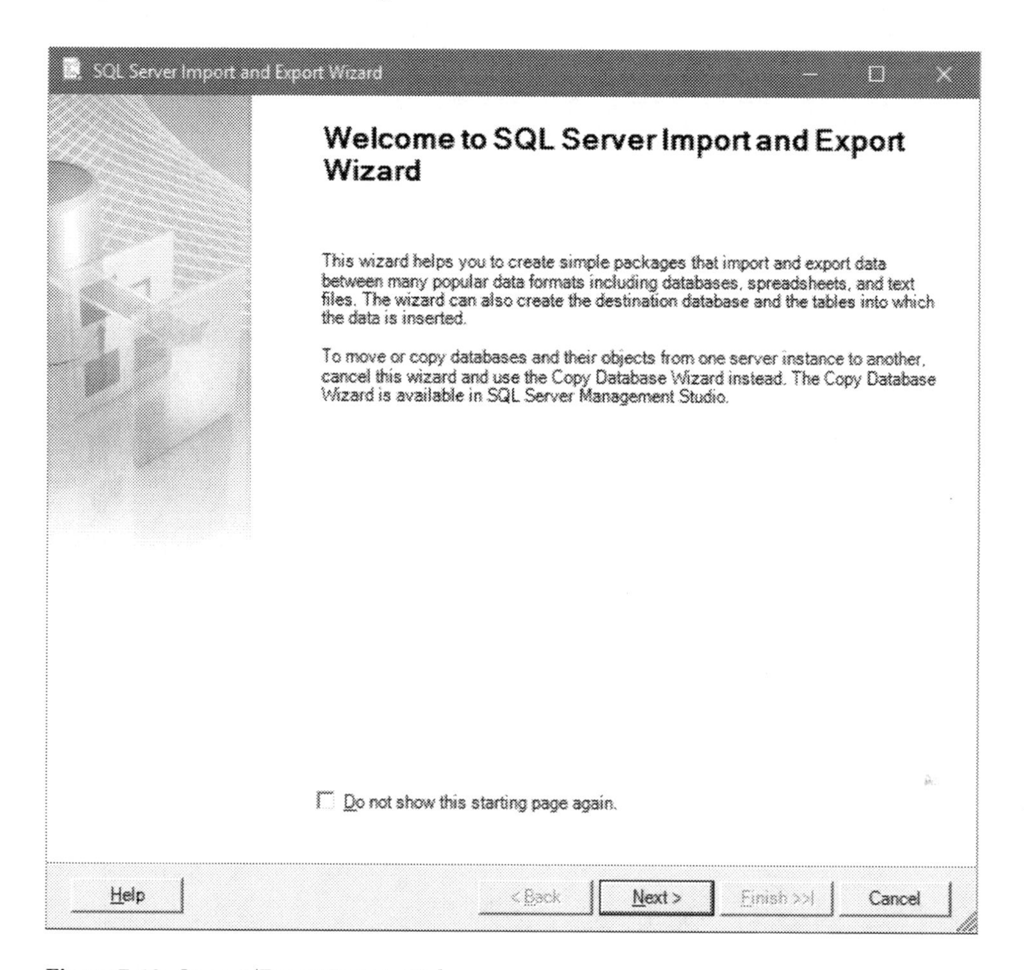

Figure 7-19. *Import/Export Data initial screen*

Click **Next** here. You see what is shown in Figure 7-20.

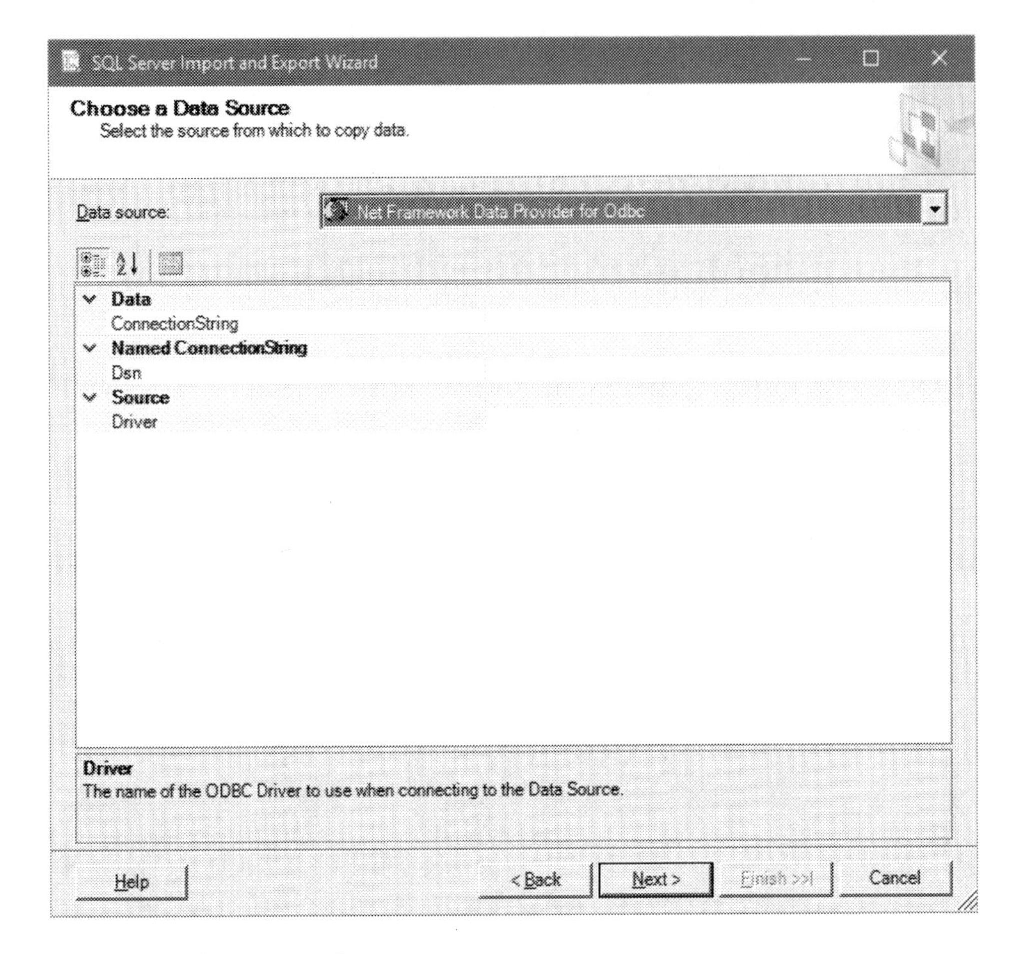

Figure 7-20. *Choose a Data Source*

Pull down the **Data Source** menu at the top and choose **Flat File Source**. You are then shown Figure 7-21, which shows the default values for this screen.

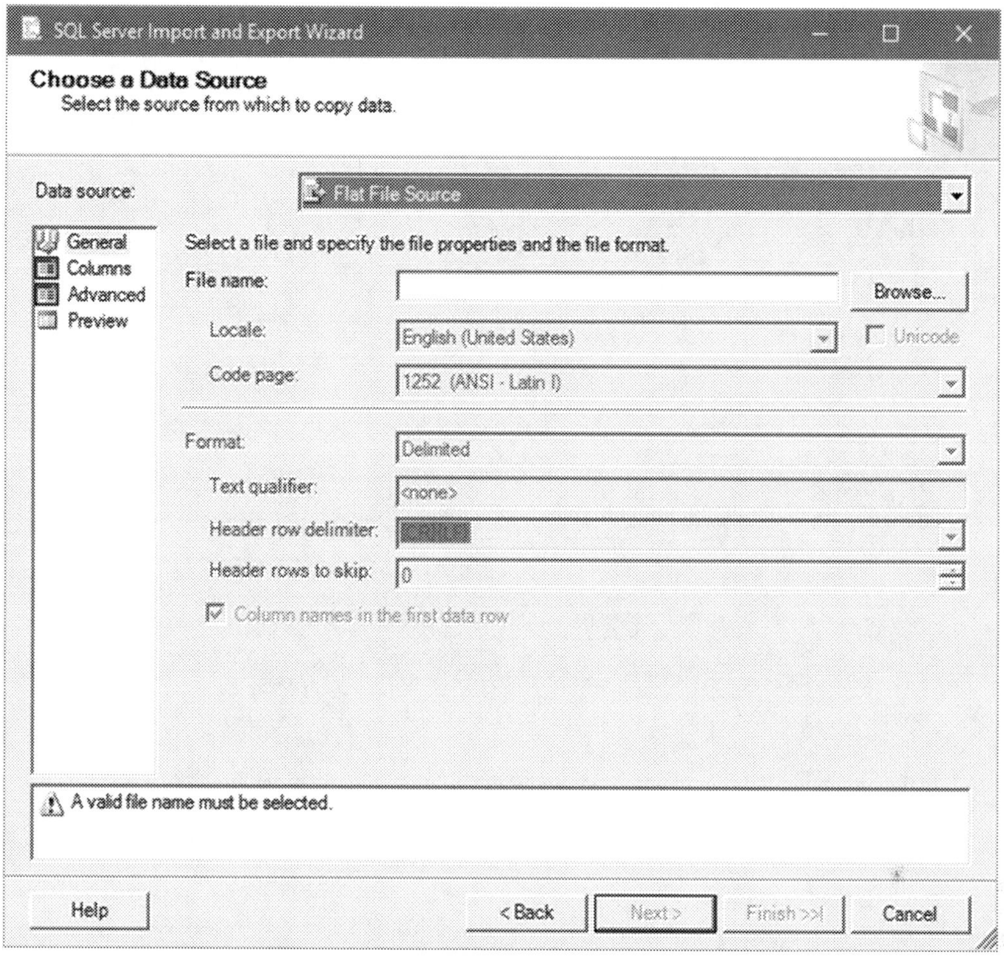

SQL Server Import and Export Wizard

Choose a Data Source

Select the source from which to copy data.

Data source: Flat File Source

General
Columns
Advanced
Preview

Select a file and specify the file properties and the file format.

File name: Browse...

Locale: English (United States) Unicode

Code page: 1252 (ANSI - Latin I)

Format: Delimited

Text qualifier: <none>

Header row delimiter: {CR}{LF}

Header rows to skip: 0

Column names in the first data row

A valid file name must be selected.

Help < Back Next > Finish >>| Cancel

***Figure 7-21.** Default values for Flat File Source*

Click the **Browse...** button and navigate to where you have saved the `Weather_Sample.csv` file. You need to pull down the file type menu and change the selected option to search for `.csv` files in order to find the file we are looking for. Once you find the `Weather_Sample.csv` file, click the **Open** button and the interface populates with information from the selected file. My example is shown in Figure 7-22.

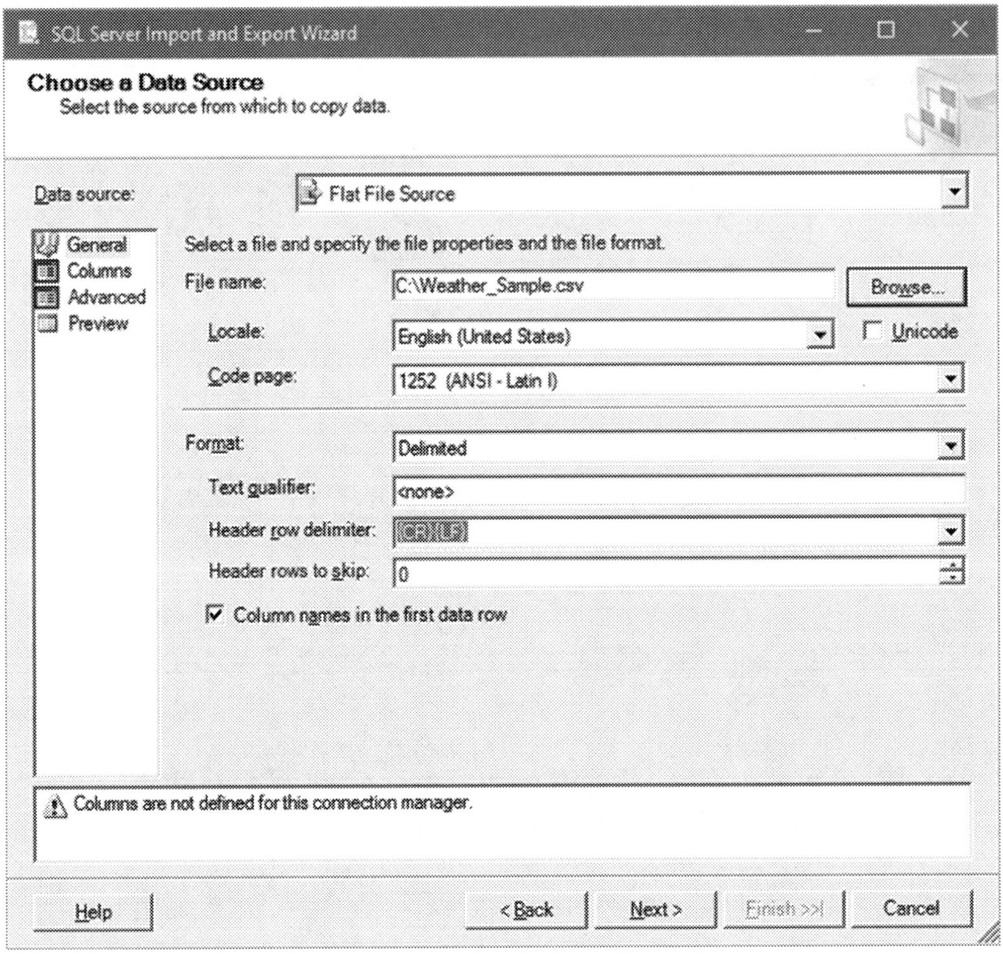

SQL Server Import and Export Wizard

Choose a Data Source
Select the source from which to copy data.

Data source: Flat File Source

General
Columns
Advanced
Preview

Select a file and specify the file properties and the file format.
File name: C:\Weather_Sample.csv Browse...
Locale: English (United States) Unicode
Code page: 1252 (ANSI - Latin I)

Format: Delimited
Text qualifier: <none>
Header row delimiter: {CR}{LF}
Header rows to skip: 0
Column names in the first data row

Columns are not defined for this connection manager.

Help < Back Next > Finish >>| Cancel

Figure 7-22. *Populated values*

Note that SQL Server automatically pulled in the values for those columns, so it knows what the data types and formats are.

Notice that yellow warning on the bottom? Click the **Columns** tab in the upper left and then click the **General** tab again. Figure 7-23 shows what happened when I did this.

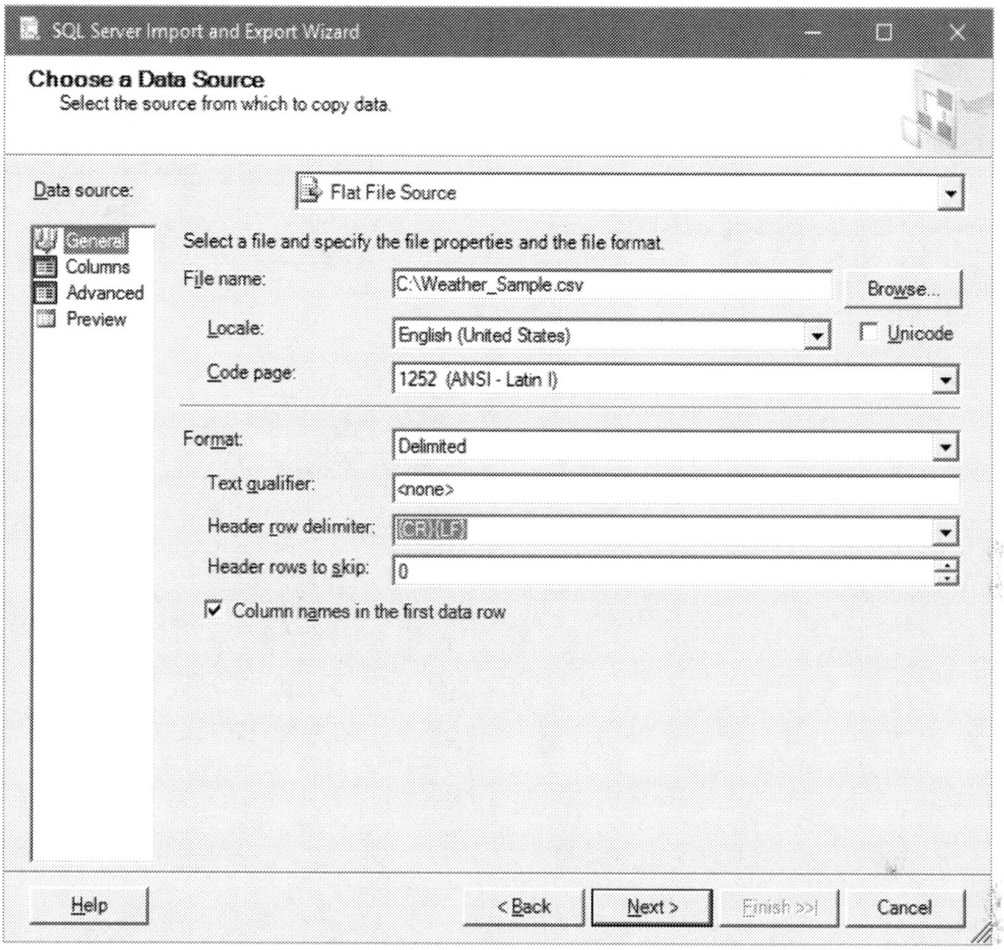

SQL Server Import and Export Wizard

Choose a Data Source

Select the source from which to copy data.

Data source: Flat File Source

General
Columns
Advanced
Preview

Select a file and specify the file properties and the file format.

File name: C:\Weather_Sample.csv Browse...

Locale: English (United States) Unicode

Code page: 1252 (ANSI - Latin I)

Format: Delimited

Text qualifier: <none>

Header row delimiter: {CR}{LF}

Header rows to skip: 0

Column names in the first data row

Help < Back Next > Finish >>| Cancel

Figure 7-23. *The disappearing warning*

No big deal, doing this action just made that warning disappear because the columns were mapped once the Columns section was selected. Everything else is the same, but I guess the interface needed to be refreshed. At any rate, we are good to go with this screen, so click **Next**. Figure 7-24 shows what you will see next.

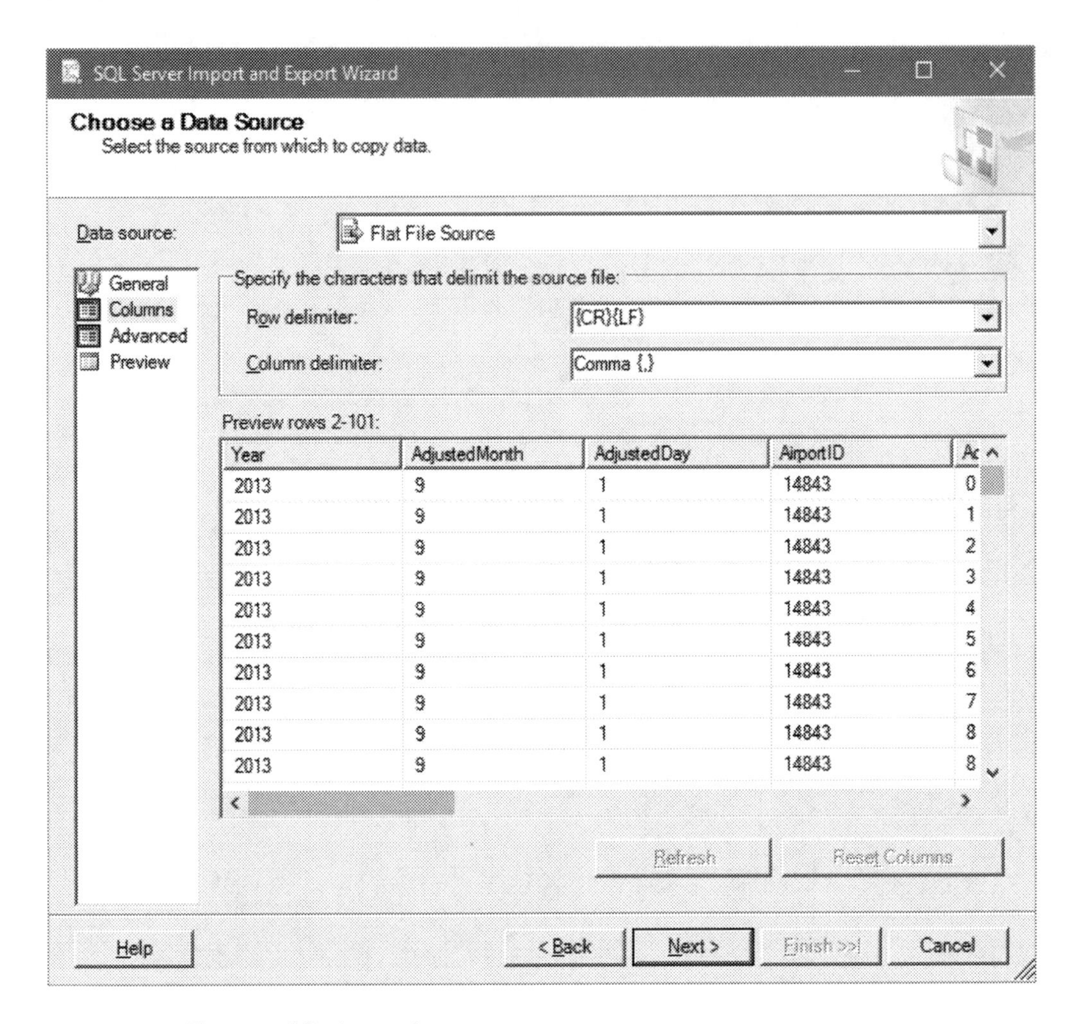

Figure 7-24. *Character delimiter section*

This section allows you to choose a delimiter for the file, if needed. We don't need one since the default works just fine, so just click **Next**.

We now come to the screen where we can choose our destination file, shown in Figure 7-25.

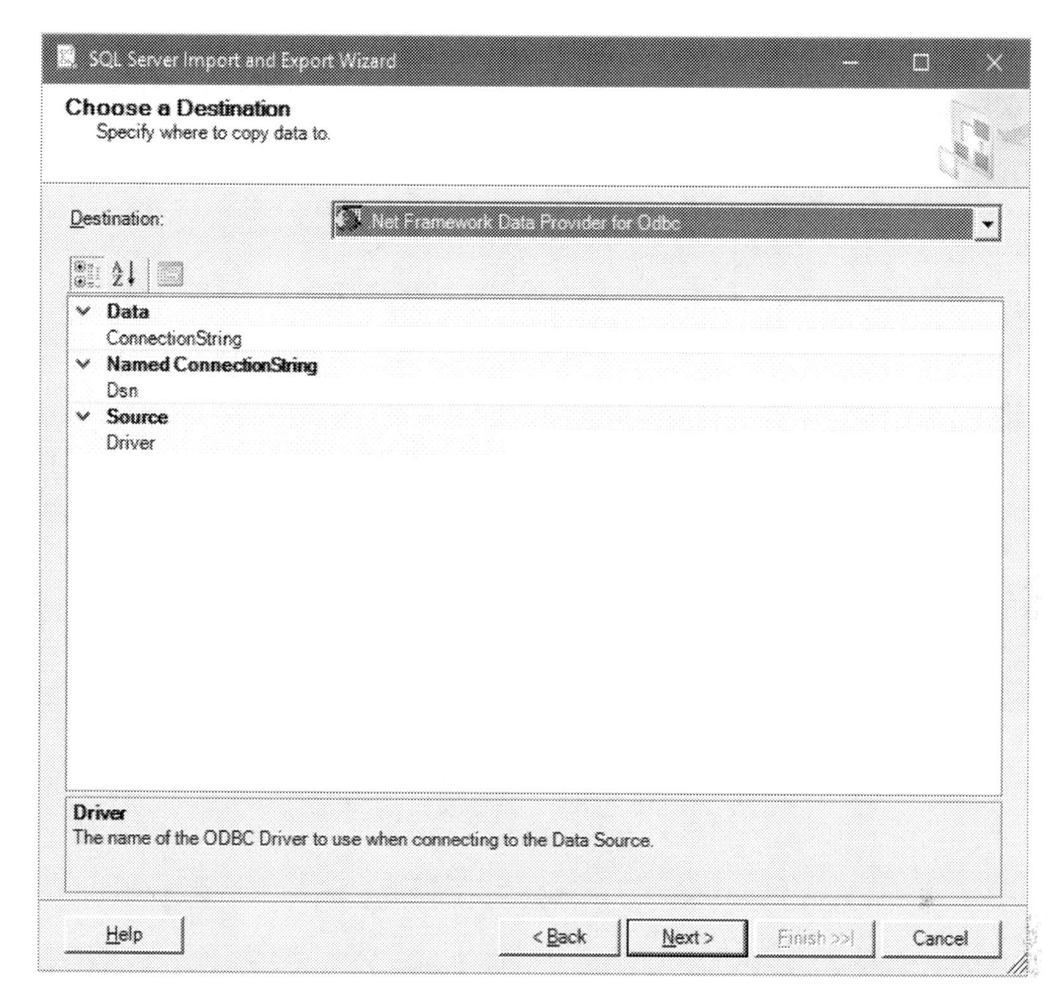

Figure 7-25. *Destination location*

We want to change the Destination to **SQL Server Native Client 11.0** here. The reason for this is because we aren't connecting over the other destination types (.Net and OLE DB). This is the native client. Make this change. You should see what is shown in Figure 7-26.

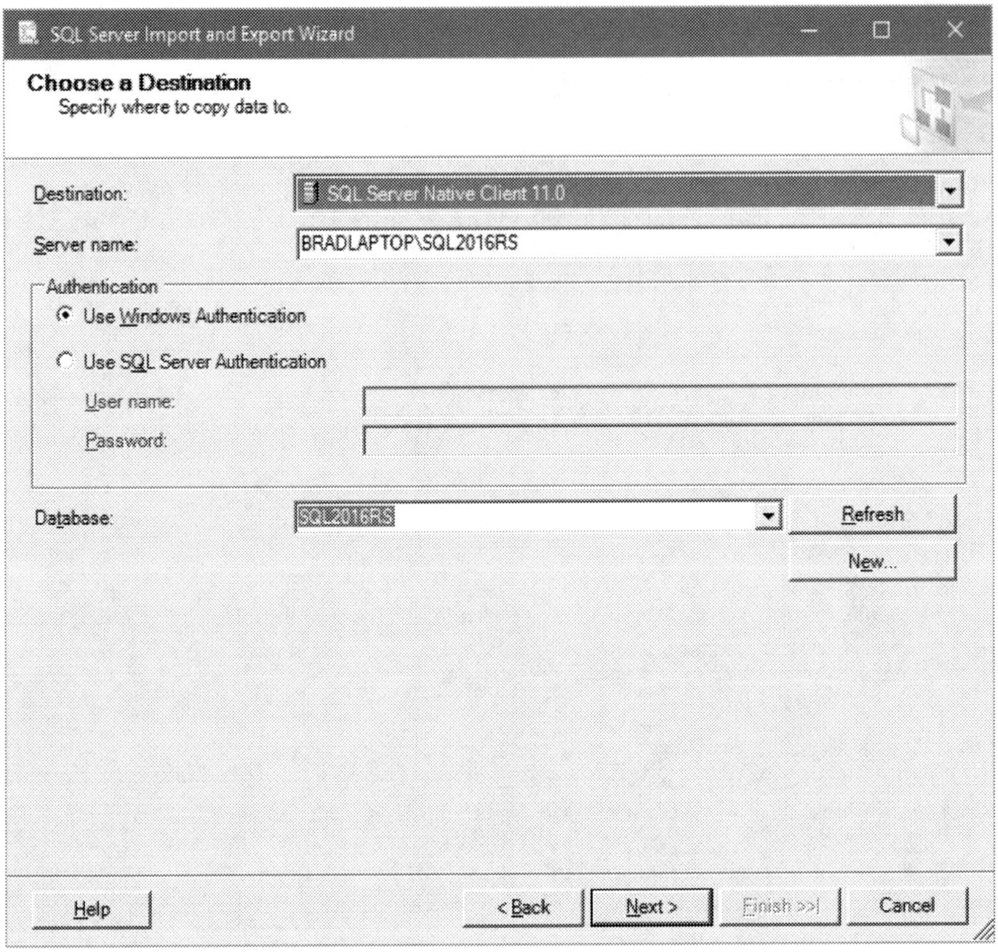

Figure 7-26. Updated Destination

We want to keep these default values, so click **Next** at this screen. You see the source and destination information shown in Figure 7-27.

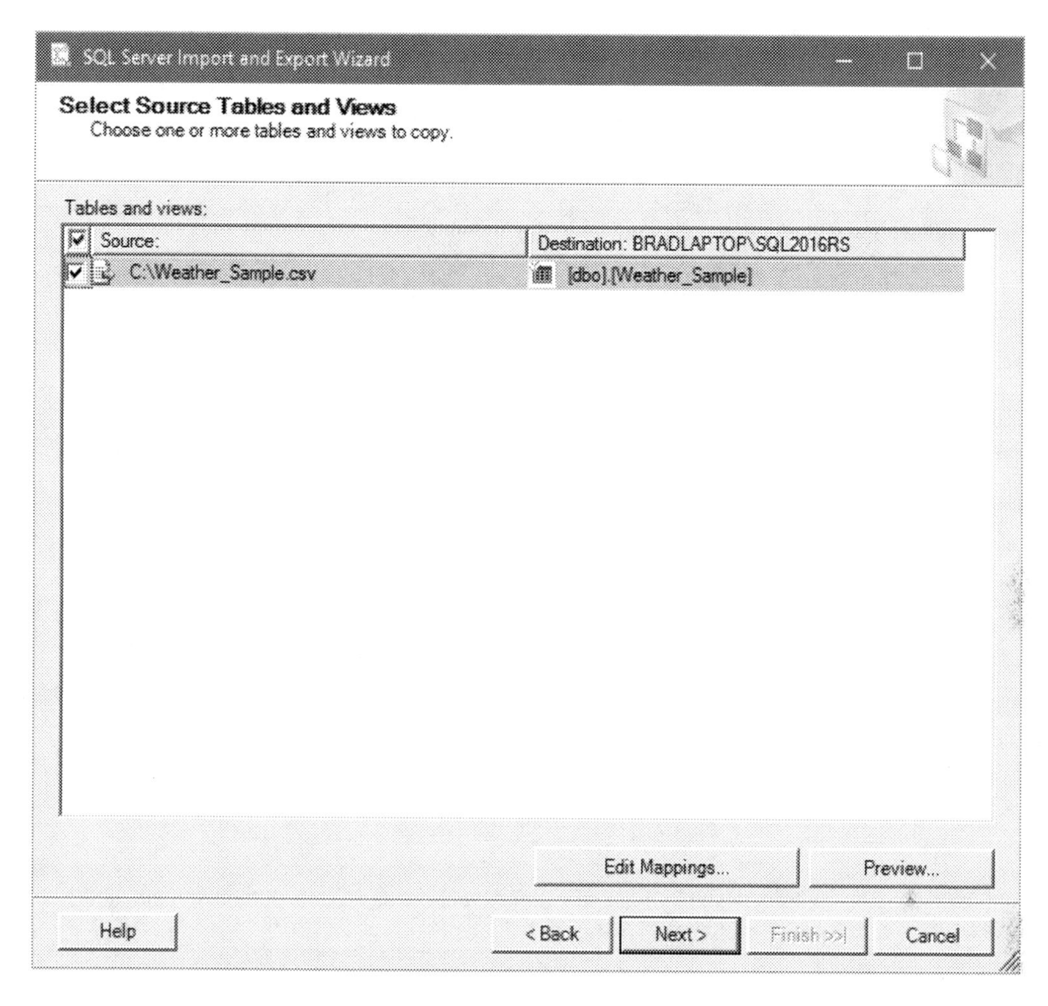

Figure 7-27. *Source and Destination information*

This basically says that we take the data represented in the Source column on the left and put it into the destination on the right.

The source points to the `Weather_Sample.csv` file we originally downloaded.

The destination points to our database named instance. A table is specified by `[dbo].[Weather_Sample]`.

Click **Next** to see the Save and Run options shown in Figure 7-28.

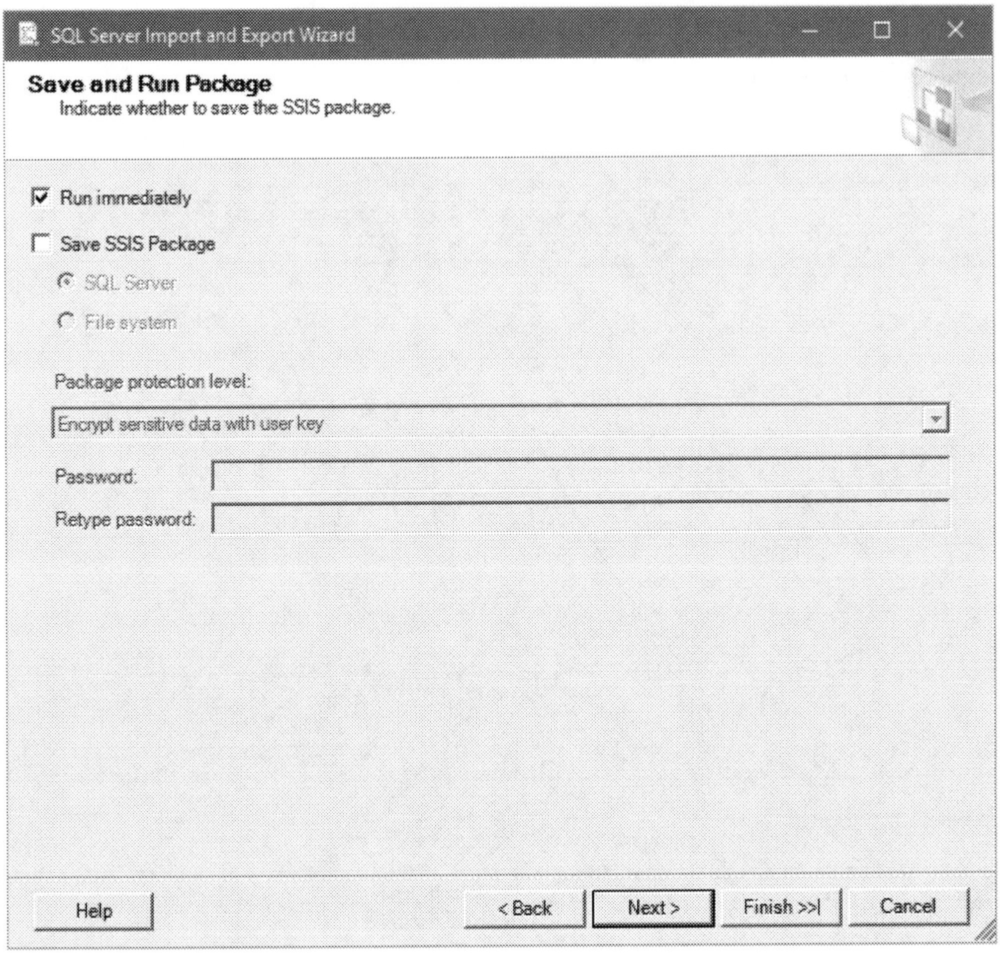

Figure 7-28. *Save and Run*

When you see this, just click **Finish**. You're then shown a screen that gives a quick recap of what we're going to do, as shown in Figure 7-29. Go ahead and click **Finish** here, too.

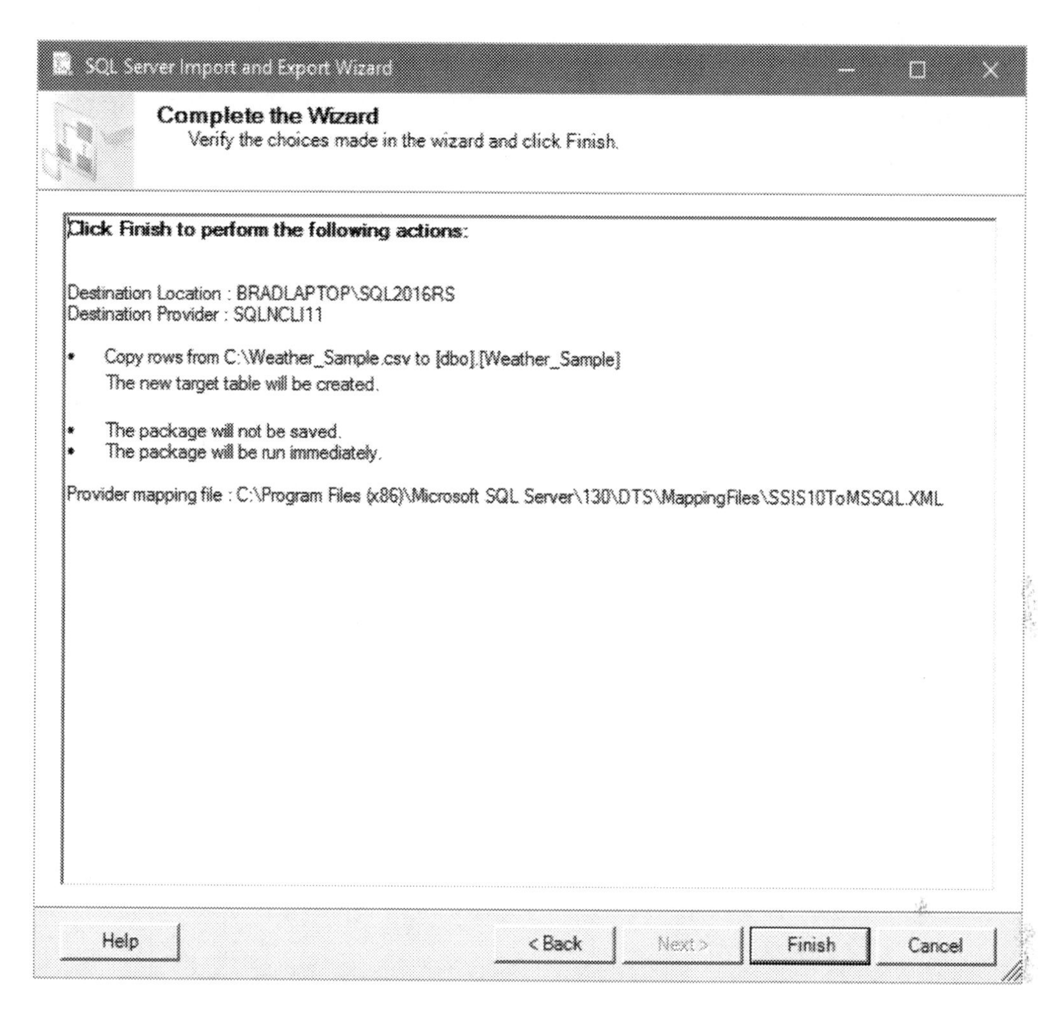

Figure 7-29. *Complete the Wizard*

The installation runs for a few seconds, but eventually, you will see what is shown in Figure 7-30.

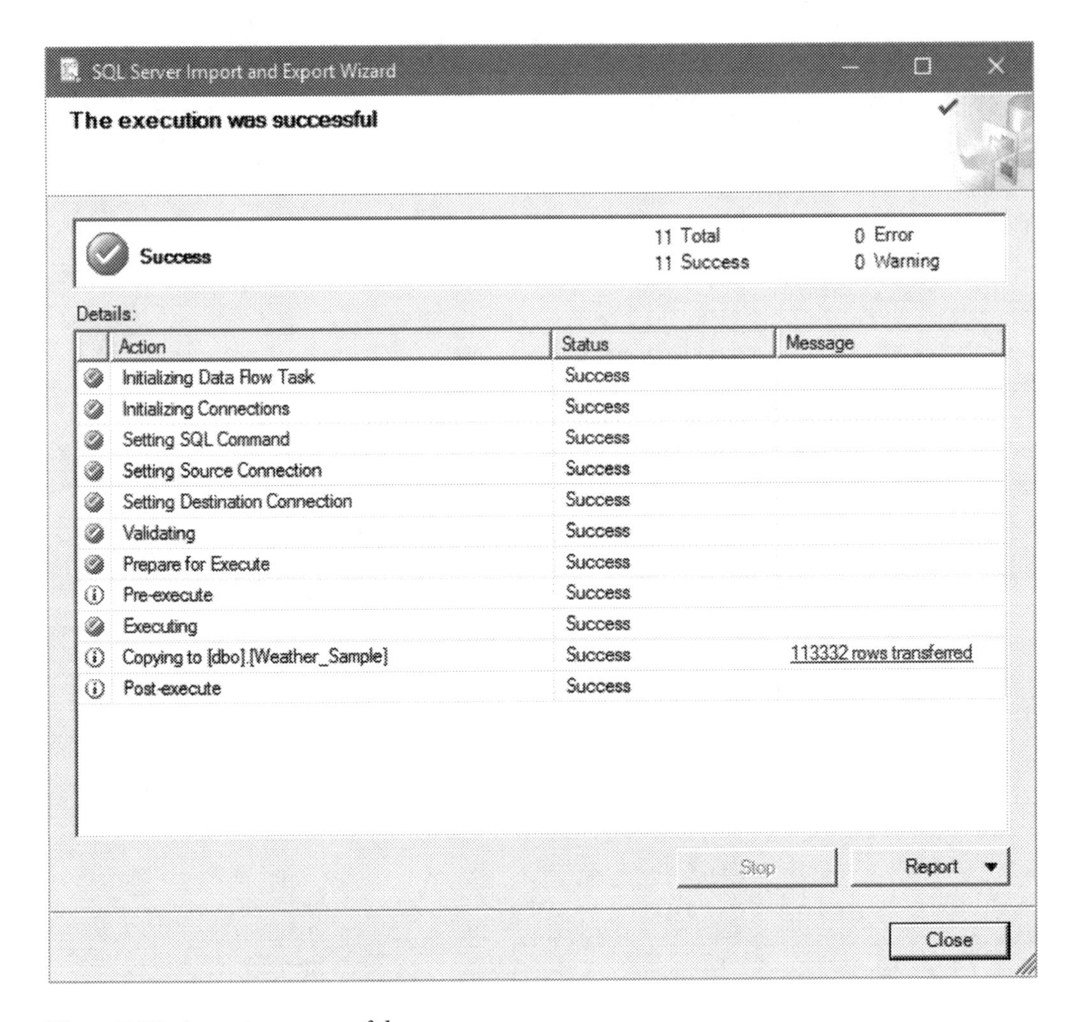

Figure 7-30. *Import was successful*

Go ahead and click **Close**. We've now got that external data brought in to SQL Server as internal data. Don't worry, we're not off track yet. We will generate that binary data very soon, but we have to take care of the leg work first.

Recall that we were getting the Average Wind Speed by Airport ID earlier. In SQL, now that we have that data in the database, the query looks like this:

```
SELECT AirportID, AVG(CONVERT(float, WindSpeed)) as WindSpeed
FROM [Weather_Sample]
GROUP BY AirportID
ORDER BY AirportID
```

Type that into a new query window and press **F5** to execute the code. You should see what is shown in Figure 7-31.

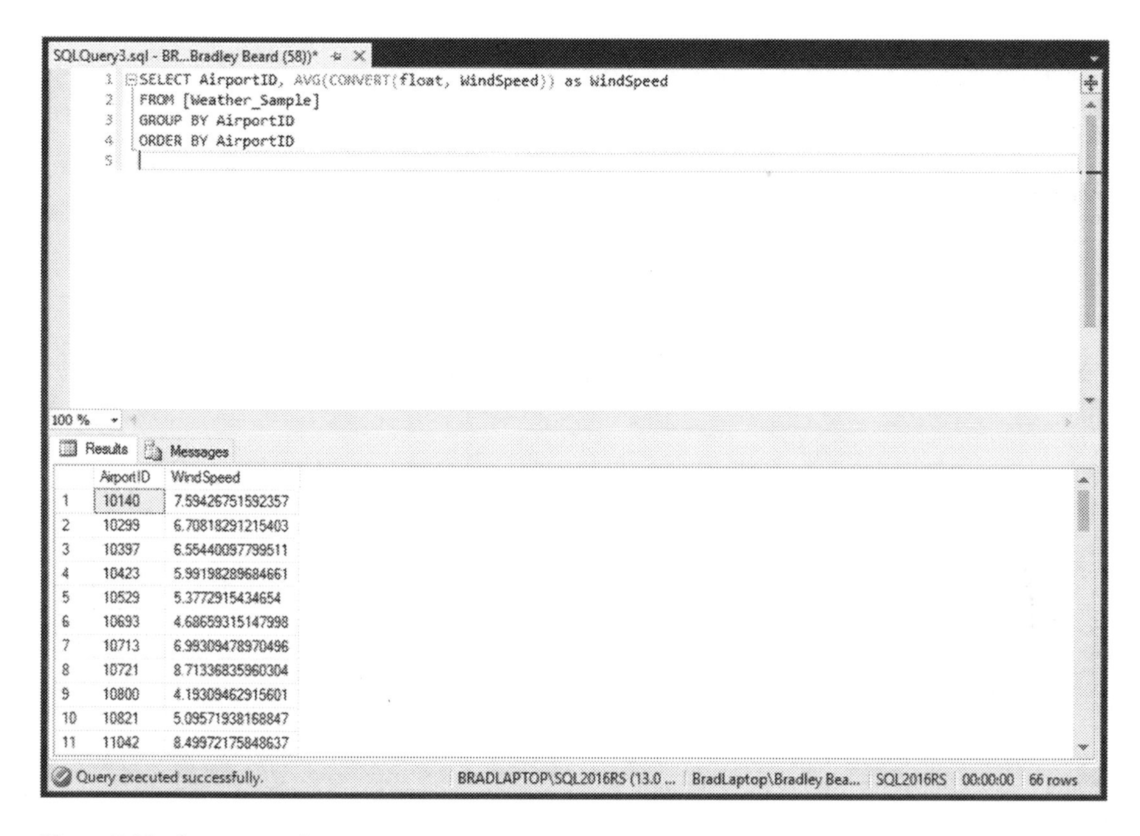

Figure 7-31. *Query execution*

Excellent! This shows us that we have the data in the correct table and that we can query it normally. That is going to be absolutely vital to generating the chart data shortly.

Generating the Binary Data

Next, we want to figure out the stored procedure syntax that we need to use to get the binary data of the chart we need to produce. Borrowing heavily from the previous work we did with creating the charts earlier, we can deduce that it is probably similar to the following.

```
EXEC sp_execute_external_script
@language = N'R',
@script = N'
library("ggplot2");
img <- inputDataSet;
image_file = tempfile();
png(filename = image_file, width=800, height=600);
print(ggplot(img, aes(x = AirportID, y = WindSpeed)) +
labs(x = "Airport ID", y = "Wind Speed") +
theme(axis.text.x = element_text(angle=90, hjust=1, vjust=0)) +
geom_point(stat = "identity") +
geom_smooth(method = "loess", aes(group = 1)) +
```

```
geom_text(aes(label = AirportID), size = 3, vjust = 1.0) +
geom_text(aes(label = round(WindSpeed, digits = 2)), size = 3, vjust = 2.0));
dev.off();
OutputDataset <- data.frame(data=readBin(file(image_file,"rb"),what=raw(),n=1e6));',
@input_data_1 = N'SELECT AirportID, AVG(CONVERT(float, WindSpeed)) as WindSpeed FROM
[Weather_Sample] GROUP BY AirportID ORDER BY AirportID;',
@input_data_1_name = N'inputDataSet',
@output_data_1_name = N'OutputDataset'
WITH RESULT SETS ((plot varbinary(max)));
```

▒ **Note** We are generating a PNG image here, not a JPG. This will be important to remember once we dynamically generate these charts using Report Builder.

Entering this into a new query window and executing it gives me a binary data result. Figure 7-32 shows what you should see in SSMS at this point.

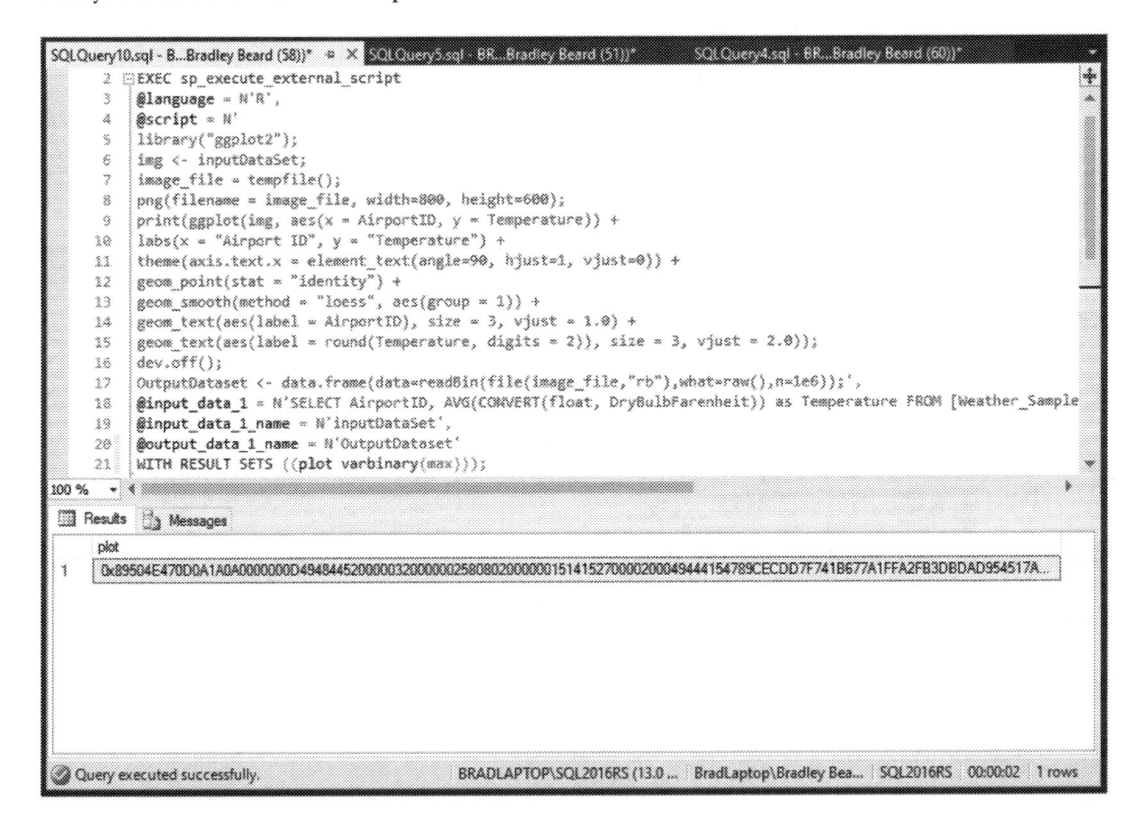

Figure 7-32. *Binary data*

Recall that we created a table to store our charts in. We are now going to insert this binary data into the table within the stored procedure by using the following code.

Now, we just need to rewrite the stored procedure to automatically insert the data. Any idea how to do this? That's right, we're going to use INSERT INTO ... EXEC.

So how do we build the query? It's actually very easy. I'm sure some of you already know how this is done, but for those of us that don't, it's looks like this:

```
INSERT INTO chartBinary (binData)
EXEC sp_execute_external_script
@language = N'R',
@script = N'
library("ggplot2");
img <- inputDataSet;
image_file = tempfile();
png(filename = image_file, width=800, height=600);
print(ggplot(img, aes(x = AirportID, y = WindSpeed)) +
labs(x = "Airport ID", y = "Wind Speed") +
theme(axis.text.x = element_text(angle=90, hjust=1, vjust=0)) +
geom_point(stat = "identity") +
geom_smooth(method = "loess", aes(group = 1)) +
geom_text(aes(label = AirportID), size = 3, vjust = 1.0) +
geom_text(aes(label = round(WindSpeed, digits = 2)), size = 3, vjust = 2.0));
dev.off();
OutputDataset <- data.frame(data=readBin(file(image_file,"rb"),what=raw(),n=1e6));',
@input_data_1 = N'SELECT AirportID, AVG(CONVERT(float, WindSpeed)) as WindSpeed FROM
[Weather_Sample] GROUP BY AirportID ORDER BY AirportID;',
@input_data_1_name = N'inputDataSet',
@output_data_1_name = N'OutputDataset';
```

That updated our binData column, but left our title column without a value. Since this is the first record in the table, the UID is set to 1; so make sure that your WHERE clause is set to specify that. To update the title value for this chart, run this UPDATE query after you run that first query.

```
UPDATE chartBinary
SET title = 'Average Wind Speed by Airport ID'
WHERE uid = 1
```

That should do it. To verify, run the following query to view your data.

```
SELECT title, binData FROM [dbo].[chartBinary] ORDER BY uid
```

The results should be what you see in Figure 7-33.

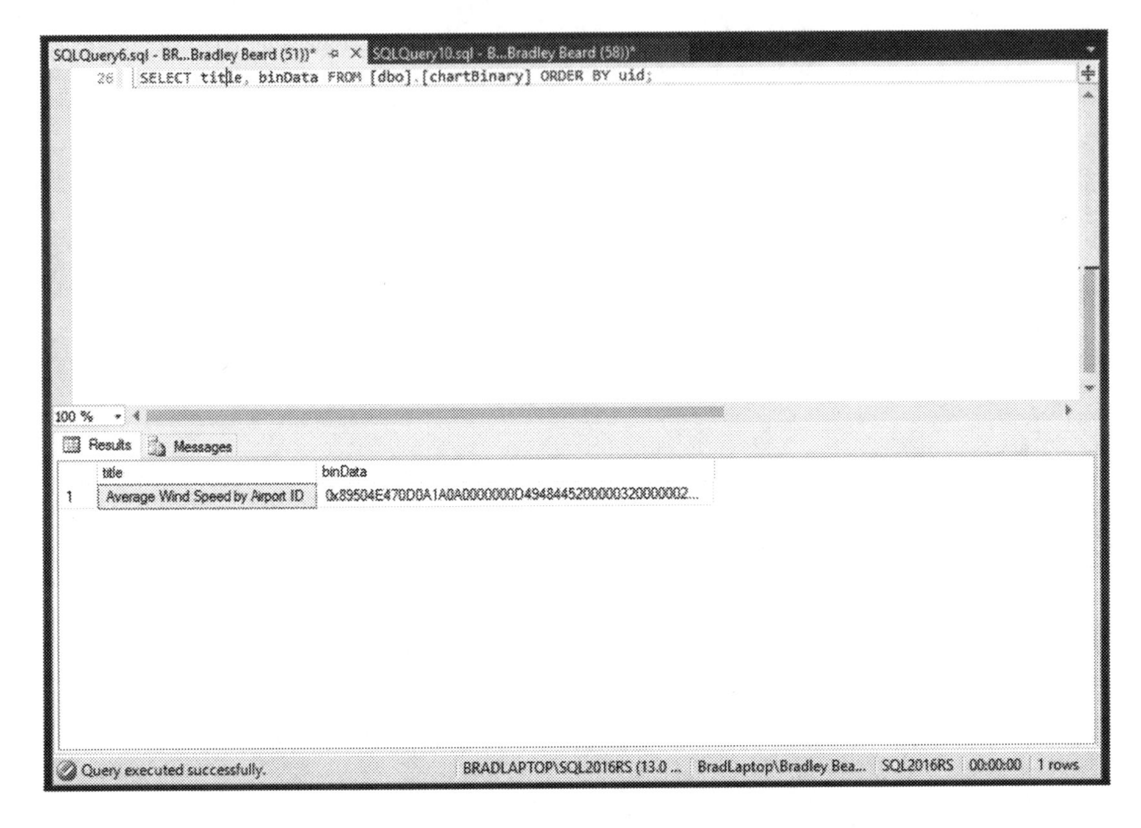

Figure 7-33. *Query results*

Next, we need to build the query for the other report, but this is actually quite easy since we've already done the hard work for it. The following code is what you need for this action.

```
INSERT INTO chartBinary (binData)
EXEC sp_execute_external_script
@language = N'R',
@script = N'
library("ggplot2");
img <- inputDataSet;
image_file = tempfile();
png(filename = image_file, width=800, height=600);
print(ggplot(img, aes(x = AirportID, y = Temperature)) +
labs(x = "Airport ID", y = "Temperature") +
theme(axis.text.x = element_text(angle=90, hjust=1, vjust=0)) +
geom_point(stat = "identity") +
geom_smooth(method = "loess", aes(group = 1)) +
geom_text(aes(label = AirportID), size = 3, vjust = 1.0) +
geom_text(aes(label = round(Temperature, digits = 2)), size = 3, vjust = 2.0));
dev.off();
OutputDataset <- data.frame(data=readBin(file(image_file,"rb"),what=raw(),n=1e6));',
```

```
@input_data_1 = N'SELECT AirportID, AVG(CONVERT(float, DryBulbFarenheit)) as Temperature
FROM [Weather_Sample] GROUP BY AirportID ORDER BY AirportID;',
@input_data_1_name = N'inputDataSet',
@output_data_1_name = N'OutputDataset';
```

And once again, to set the title column:

```
UPDATE chartBinary
SET title = 'Average Temperature by Airport ID'
WHERE uid = 2
```

Run those two code blocks, and then run the following query to check and make sure that everything was inserted correctly.

```
SELECT title, binData FROM [dbo].[chartBinary] ORDER BY uid
```

At this point, you should see what is shown in Figure 7-34.

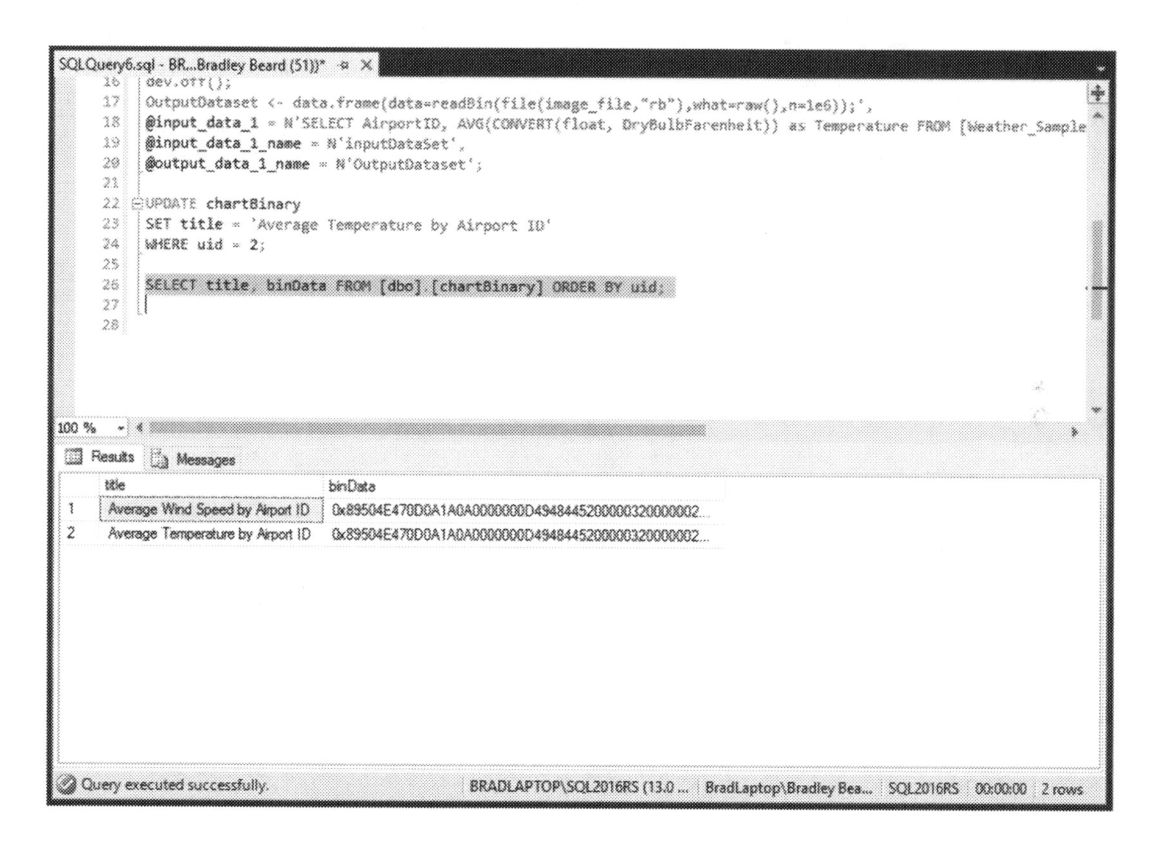

Figure 7-34. *Query results*

There we go! Two results, just like we wanted. Now that this data is loaded into the table correctly, we can start building our reports to be delivered to the customer. Don't forget to verify our progress with what we have in the software requirements document periodically.

Summary

Let's review what we did in this chapter, because it was actually quite a lot.

- Downloaded and installed Report Builder
- Loaded our weather data into SQL Server
- Generated charts as binary based on this weather data

That doesn't seem like a lot, but it is. Be sure that you have gone through this chapter very carefully, because you aren't going to get very far in the next chapter without that binary data.

Next, we tie this all together into the actual reports to be delivered to the customer. If you made it through this chapter without a headache, you are probably doing it wrong. Seriously, there was an awful lot of information to digest. I highly recommend going back and re-reading the steps that generated the binary data in SQL Server. The individual attributes of the `ggplot2` function, in particular, are actually really fun to play around with. You can customize your charts just about any way you can dream of. There is even information on the Internet about how to create your own R packages for private use or general distribution.

We're almost done now! Keep going—we will get this wrapped up shortly. If you've made it this far, I congratulate you and encourage you to go for just a bit longer. I promise that it will be worth it!

CHAPTER 8

Building Reports Using Report Builder

Up until now, we've dealt with an awful lot of information in this book. We've performed a new installation of SQL Server R Services, installed R Tools for Visual Studio, configured Reporting Services, and installed and configured Report Builder. We learned quite a bit about R and how it works and we generally got a lot of experience with writing code in general. If you have gotten this far, nice work! This is really sort of advanced stuff, since it's basically brand-new functionality. It's pretty safe to assume that R will continue to be offered as part of SQL Server for future releases, so it's best to go ahead and get acquainted with it now instead of having to deal with the learning curve later.

In this chapter, we create a report using the binary data we created in Chapter 7. This binary data is going to be converted into images dynamically. That sounds pretty complicated, but luckily, Report Builder makes it actually pretty easy for us.

Here's what we're doing in this chapter:

- Build the Average Wind Speed by Airport ID report.
- Build the Average Temperature by Airport ID report.

We're only going to do two things? That's correct. This chapter is sort of quick compared to the other chapters, which are a lot more in-depth. Consequently, this chapter is much shorter than the others, but hopefully you can still find this information useful.

Report 1: Average Wind Speed by Airport ID

Recall that the software requirements document specifically called out two reports that the customer wanted delivered. The first of these reports was the average wind speed by airport ID report. This particular report is important to the customer, so we are going to put a little time into setting it up, and then use that same format to create our second report. Keep in mind that, once the first report is written, it really is much faster to create any subsequent reports because we already have a general idea of how it is done. All we really need to do from that point is apply any specific formatting to the report, disregarding the image itself since that is generated from the database using pre-compiled binary data.

First of all, start up Report Builder and run it as Administrator. The initial interface is shown in Figure 8-1.

B. Beard, *Beginning SQL Server R Services*, DOI 10.1007/978-1-4842-2298-0_8

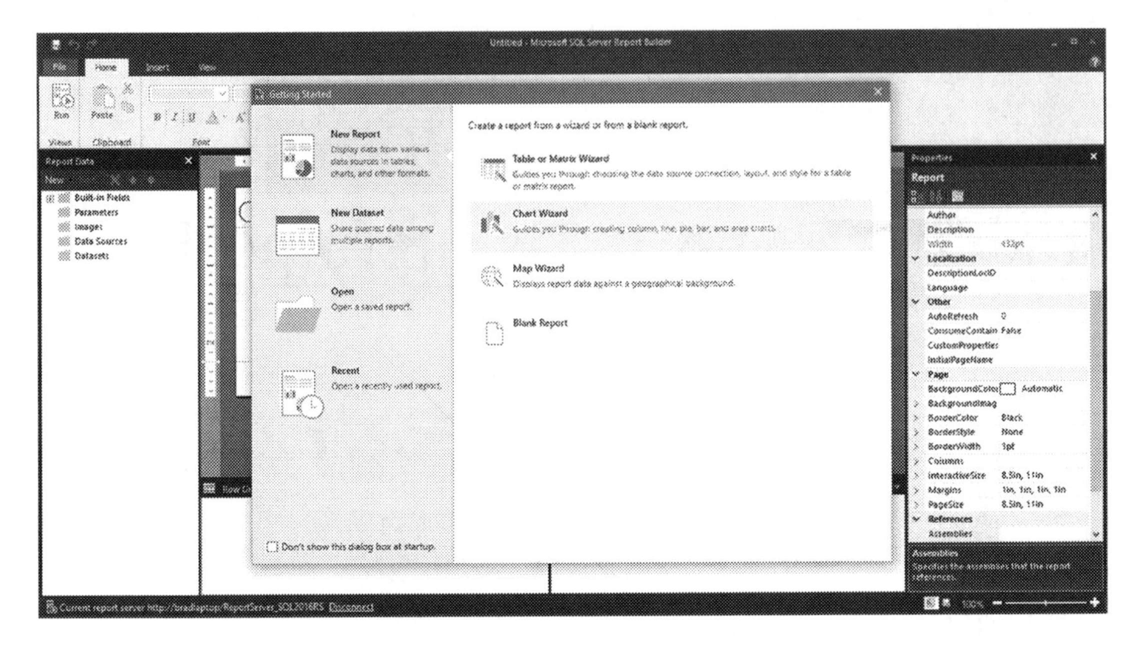

Figure 8-1. *Report Builder initial interface*

Recall that we have already stepped through the features of this interface, so let's just get right into it.

Setting up the Report Layout

Click **Blank Report** at the bottom of the Getting Started screen. The layover window should disappear, so that you are left with what is shown in Figure 8-2.

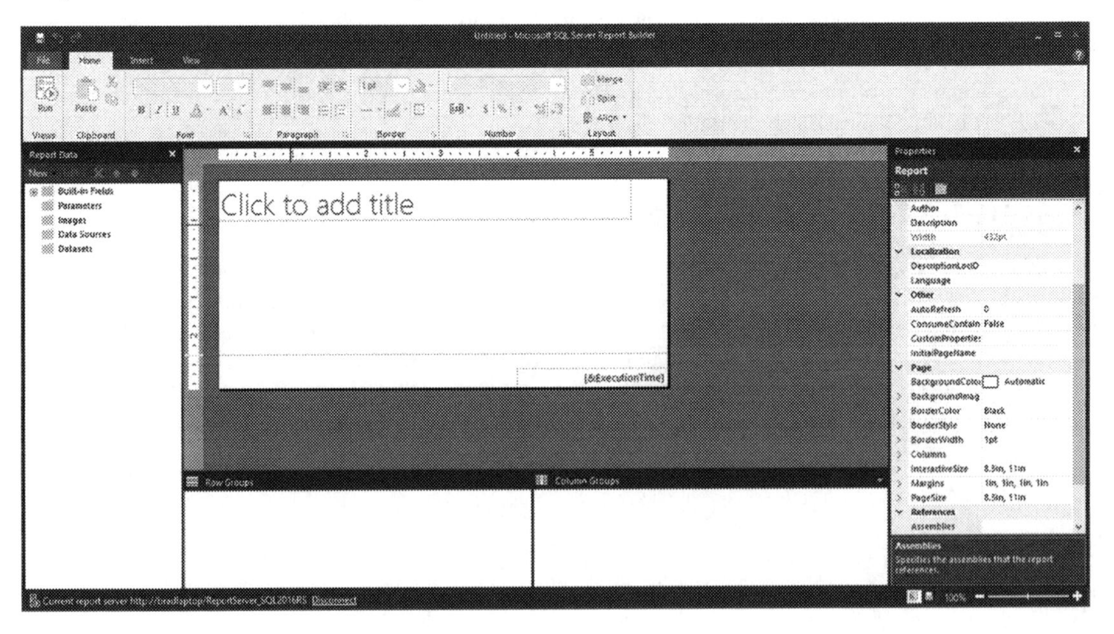

Figure 8-2. *Blank Report*

Click inside of the box labeled **Click to add title** and type **Average Wind Speed by Airport ID**. And then stretch the box to the height and width of the text. After that, center the text using the controls at the top of the screen. Figure 8-3 shows what your report should now look like.

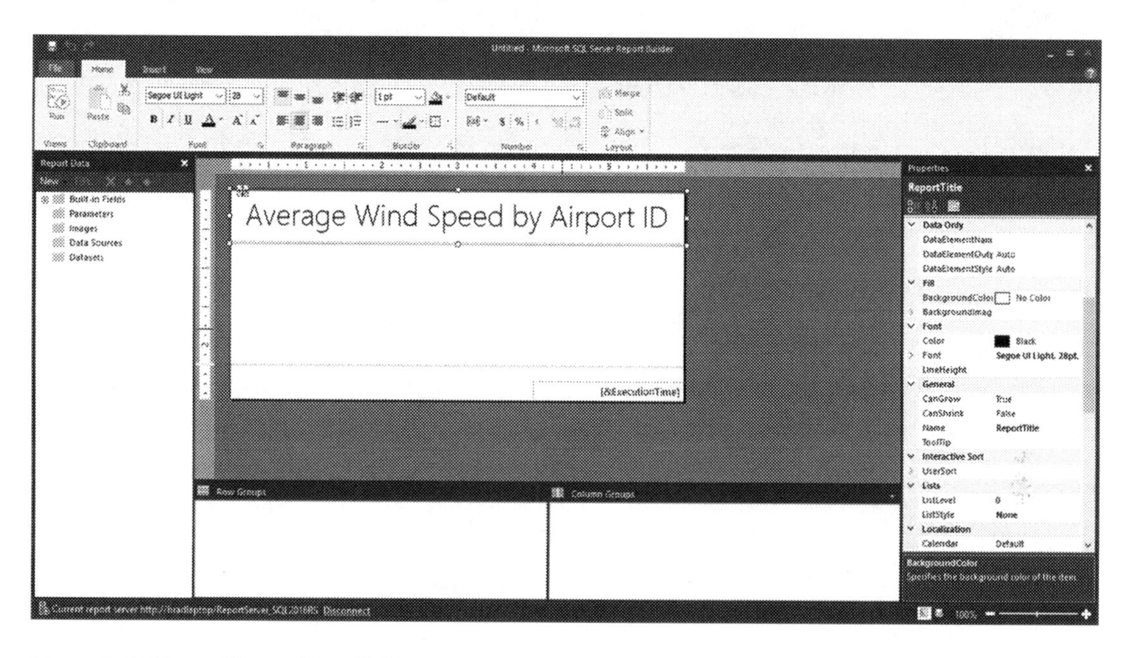

Figure 8-3. General formatting of Title

Next, we want to just do some really basic formatting for the body, so right-click anywhere in the middle part of the report and choose **Body Properties**. The location of this option is shown in Figure 8-4.

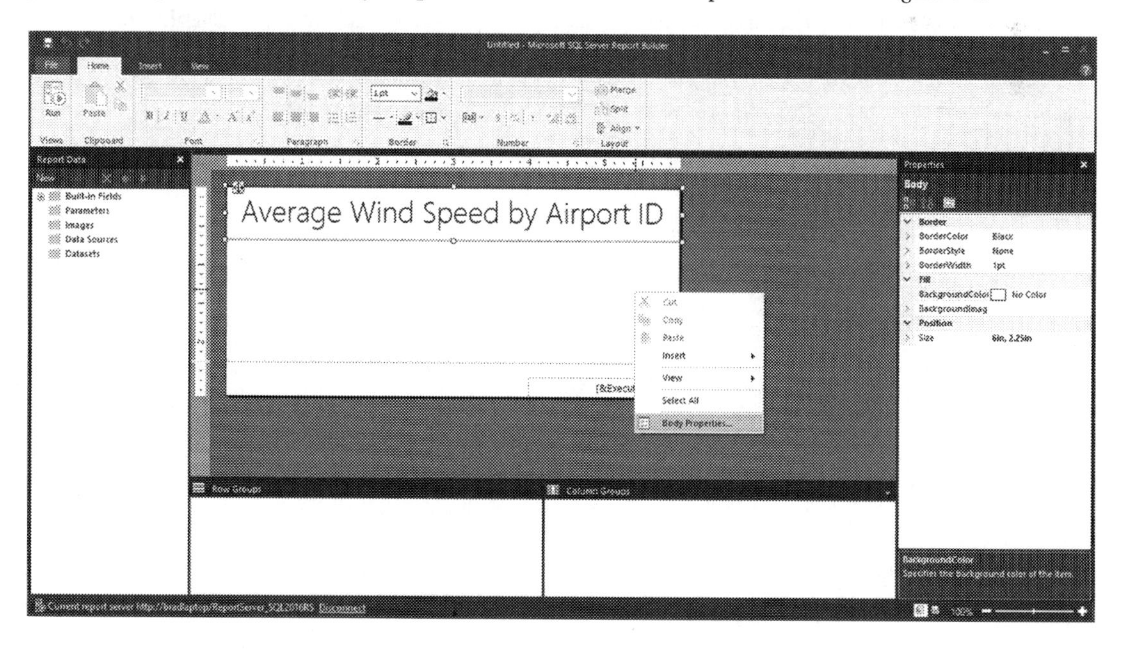

Figure 8-4. Option location

Selecting that option opens another screen, shown in Figure 8-5.

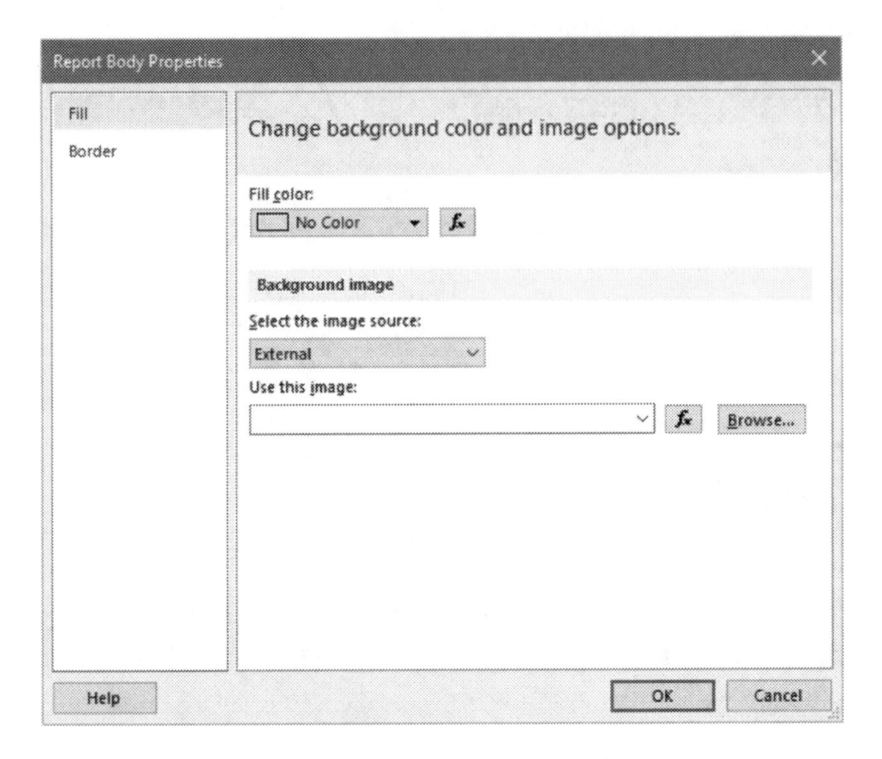

Figure 8-5. *Report Body Properties*

Now, we don't want to have a fill color except for plain white, but if you want to get adventurous, go right ahead. We also don't want a background image, so leave that blank for now. Like I said, if you want to change that, go right ahead. It's your report!

Click the **Border** option on the left; you should see what's shown in Figure 8-6.

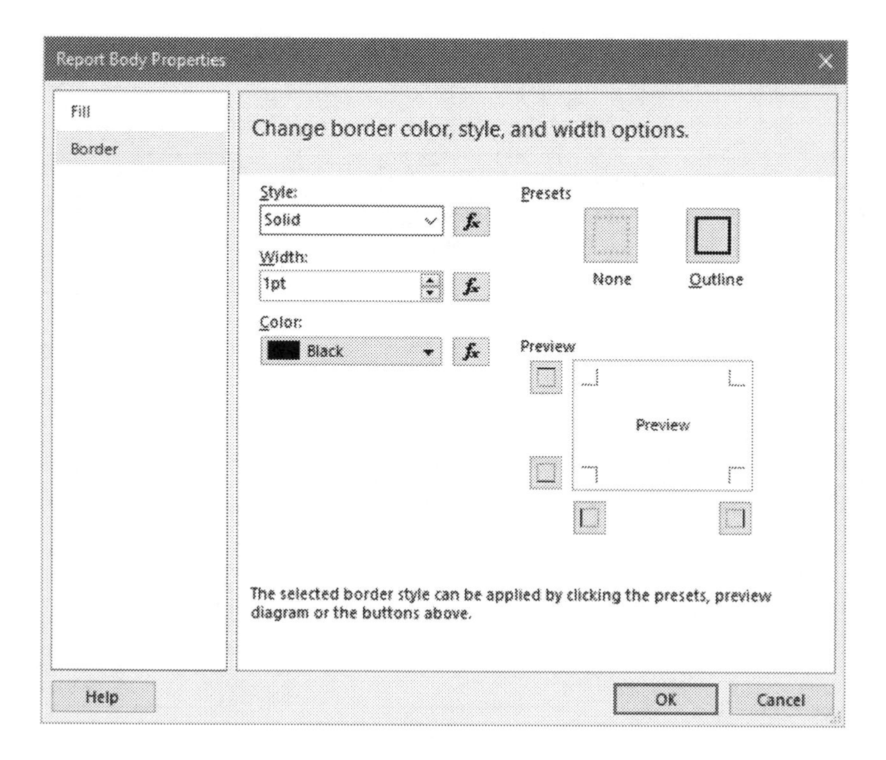

Figure 8-6. *Border properties*

Just click the **Outline** option in the top-right corner and then click **OK**. That screen closes. You can't really see a border on the page, but it's there. You will see it when you preview the report.

Data Configuration

Now we need to set up our data source and our dataset. Isn't that sort of the same thing? In this context, a data source (connection to the data in the database) feeds a dataset (a query to the data source), much like a lot of other examples dealing with these concepts.

You can't have a dataset without a data source and a data source is useless without a dataset.

First, we have to set up the data source. To do this, right-click the **Data Sources** option on the left of the screen and select **Add Data Source...**. Figure 8-7 shows the location of this menu option.

A screen appears, as shown in Figure 8-8.

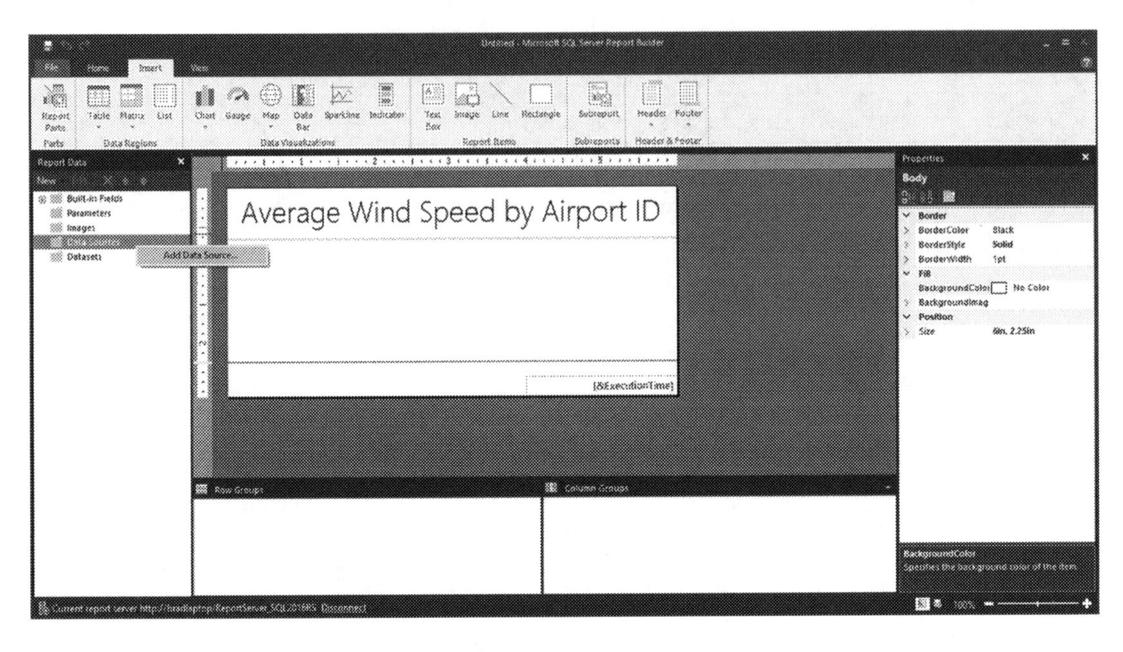

Figure 8-7. *Menu option*

Update that screen to what is shown in Figure 8-9.

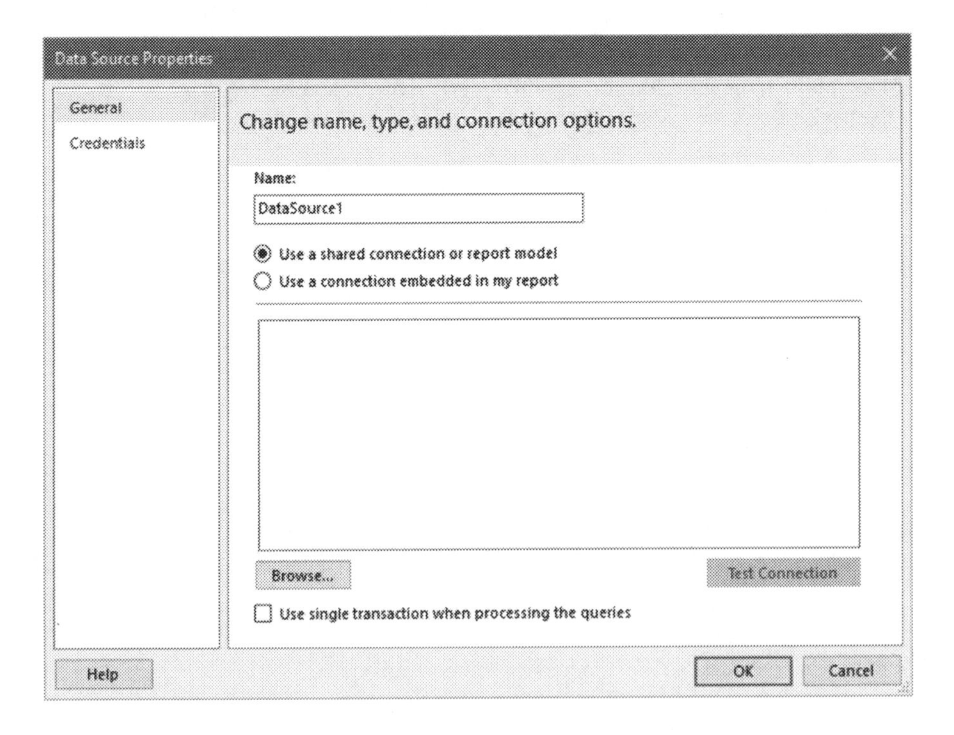

Figure 8-8. *Data Source Properties*

We need to configure that connection string, so click the **Build...** button; you should see what is shown

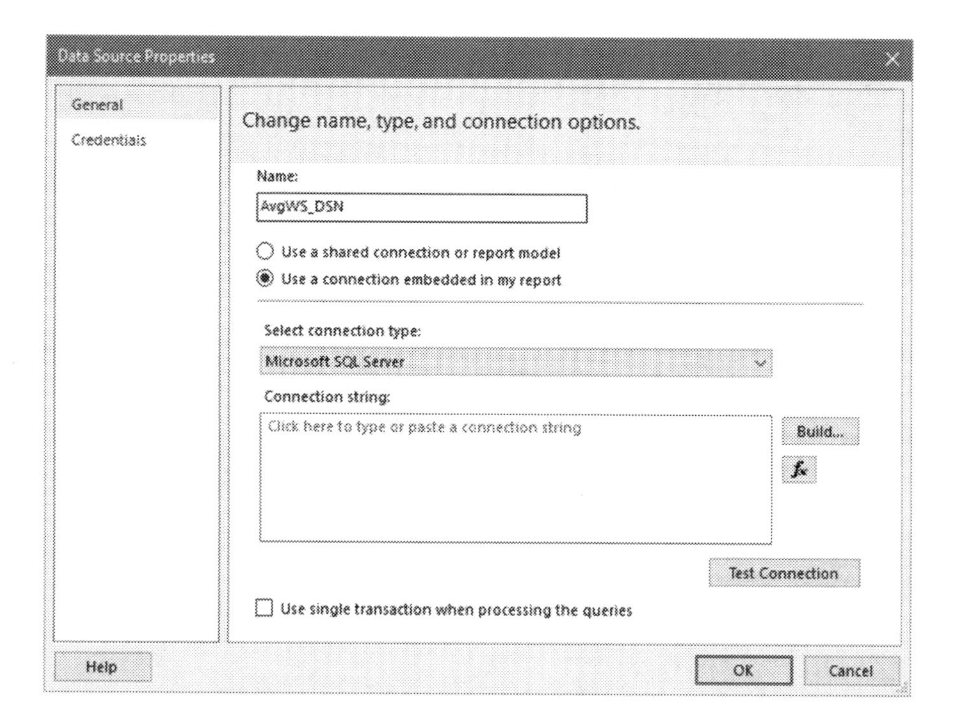

Figure 8-9. *Updated values*

in Figure 8-10.

Update that screen to what is shown in Figure 8-11.

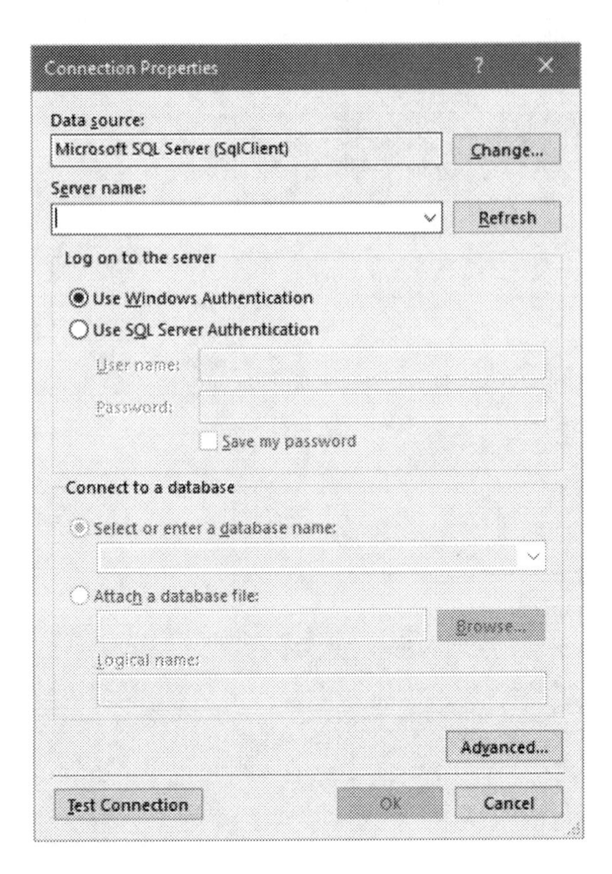

Figure 8-10. *Connection Properties*

Obviously, you want to keep in mind that your settings are probably different than mine.

Figure 8-11. *Updated values*

Once you get here, you can click the **Test Connection** button to verify that you're talking to the database. Figure 8-12 shows the result of clicking this button.

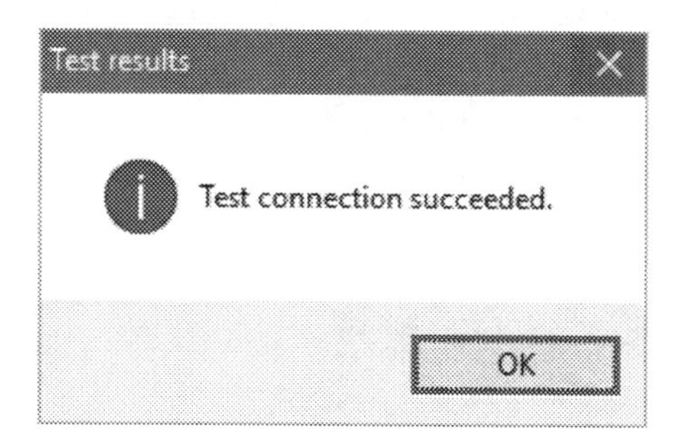

Figure 8-12. *Test connection succeeded*

Click **OK** here and then click **OK** again to save the Connection information. Figure 8-13 shows what you should see now.

Data Source Properties

General
Credentials

Change name, type, and connection options.

Name:
AvgWS_DSN

Use a shared connection or report model
Use a connection embedded in my report

Select connection type:
Microsoft SQL Server

Connection string:
Data Source=BRADLAPTOP\\SQL2016RS;Initial Catalog=ReportServer
$SQL2016RS

Build...

Test Connection

Use single transaction when processing the queries

Help
OK
Cancel

***Figure 8-13.** Updated values*

So there's our connection string value all done. Again, you can click the **Test Connection** button here to verify connectivity, but I think we're pretty much good for this area. Click **OK** to close this window.

Notice that Figure 8-14 shows that we now have a data source available.

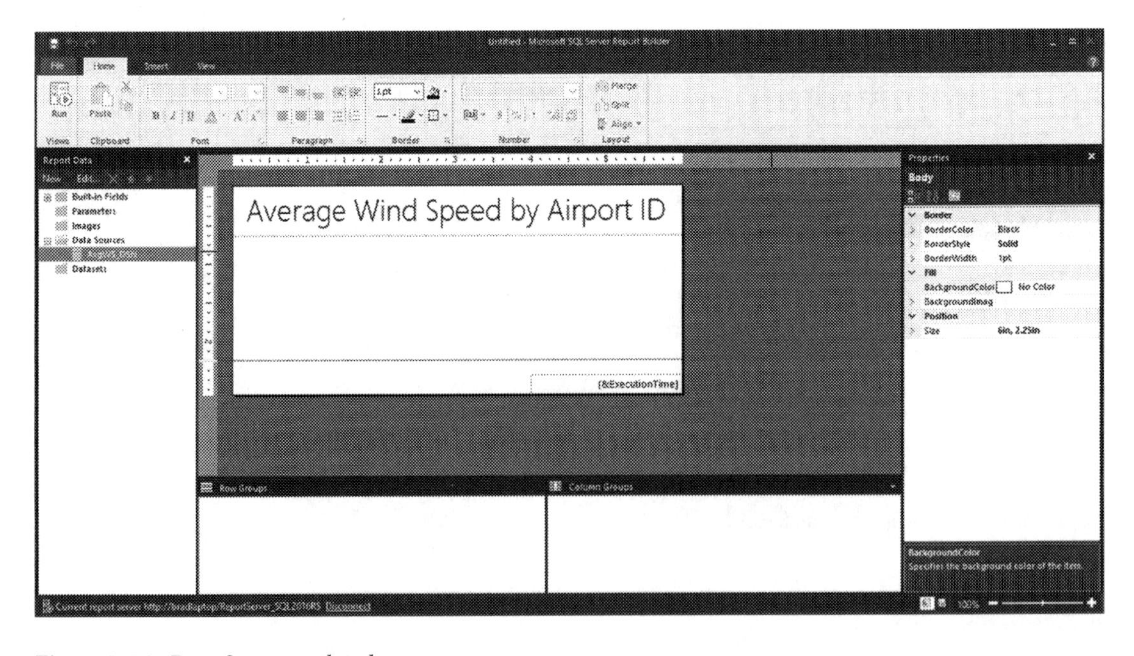

***Figure 8-14.** Data Source updated*

Next, we need to add our dataset. To do this, right-click the **Datasets** option on the left and select **Add Dataset...**, as shown in Figure 8-15.

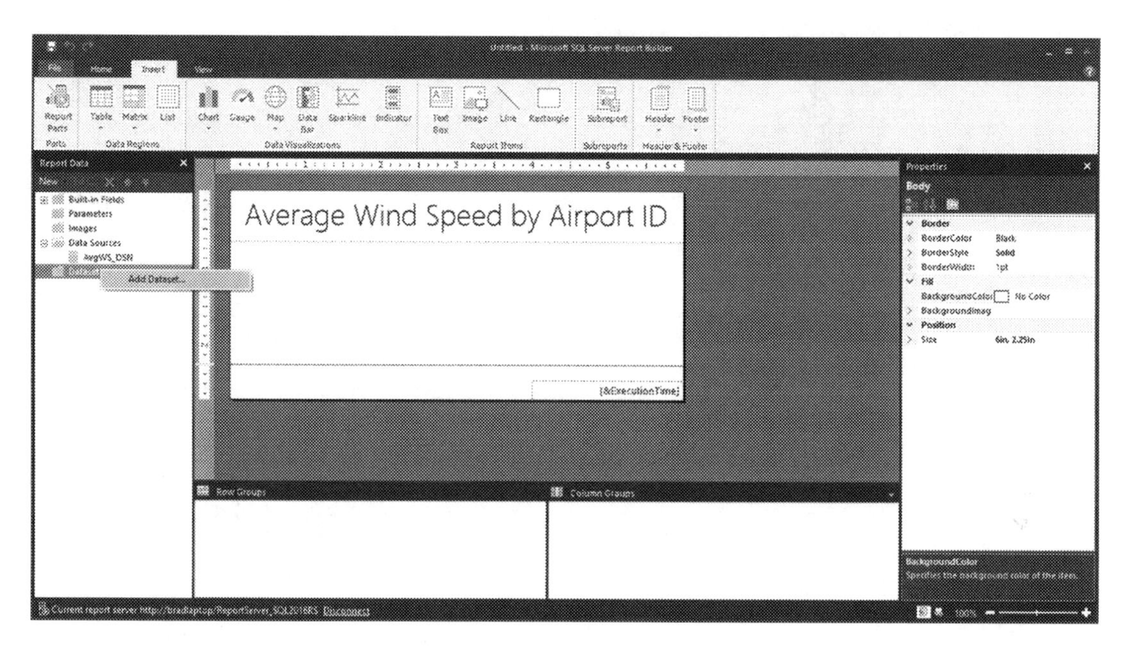

Figure 8-15. *Add Dataset*

This opens the interface shown in Figure 8-16.

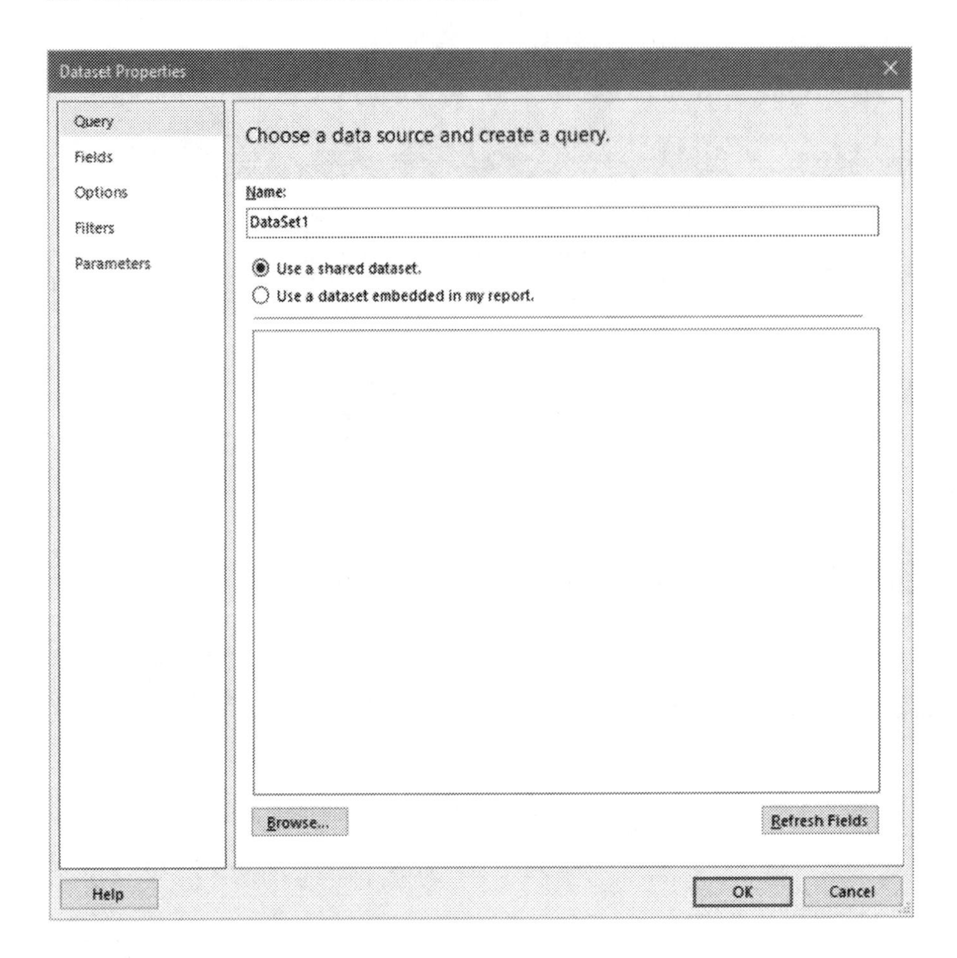

Figure 8-16. *Dataset Properties*

Update that interface to match what is shown in Figure 8-17.

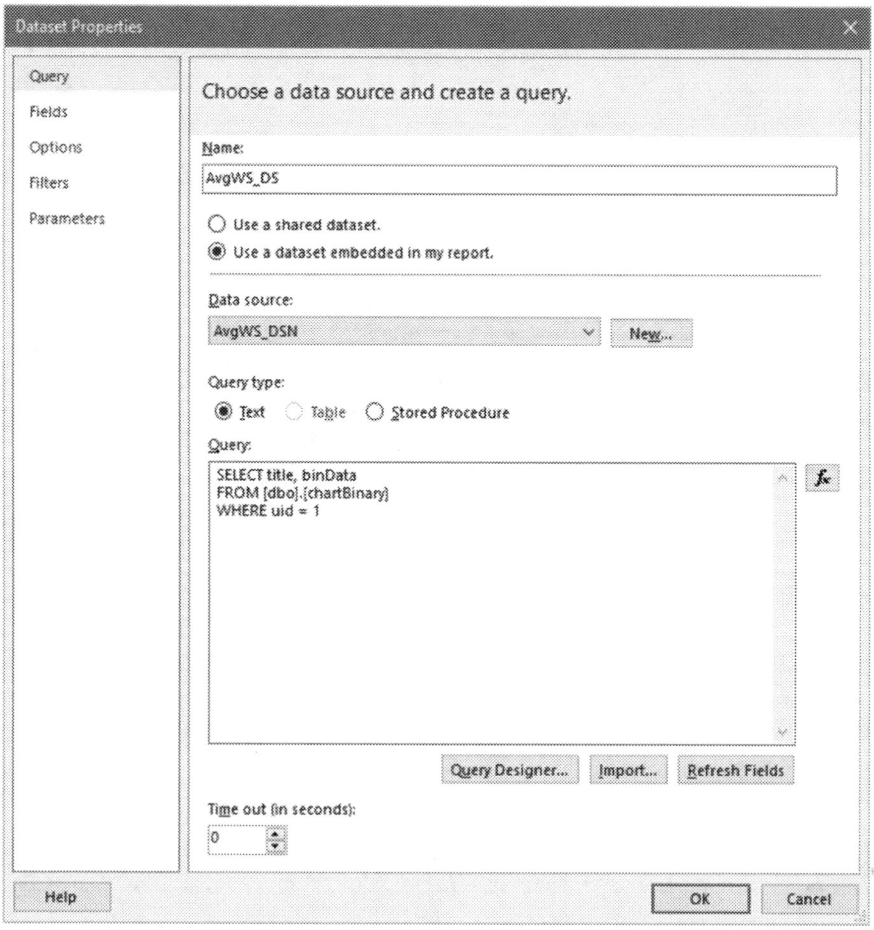

Figure 8-17. *Dataset Properties updated*

The query I wrote to get the data for this report is:

```
SELECT title, binData
FROM [dbo].[chartBinary]
WHERE uid = 1
```

Super simple, yet effective.

The options on the left don't really apply to this section, but you can play with those later. Click **OK** to set up the dataset. You should be returned to the blank report screen shown in Figure 8-18.

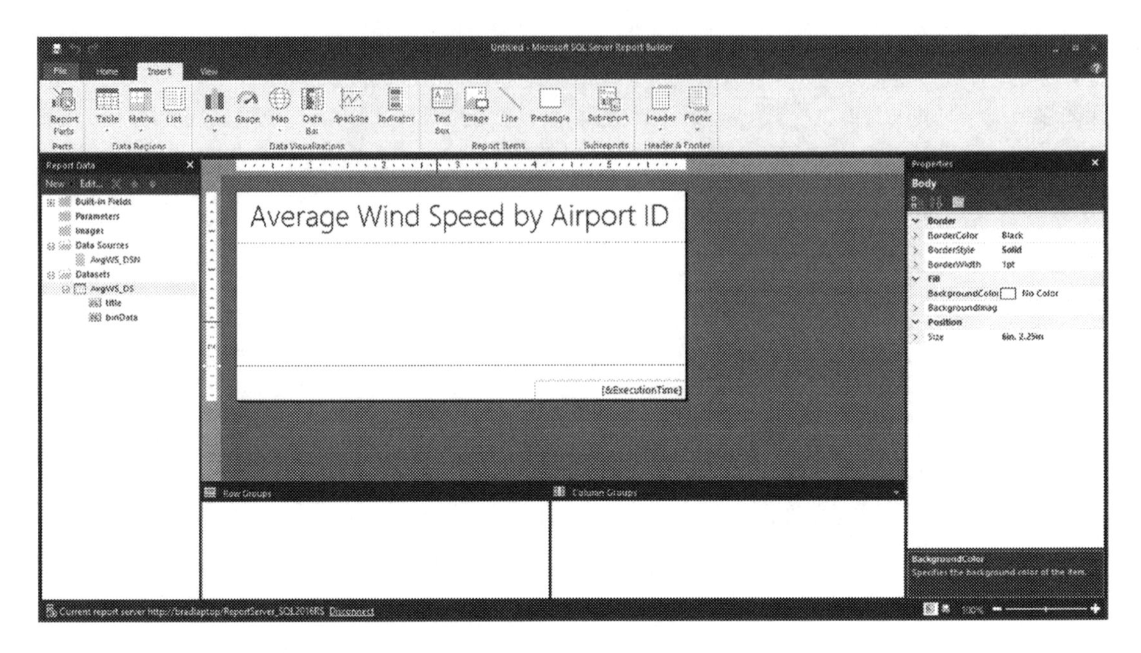

Figure 8-18. *Main screen*

Notice that there are now entries under the Data Sources and Datasets folders, respectively.

Let's change that title text to the title we entered in our query. Go ahead and delete the title that you entered earlier (Average Wind Speed by Airport ID). Just highlight the text and then press **Delete**. You should see what is shown in Figure 8-19.

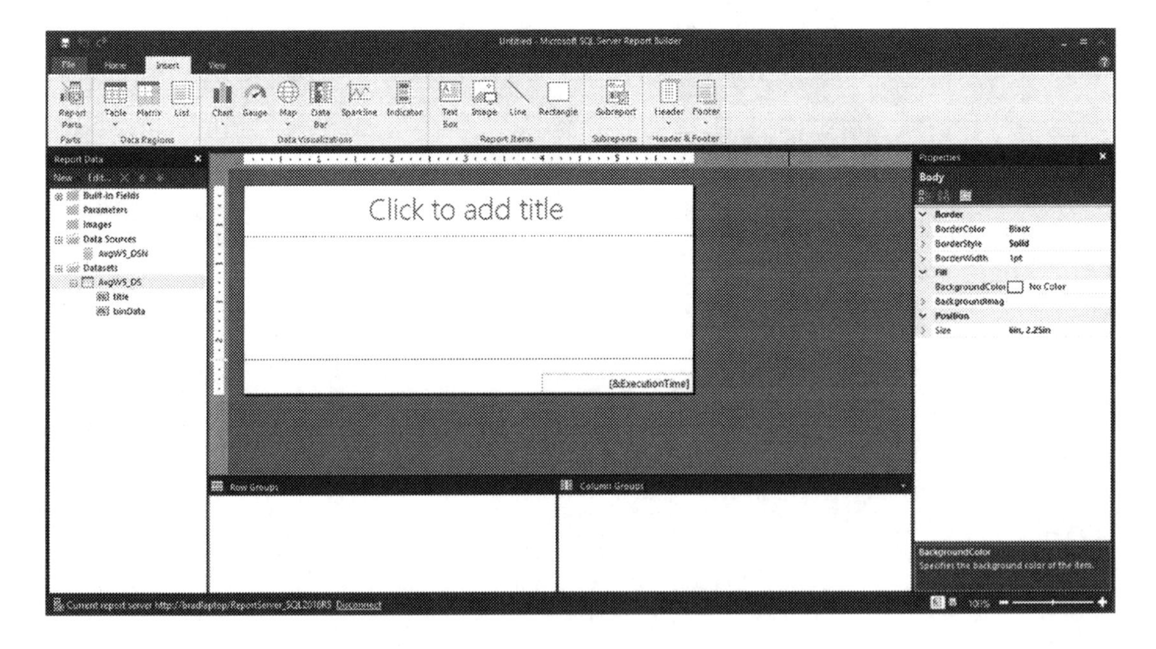

Figure 8-19. *Deleted initial title value*

Next, click and drag the **title** value under the **AvgWS_DS** option on the left to inside the title box on the main screen. Figure 8-20 shows what you should see now.

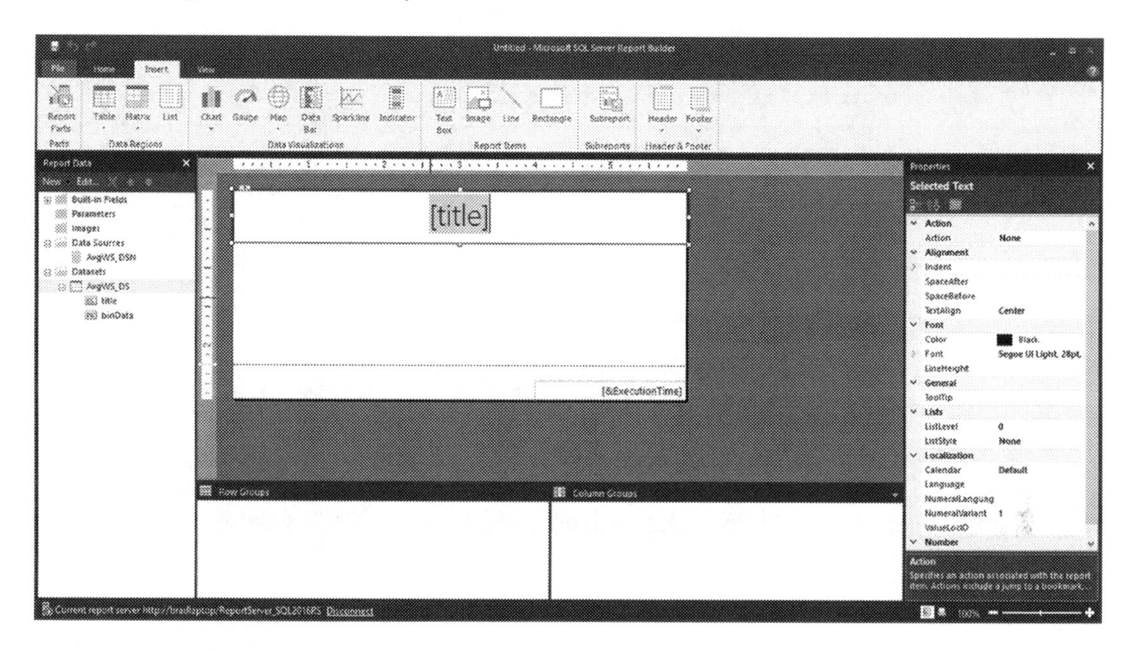

Figure 8-20. *Title addition*

That's perfect! So now, when the report is run, the value of the title is displayed.

Adding the Dynamic Image

Next, we need to add our image. Right-click anywhere in the body and hover the **Insert...** selection until a submenu appears. This submenu contains an Image option. Figure 8-21 shows the location of this menu option.

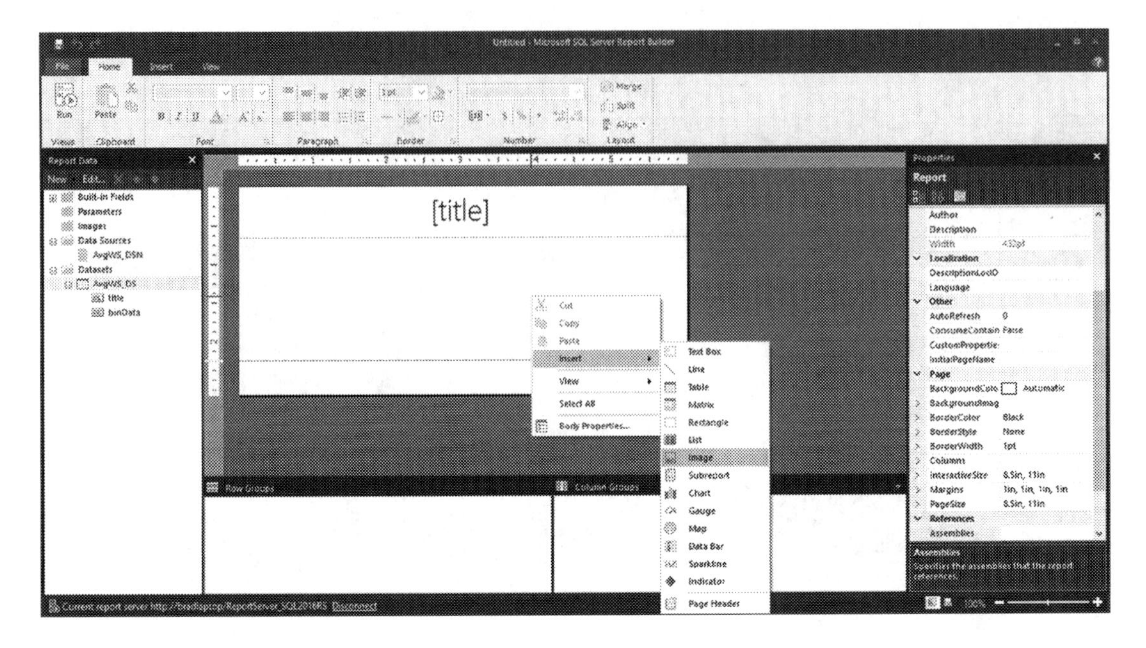

Figure 8-21. *Image option*

Consequently, you could also click the **Insert** menu option at the top of the window and then click the **Image** option. However is more comfortable for you is fine, since they both accomplish the same task.

Once you have selected to put an image on the report, an interface appears, as shown in Figure 8-22.

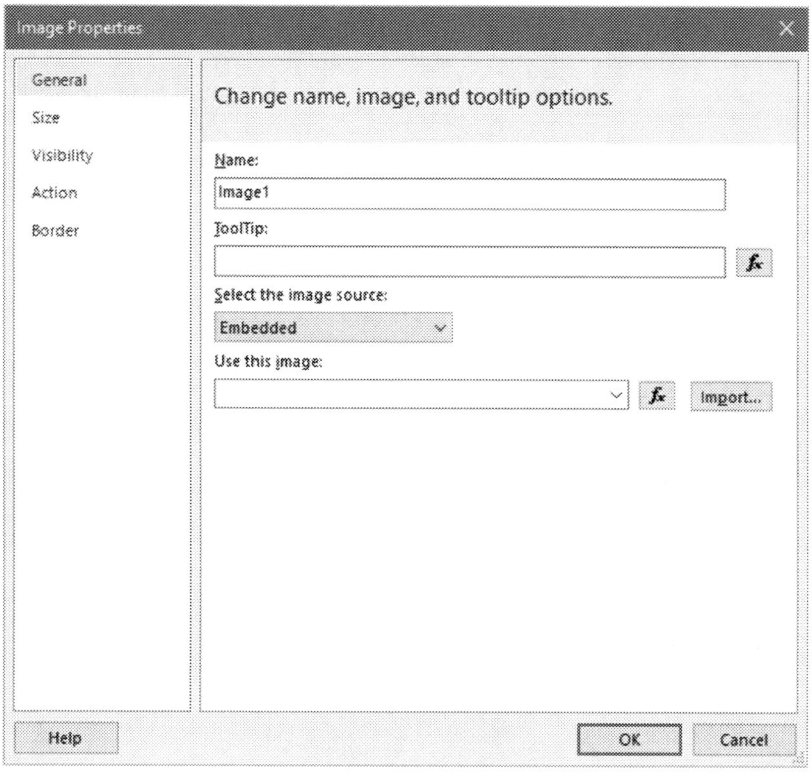

Figure 8-22. *Image Properties*

We need to update these values to match what is shown in Figure 8-23. Notice that I only updated the Name, ToolTip, and image source fields.

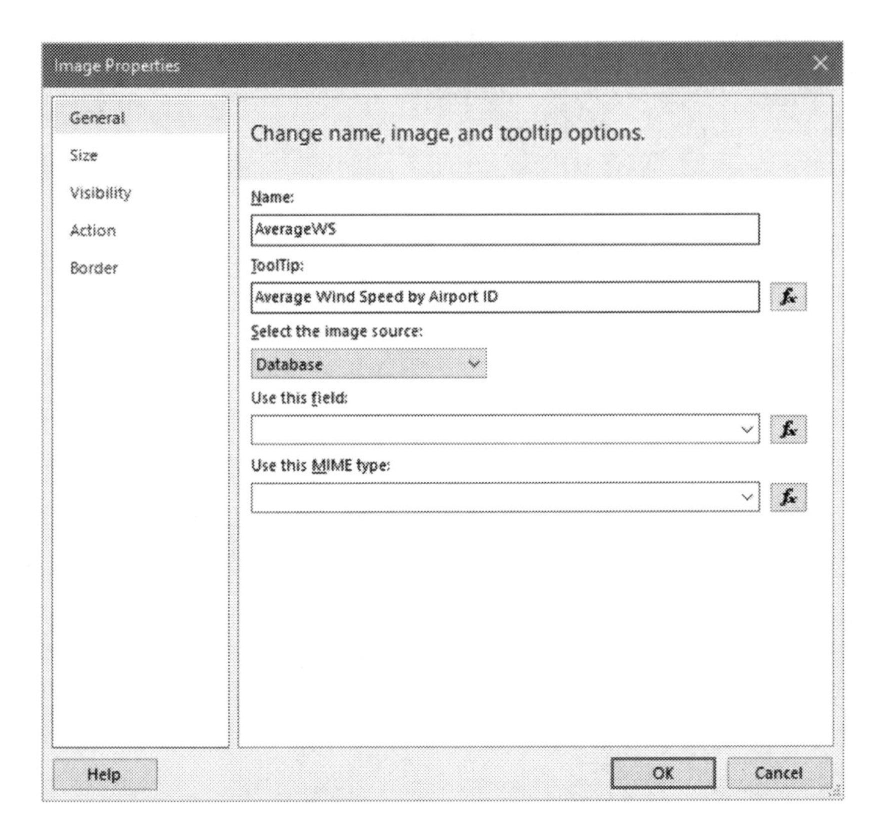

Figure 8-23. *Updated Image Properties screen*

Pull down the **Use this field** menu option and select the value =**First(Fields!binData.Value, "AvgWS_DS")**. It is easy to select the wrong value, so be sure you're choosing the **binData.value** option.

For MIME type, we want to choose **image/png**.

The finished product is shown in Figure 8-24.

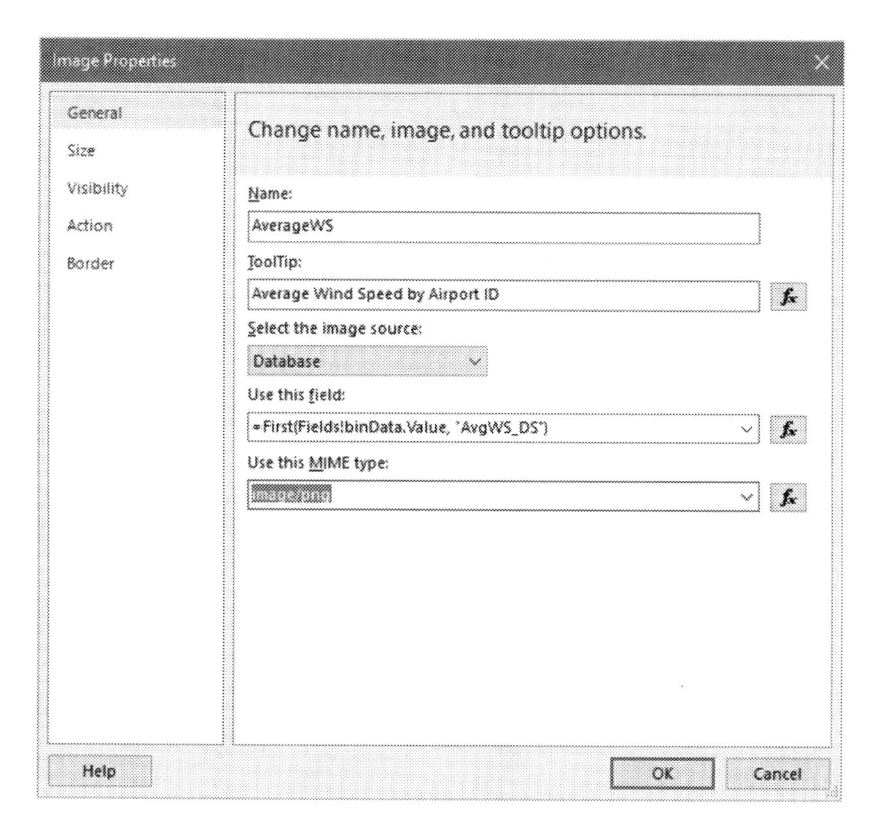

Figure 8-24. *Updated values*

Next, click the **Size** option on the left. Figure 8-25 shows this interface.

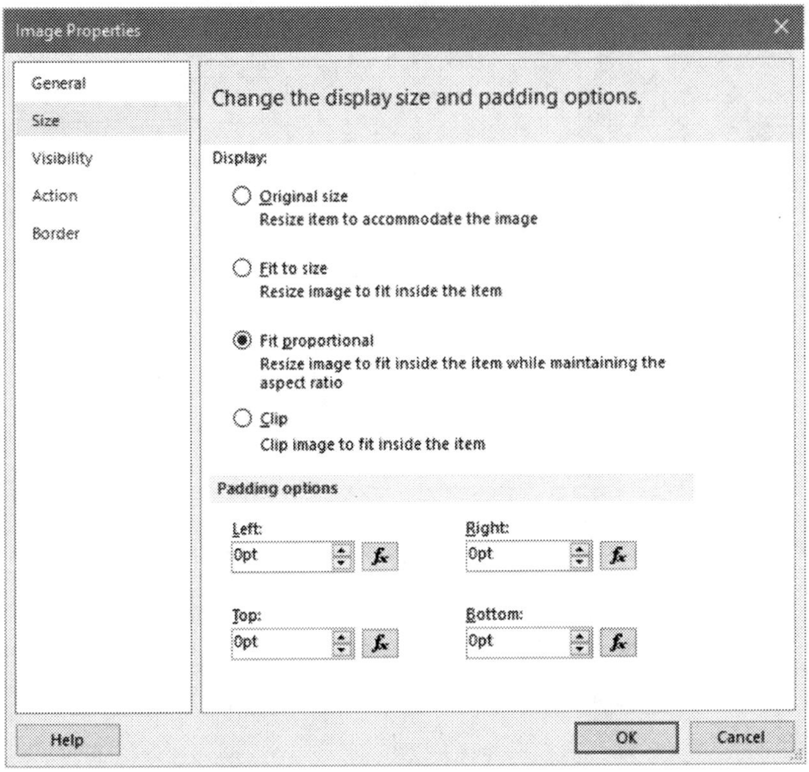

Figure 8-25. *Size options*

Choose the **Fit to size** option here. That keeps the image inside the printable area. Click **OK**. You should see something very similar to what is shown in Figure 8-26.

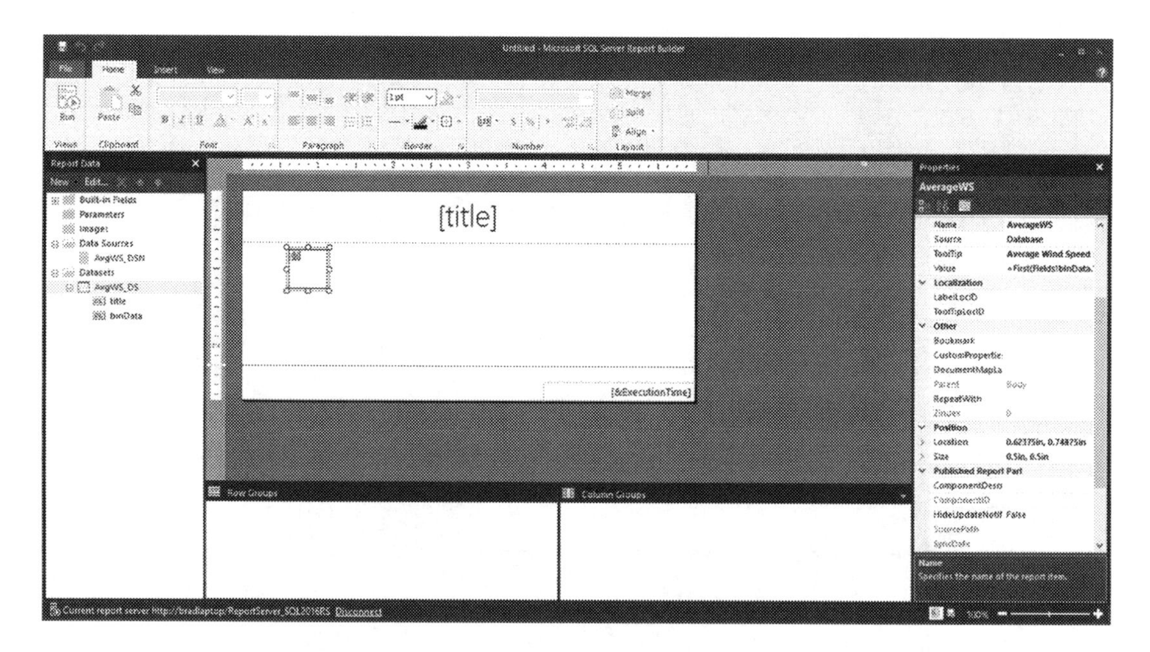

Figure 8-26. *Main screen, updated*

At this point, we only see the image placeholder, since the image is generated dynamically from the database. Once the report is run, the image is shown, as you would expect.

Before we move on, just grab the image on the screen and manually align it to the left of the screen inside the white space. Figure 8-27 shows the result of this action.

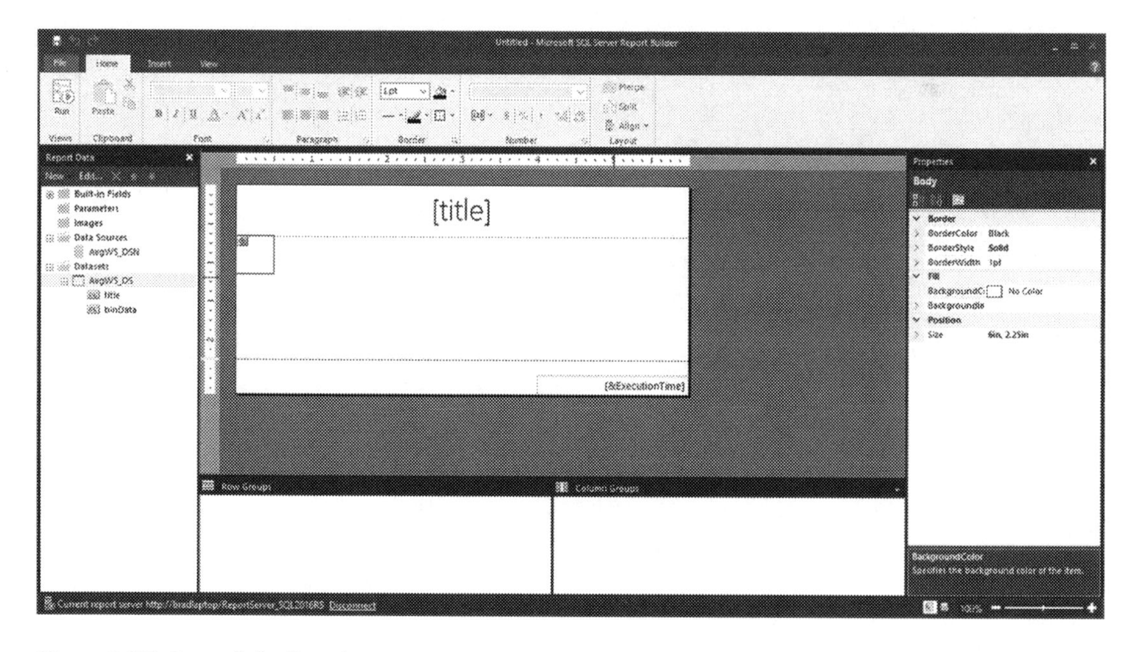

Figure 8-27. *Image left-aligned*

Report Body Properties

Recall that the image we created was 800×600, so let's update the style of the report. Right-click the gray area and choose **Report Properties...** from the menu. Figure 8-28 shows this action.

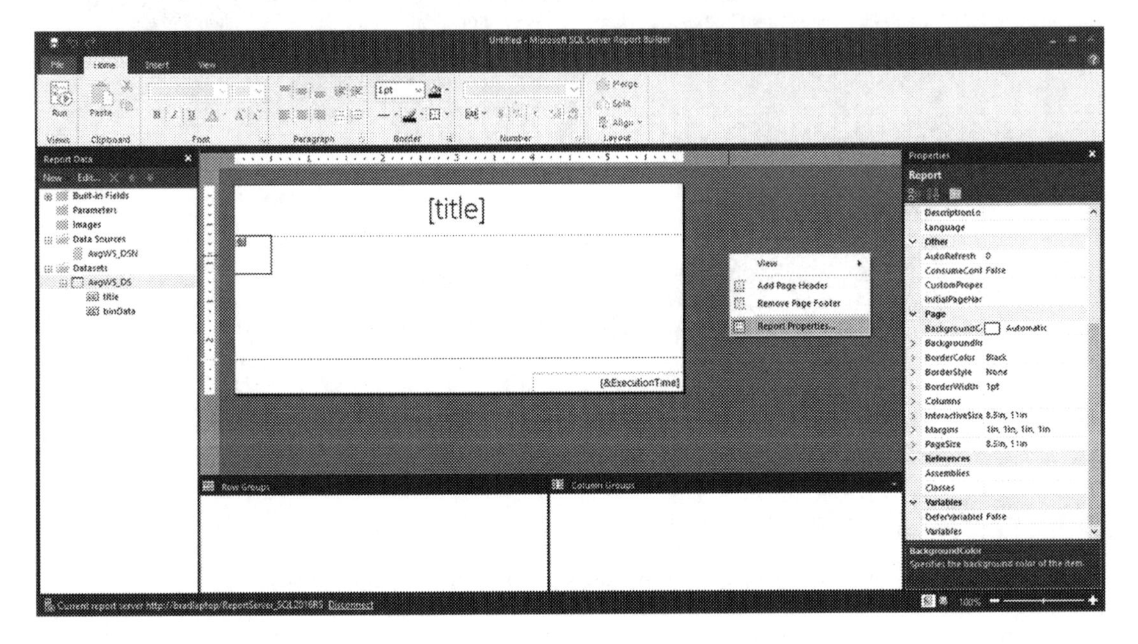

Figure 8-28. *Report Properties location*

This opens up the interface shown in Figure 8-29.

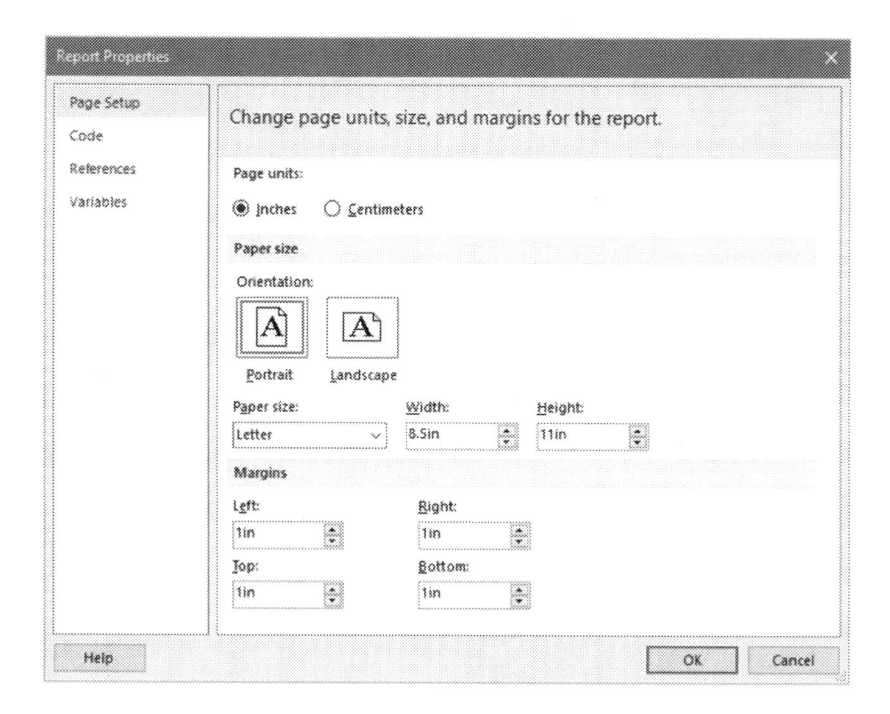

Figure 8-29. *Report Properties*

All that we need to do is change the **Orientation** value to **Landscape** and then click **OK**. We need to check the size of the page to adjust the image. A little experimenting shows that we want to be right around 9×5 for the stage dimensions (just click and drag them to that size) and the Figure 8-30 shows what this should look like.

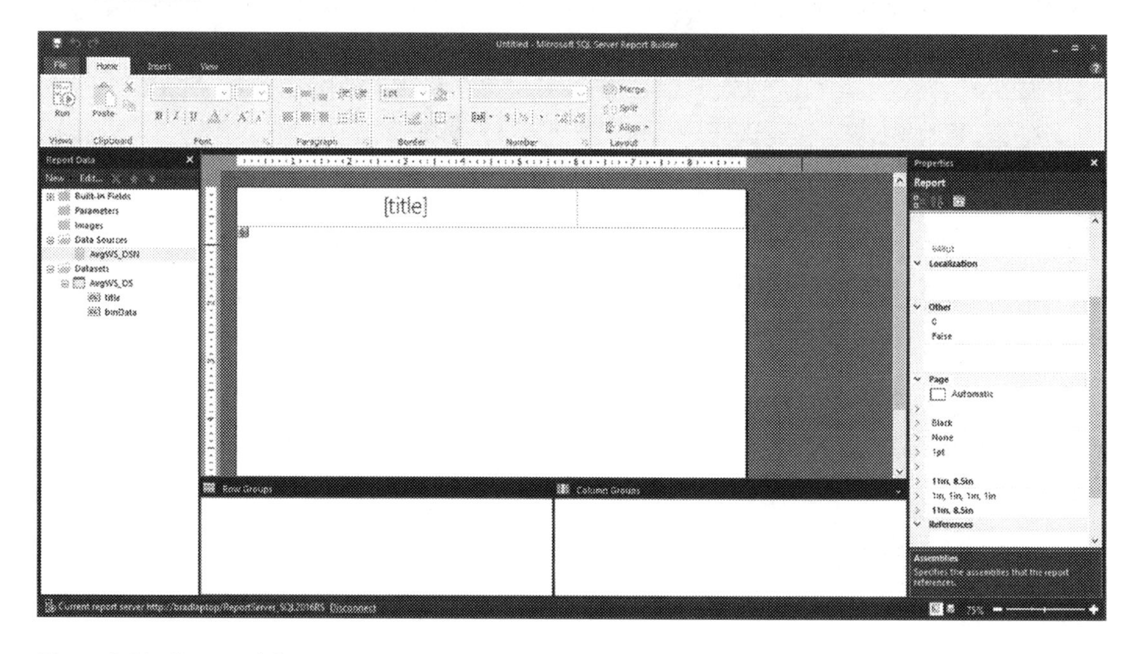

Figure 8-30. *Stage resizing*

I colored my footer blue with white text, but you can do whatever you want as far as styling to make this report your own.

Running the Report

Go ahead and click the **Run** button in the top-left corner. Get ready for a big surprise. Figure 8-31 shows what you should see after this happens.

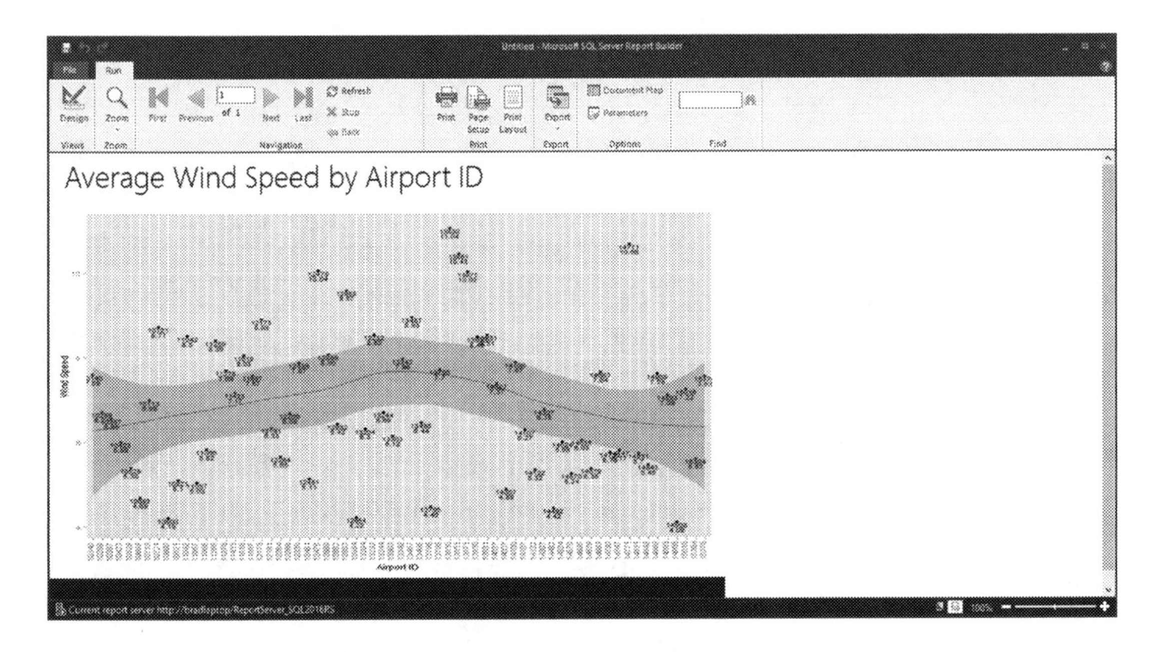

Figure 8-31. *Completed Report*

Not bad! Now click the **Print Layout** button in the menu bar. Figure 8-32 shows this result.

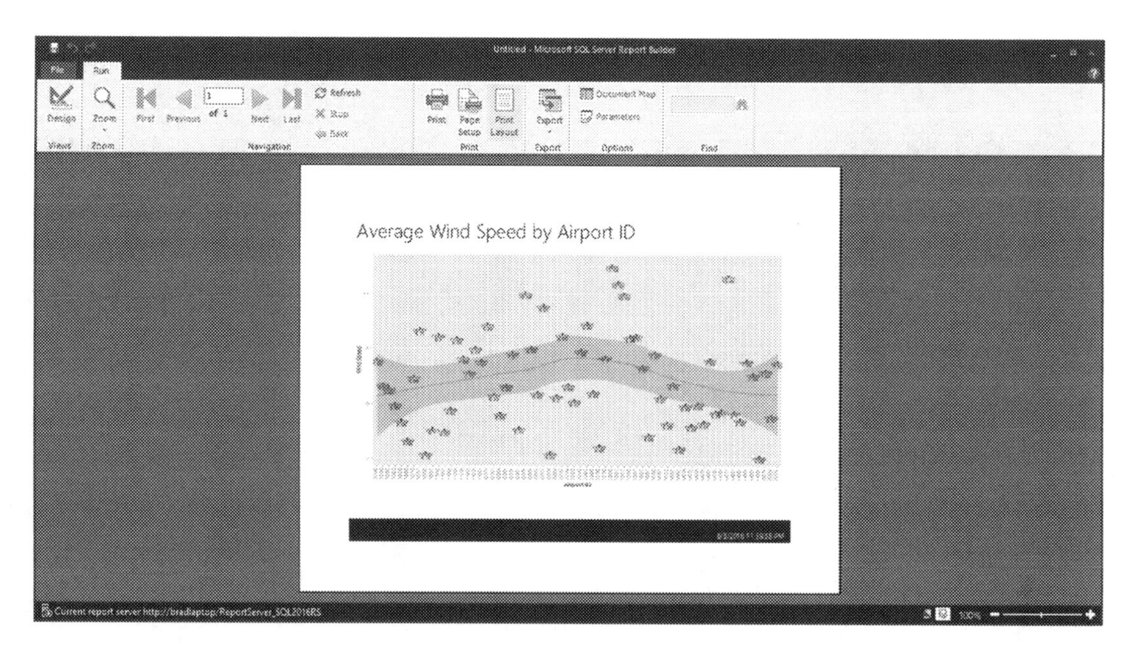

Figure 8-32. *Print Layout view*

There it is, nicely laid out for us on one page, with our rendering time displayed in the blue footer.

Click the **Design** button in the top-left corner and then click **Ctrl+S** to save the report. The default location and values are shown in Figure 8-33.

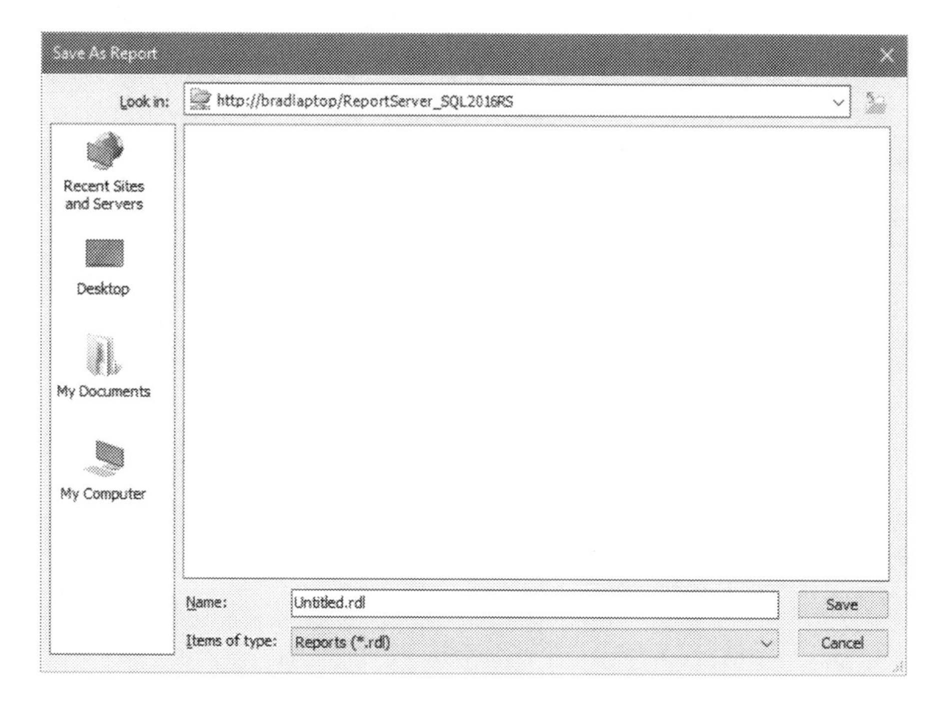

Figure 8-33. *Default report location*

We can see that the default location is on our report server, so that's perfect. Update the **Name** of the file to **AverageWSbyID.rdl**, as shown in Figure 8-34.

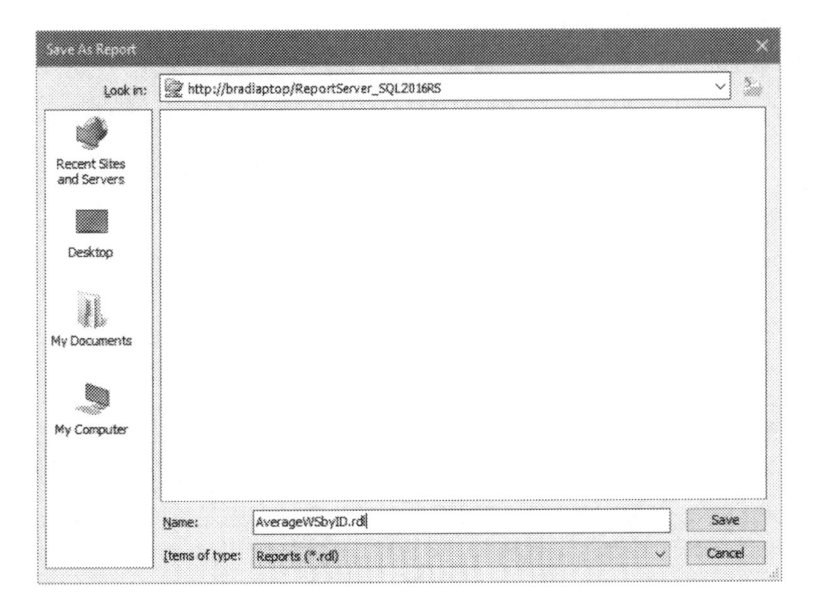

Figure 8-34. *Updated name information*

Press the **Save** button here to save the report.

This report is now saved to the report server. We will view it right after we get the other report going.

Report 2: Average Temperature by Airport ID

Watch how easy it is to create the second report. I hope that you are starting to see the flexibility that this technology gives you!

Press **Ctrl+N** to open a new report (be sure to save the old one). You are immediately shown a blank report, just like before. You can also go to the File menu and choose **New...** to choose a different type.

We need to essentially follow the exact same instructions as before, except this time, when we create the dataset, we need to change the `WHERE` clause to show where the UID is equal to 2 instead of 1. I am going to leave making that change as an exercise for you to complete on your own. It's a trivial change to make.

When you are done setting it up, your screen should look similar to Figure 8-35.

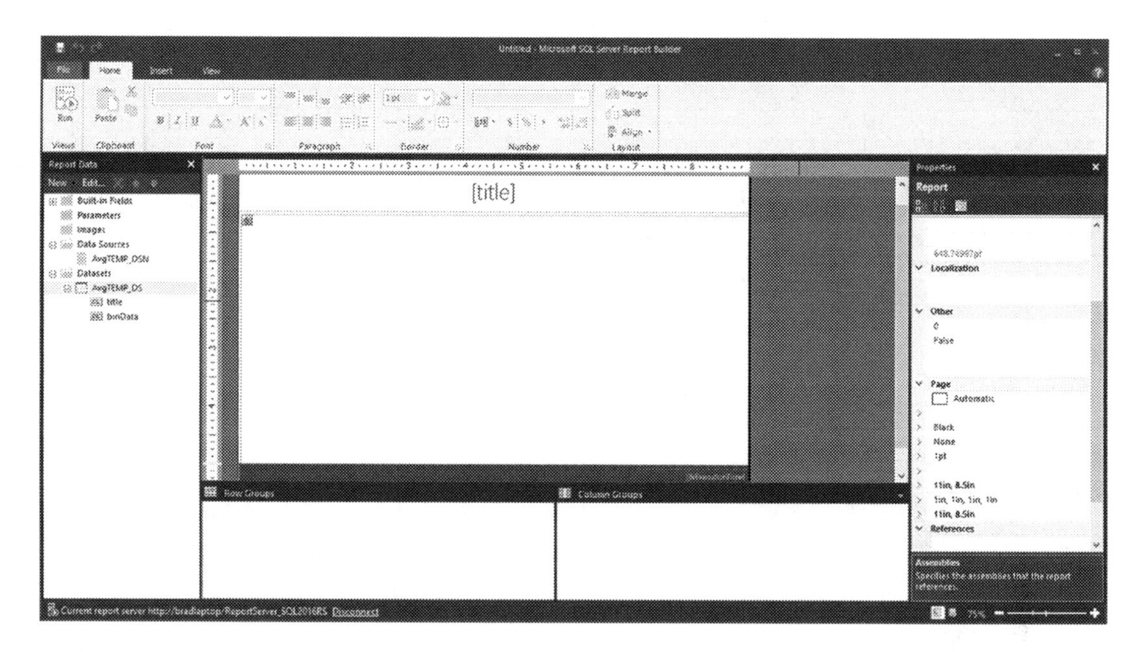

Figure 8-35. *Average Temp report layout*

And your report should be very similar to what is shown in Figure 8-36.

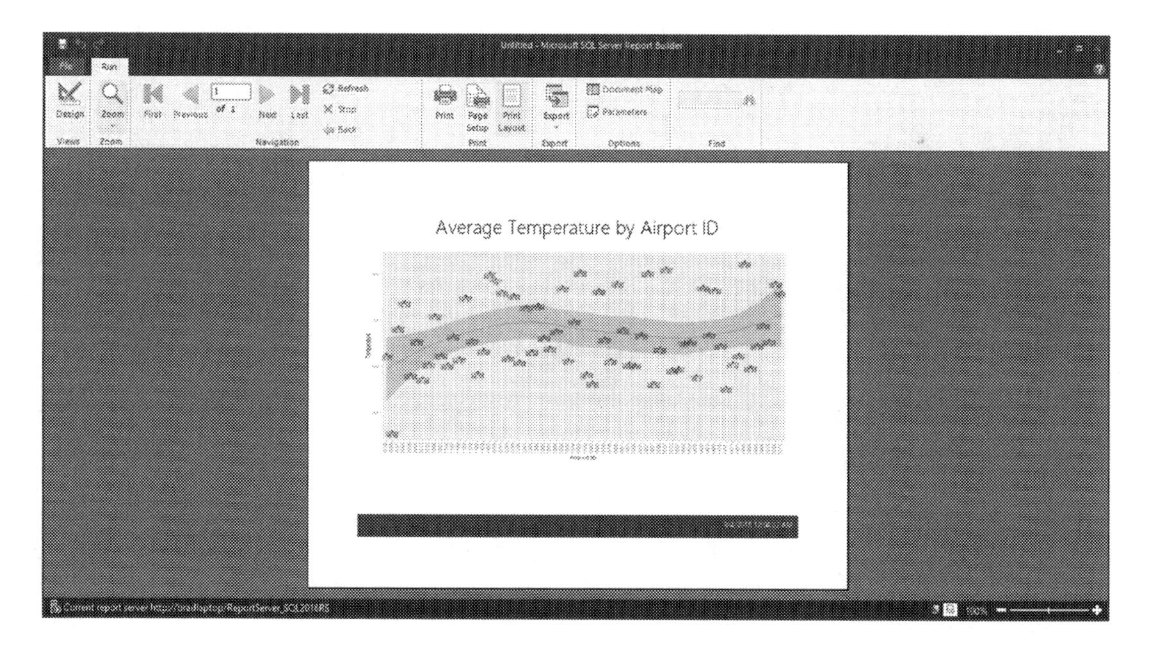

Figure 8-36. *Second report*

Excellent work! Be sure to save that report as `AvgTempByID.rdl`.

Summary

I hope you learned a lot in this chapter. I encourage you to continue your journey as data scientists by learning as much as you possibly can about R and business intelligence as you can. This is guaranteed to be a very lucrative career field in the future and I am very excited to be a part of it.

If any part of this chapter didn't make sense, as I have said in previous chapters, please be sure to go back and redo the examples.

Chapter 9 shows you how to access these reports from the Report Server. It is going to be a fun chapter, not only because it's the last chapter, but because it's the culmination of our work in this book.

CHAPTER 9

Viewing the Reports in Report Server

Now that we have built the required reports in Report Builder, we can view them through the report server web portal. Later in this chapter, we get into other aspects of Report Server, but for now, let's take a look at our reports.

Viewing Reports

Recall back in Chapter 6 when we set up the report server, we specified the URL as `http://bradlaptop:80/Reports_SQL2016RS`. Type your web portal address into your browser and click **Enter**. You should see something similar to what is shown in Figure 9-1.

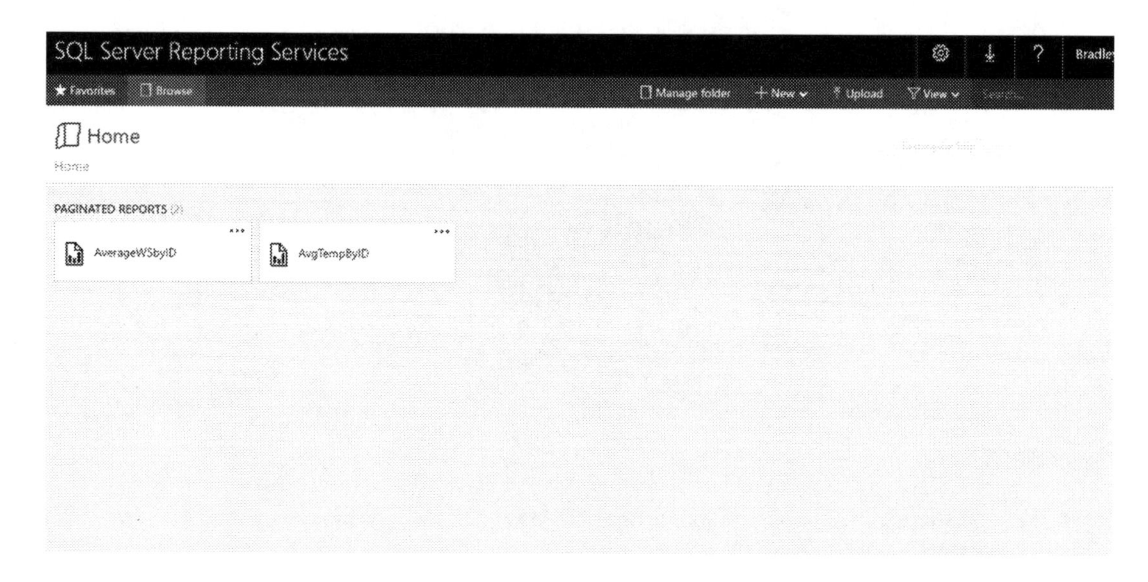

Figure 9-1. *Web portal*

Clicking the **AverageWSbyID** report on the left shows what you see in Figure 9-2.

B. Beard, *Beginning SQL Server R Services*, DOI 10.1007/978-1-4842-2298-0_9

Figure 9-2. *Generated report*

The same happens when you click the **AvgTempByID** link on the home page. Figure 9-3 shows this.

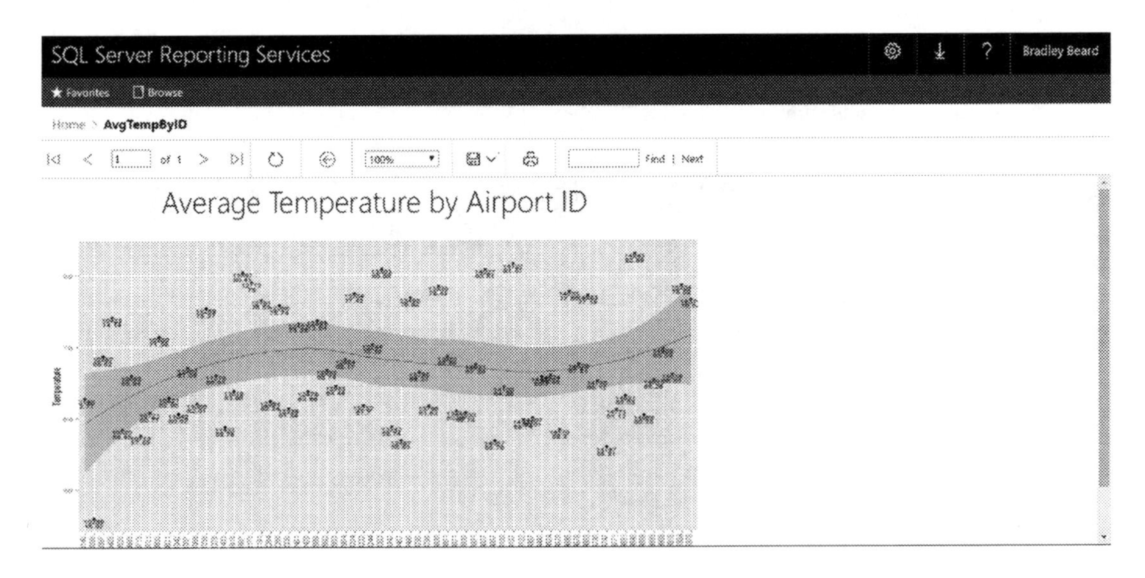

Figure 9-3. *Generated report*

So there you have it… a complete report using binary data to create images from original R code.

Managing Reports

With this release of Report Server, as with previous versions, there is the capability to manage the reports from the Report Server interface. Managing reports in Report Server requires that the user is an administrator (which we are, or should be) and belongs to the Content Manager group. The following security roles are available in Report Server:

- **Browser**: This role can view folders and reports in specified folders, and it can subscribe to reports. This role cannot create reports, though. *This would be a regular read-only user role.*
- **Content Manager**: This role manages the content in Report Server. *This is the administrator account for Report Server and can do anything within the confines of Report Server.*
- **My Reports**: This role can publish reports and manage user folders they are specifically designated to access. *This would be a regular read/write "power user" account type.*
- **Publisher**: This role can only publish reports and linked reports to Report Server. *This would be a write-only user role.*
- **Report Builder**: This role can manage and view the report definitions and attributes. *This would be a role reserved for specific users that need to configure the reports, but don't need to be full administrators (in the Content Manager role).*

Keep in mind that the Content Manager role can do anything we need in this instance, so the principal of least privilege is really overlooked here.

Click the ellipse, as shown in Figure 9-4. Select the **Manage** option from the pop-up menu.

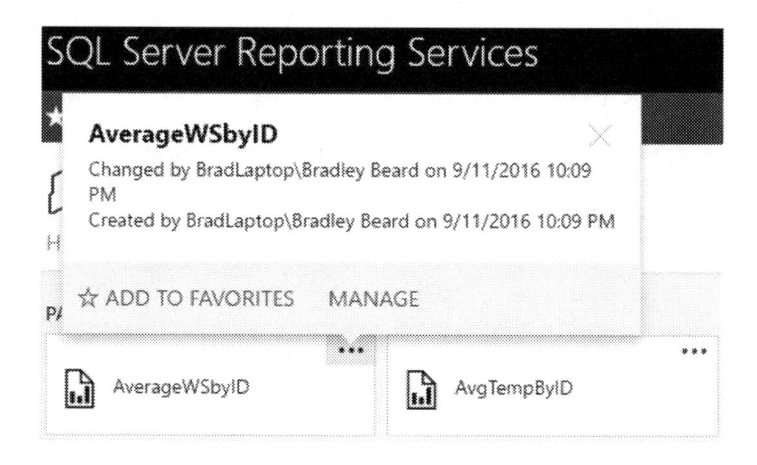

Figure 9-4. *Manage menu*

A page appears, as shown in Figure 9-5, which allows you to edit some of the properties of the report.

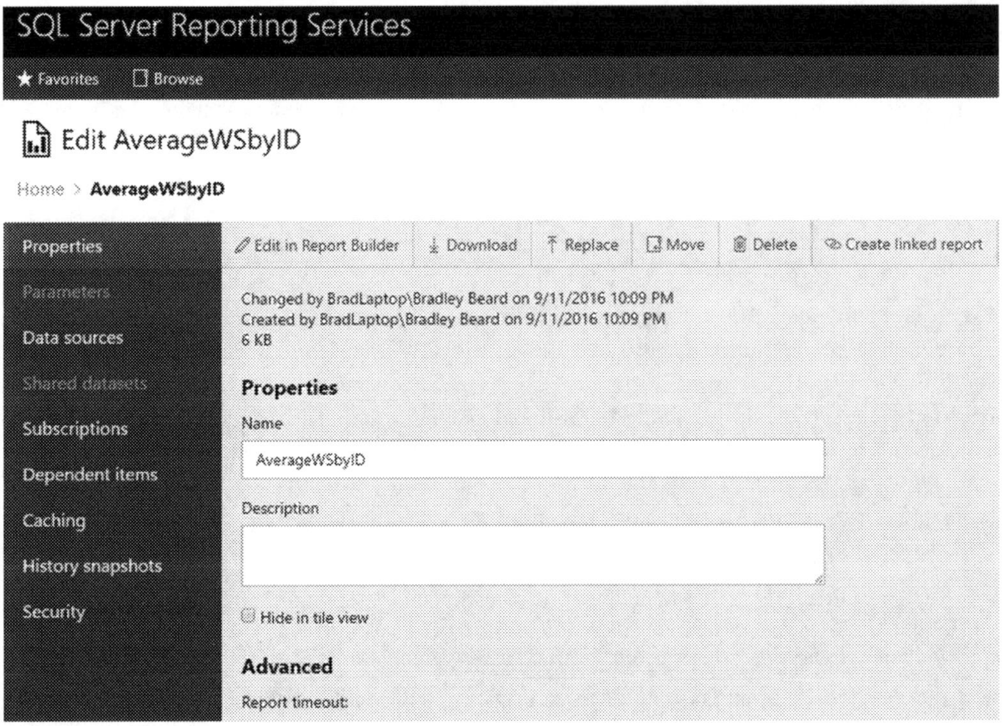

Figure 9-5. *Properties*

Properties

Notice that we are in the Properties section on the left-hand side. Across the top, there are the following options: **Edit in Report Builder**, **Download**, **Replace**, **Move**, **Delete**, and **Create linked report**. Most of these are self-explanatory, except for **Create linked report**. The **Create linked report** option allows you to create a report that retains the layout and data source information of an existing report, but still allow for the editing of the other parameters of the report, such as subscriptions and report parameters. Think of this sort of like a template creator. You essentially create a template of one report and have an entirely new report created from the basic skeleton of that report (data source and report layout), but then you can populate the body of the skeleton with entirely different report information.

On this page, we can add a description to the report in the **Description** box shown in Figure 9-5. If you were to scroll down, you would see there is another area titled **Advanced**. This area allows you to change the report timeout period, but since this is a dynamically generated report, I am going to leave this alone. There could be an instance where network traffic is keeping my report from being generated quickly. I don't want to cause the report to fail just because of that.

Data Sources

The next section, shown on the menu on the left, is Data Sources. This section allows us to update or change our existing data sources. The keyword here is *existing*. We cannot create new data sources from Report Server; they must be created in Report Builder first.

Figure 9-6 shows the top part of this screen, which has the data source information we specified back in Report Builder. Notice that we have the option here to change *anything*, because we are in the **Content Manager** role.

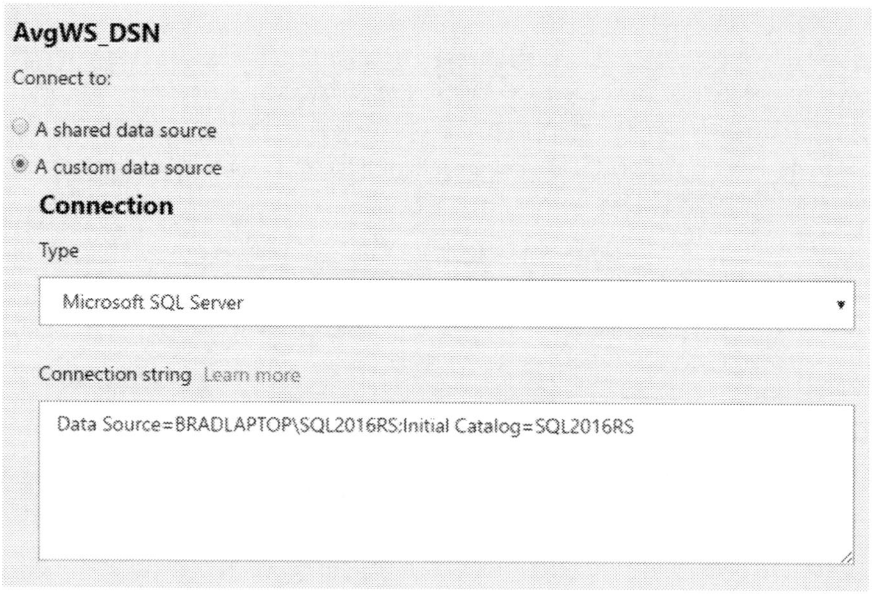

Figure 9-6. *Data source information*

The lower part of this screen is shown in Figure 9-7. This section allows us to define or edit the credentials necessary to connect to the data source. Clearly, this Credentials section already works as we have it set up, or we wouldn't have been able to see the reports we created in this Report Server interface. Figure 9-7 shows our options here.

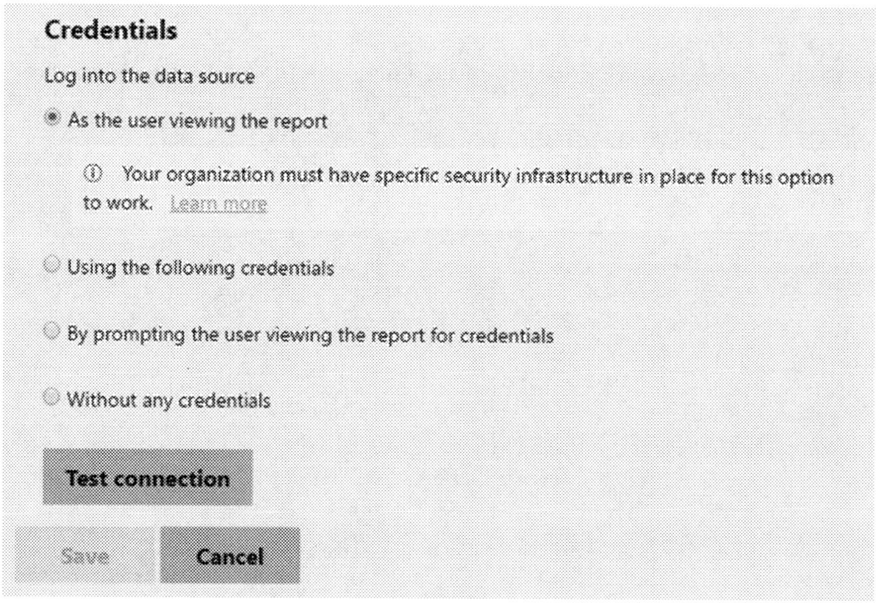

Figure 9-7. *Credentials information*

Notice that we can choose from four different login types.

■ **Note** This option may change for your particular installation, depending on your individual requirements.

On the Credentials screen, we want to scroll to the bottom and click the **Test Connection** button. Figure 9-8 shows what you should see at this point.

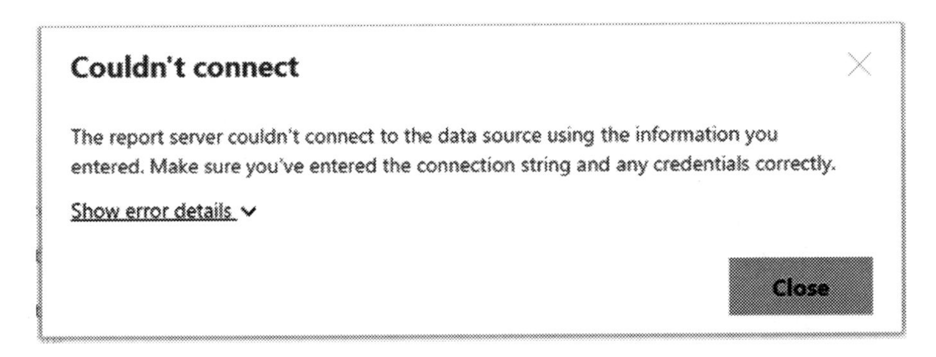

Figure 9-8. *Test Connection unsuccessful*

Additionally, an error message appears next to the Test Connection button, shown in Figure 9-9.

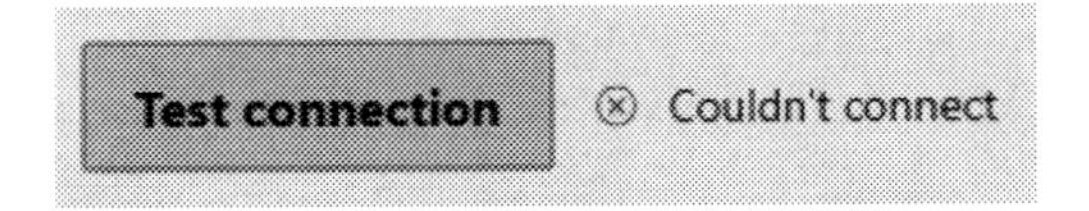

Figure 9-9. *Error message*

Next, we want to click the **Using the following Credentials** radio button and update the interface to what is shown in Figure 9-10. Your information is more than likely different than the information shown in the figure.

Using the following credentials
Type of credentials
Windows user name and password
User name
Bradley Beard
Password
****
Log in using these credentials, but then try to impersonate the user viewing the report Learn more

Figure 9-10. *Credentials information, updated*

Notice that the check box shown on the screen is not checked; this is essential to creating subscriptions in the next section. Once you have the information updated, click the **Test Connection** button at the bottom of the screen. Figure 9-11 shows the result of this operation.

Figure 9-11. *Connection successful*

The last thing we need to do is click the **Save** button, which updates the report.

▒ **Note** At this point, you may want to run the report again and make sure that it still renders correctly.

Once the connection information is saved, it becomes immediately available to the users that have subscribed to the report.

Subscriptions

Click the **Subscriptions** link in the left menu. An interface opens, as shown in Figure 9-12; it can be used to create subscriptions to reports.

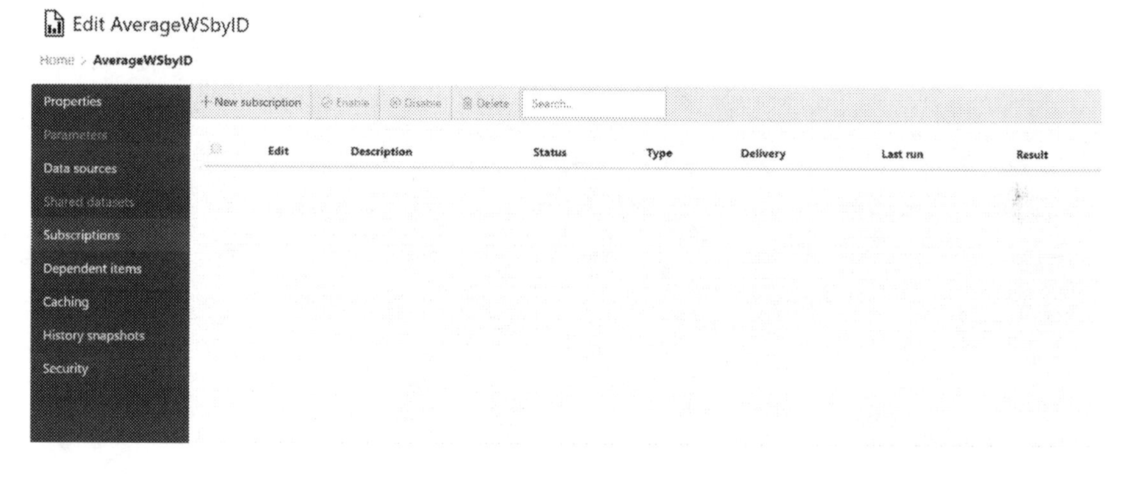

Figure 9-12. *Subscriptions*

Notice that there is a button labeled **New subscription** with a plus sign to the left of the text. Go ahead and click the **New subscription** button. You should see what is shown in Figure 9-13.

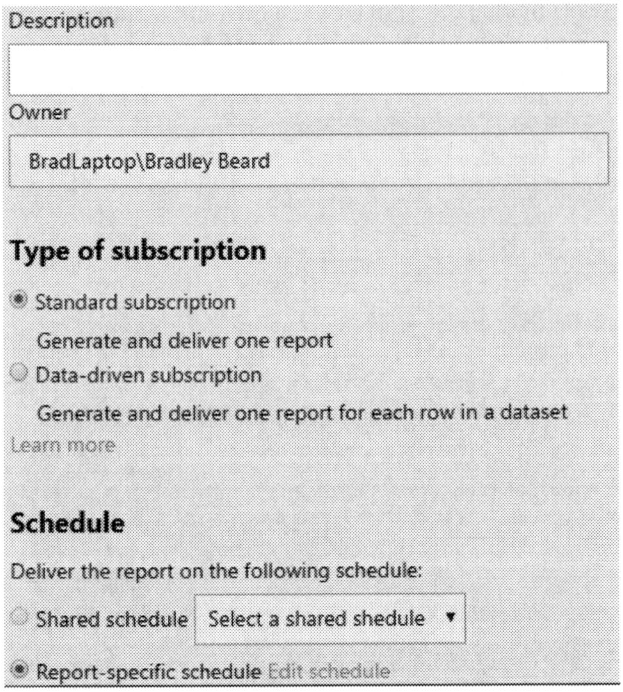

Description

Owner

BradLaptop\Bradley Beard

Type of subscription

Standard subscription

Generate and deliver one report

Data-driven subscription

Generate and deliver one report for each row in a dataset

Learn more

Schedule

Deliver the report on the following schedule:

Shared schedule Select a shared shedule

Report-specific schedule Edit schedule

Figure 9-13. *New subscription*

First, we want to update the Description field to say **Wind Speed Report**. Next, we are going to edit the schedule, but leave the subscription type alone. Clicking the blue **Edit schedule** link allows you to change the current schedule. Figure 9-14 shows the initial interface when clicking the **Edit schedule** option.

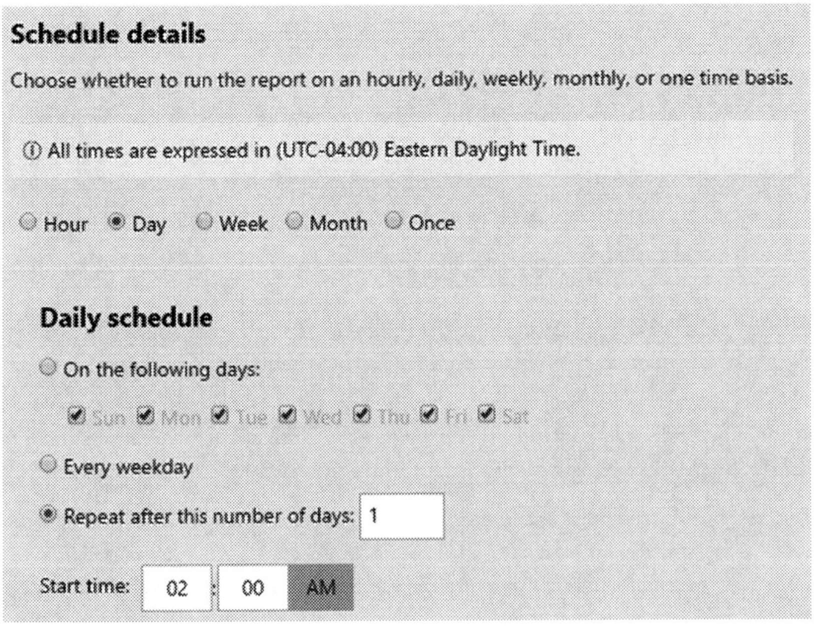

Schedule details

Choose whether to run the report on an hourly, daily, weekly, monthly, or one time basis.

ⓘ All times are expressed in (UTC-04:00) Eastern Daylight Time.

Hour Day Week Month Once

Daily schedule

On the following days:

Sun Mon Tue Wed Thu Fri Sat

Every weekday

Repeat after this number of days: 1

Start time: 02 : 00 AM

Figure 9-14. *Edit schedule (top)*

Notice that this is the top part of the **Edit Schedule** interface. Click the **Once** radio button and then enter a time that is just a few minutes away from your current time.

The bottom part of this interface is shown in Figure 9-15. In this area, we need to just click the **Apply** button.

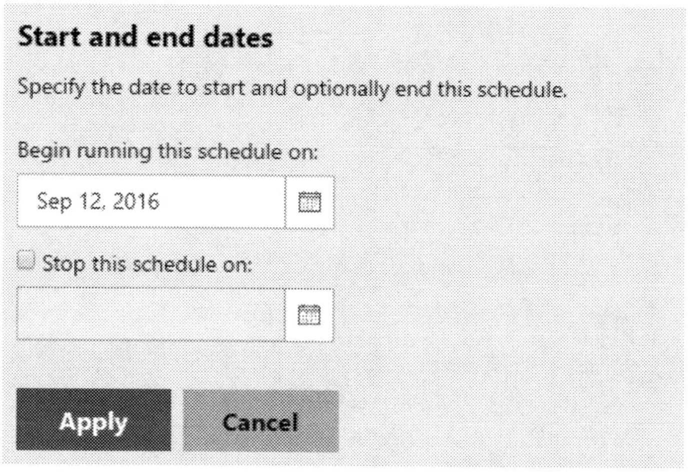

Figure 9-15. *Edit schedule (bottom)*

At this point, the subscription interface updates to show the new schedule information. Figure 9-16 shows this updated schedule information, as well as the next part that we need to update.

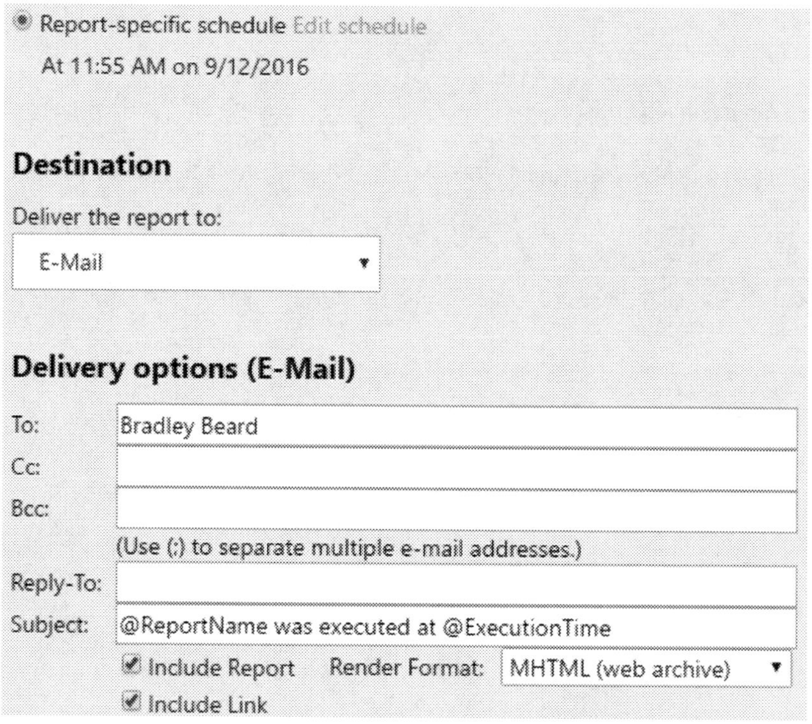

Figure 9-16. *Destination and E-mail options*

We want to leave the destination set to **E-mail** and to update the rest of the information, as shown in Figure 9-17.

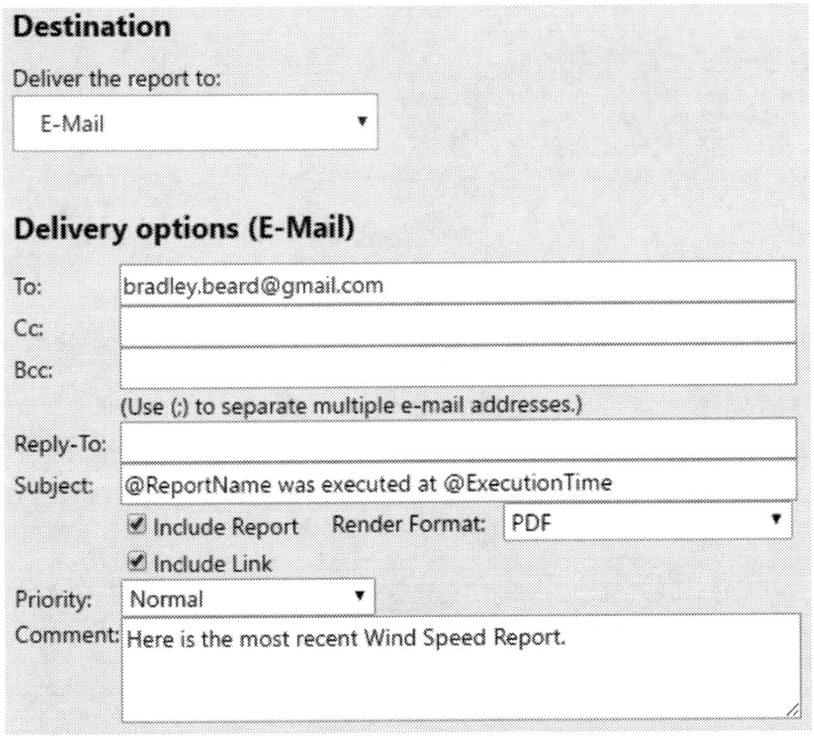

Destination

Deliver the report to:

E-Mail

Delivery options (E-Mail)

To: bradley.beard@gmail.com

Cc:

Bcc:

(Use (;) to separate multiple e-mail addresses.)

Reply-To:

Subject: @ReportName was executed at @ExecutionTime

Include Report Render Format: PDF

Include Link

Priority: Normal

Comment: Here is the most recent Wind Speed Report.

Figure 9-17. *Destination and E-mail options, updated*

Notice that I entered my actual e-mail address into the **To:** field, as well as changed the **Render Format to PDF**.

■ **Note** Available render formats (currently) are Word, Excel, PowerPoint, PDF, TIFF, MHTML, CSV, XML, and Data Feed.

Next, just add some text in the Comment field, as shown in Figure 9-18.

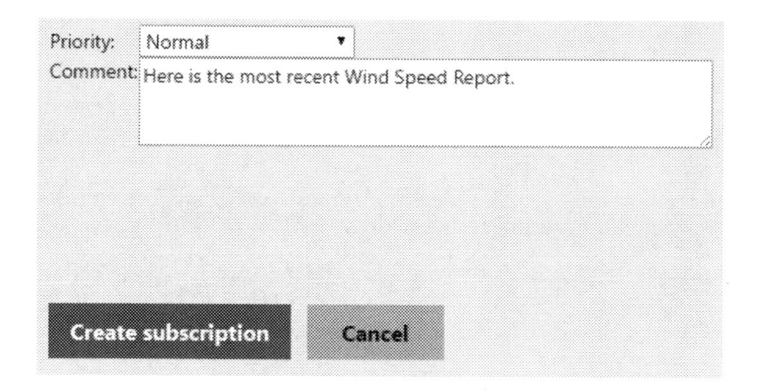

Figure 9-18. *Added comment information*

Click the **Create Subscription** button and we are done with this area. Wait a few minutes. You should see an e-mail appear in your inbox. Opening it displays something similar to what is shown in Figure 9-19.

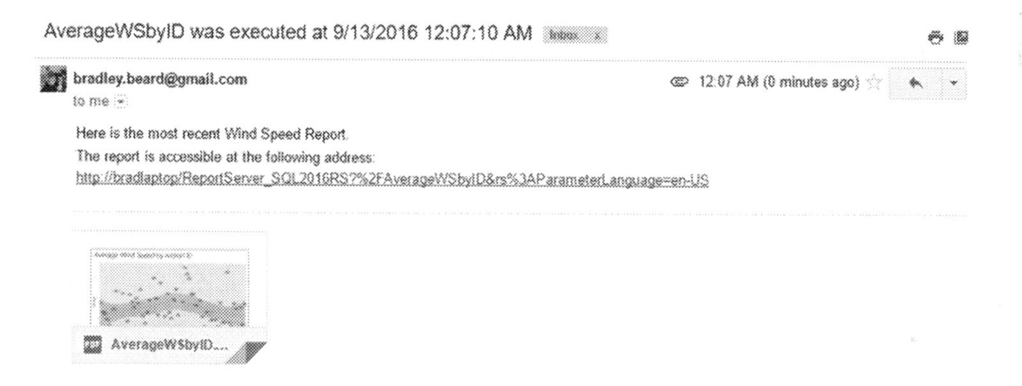

Figure 9-19. *E-mail received*

You have now successfully created an e-mail subscription for this report.

Dependent Items

The Dependent Items section has no configurable areas, so we'll just leave this area alone.

Caching

The Caching section allows you to choose whether or not you want your reports cached or not. Figure 9-20 shows the initial interface for the Caching section.

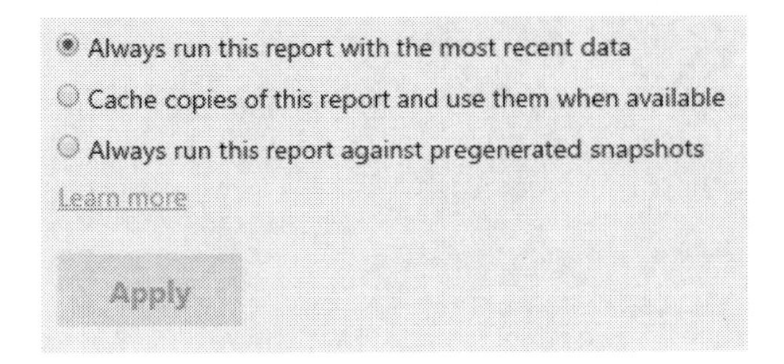

Figure 9-20. *Caching interface*

Personally, I would choose the **Always run this report with the most recent data** option, just in case the report is run out of sequence or outside of the normal scheduled run (in the case of a subscription). There are instances where the **Cache copies of this report and use them when available** option would be a better option; for instance, if this were strictly a subscription-based server and there were no contingencies for users to run reports at their leisure, then it would make sense to generate the report once and then server can serve the cached copy of the report when requested. The last option, **Always run this report against pregenerated snapshots**, means that the report is generated from a specific point in time, referred to as a snapshot, which is discussed in the next section.

History Snapshots

This section allows you to generate snapshots of the report. This enables the option discussed earlier to become in this context. Clicking the **New history snapshot** button shown in Figure 9-21 creates a snapshot of the report based on data from the current date and time.

+ New history snapshot | Delete | Schedule and settings | Search...

	View	Created	Size (Total: 124 KB)
	View history snapshot	Sep 13, 2016 12:21:05 AM	124 KB

Figure 9-21. *History snapshots*

Referring back to the Caching section, if you were then to select the **Always run this report against pregenerated snapshots** option after generating a snapshot, then any subsequent reports generated would be run against this snapshot.

▒ **Note** If you created a snapshot at this point, go ahead and delete it unless you are planning on using it in the future.

Security

The Security section allows you to customize your users and roles. The **Group or user information** comes from the Windows subsystem. The **Roles** are based in SQL Server Reporting Services.

Figure 9-22. *Security options*

Note that we don't need to change anything here either.

Saving Reports

Apart from subscriptions, there is still a way to save a generated report in Reporting Services. To do this, just run the report by clicking the name of the report on the Home screen, as shown in Figure 9-23.

Figure 9-23. *Home screen*

This opens the report that we have seen before. Figure 9-24 shows the Save dialog, which is shown by clicking the disk icon in the report toolbar.

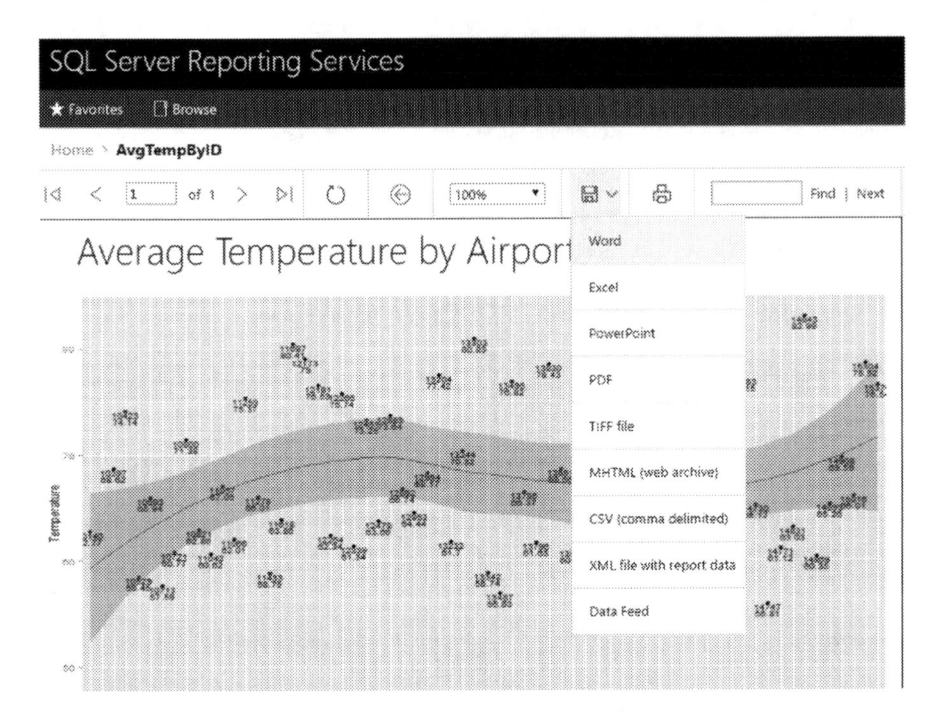

Figure 9-24. *Save options*

So again, we have the same render options as we did when we configured the E-mail settings earlier. You can save the report in any format you wish, at this point.

And with that, we are done.

Summary

Congratulations! We have reached the end of this journey into the beginnings of data science. Understand that this is just the very tip of the iceberg as far as what R can do. In no way is this book everything that you ever need to know about the language in order to function in an R environment. Quite the opposite, in fact; the purpose of this book is to introduce you into the now-blended world of R and SQL Server, in hopes that you continue your journey of discovery, if you haven't already. I encourage you to press onward and make future versions of SQL Server R Services even better through greater community involvement.

Until the next time, thank you so much for taking the time to read this book. I hope you enjoyed reading this book as much as I enjoyed writing it!

PART IV

Appendices

APPENDIX A

Installing a SQL Server 2016 Instance in a SQL Server 2014 Installation

These instructions closely follow the instructions in Chapter 1, but instead of installing an instance of SQL Server 2016 on a server by itself, we install an instance of SQL Server 2016 on an existing SQL Server 2014 installation.

We are also not going to install absolutely everything like we did in Chapter 1 either. Instead, we are only installing R Services, the SQL Server 2016 database engine, and SQL Server 2016 Reporting Services. We need Reporting Services to provide us with the R content later on.

If you choose to install SQL Server 2016 on top of SQL Server 2014, understand that there are implications to this, such as relying on older versions of Analysis Services or Integration Services. If you are okay with this, then that is why this Appendix was created.

Getting Started

Let's begin the install. Double-click the **setup.exe** file in the download folder. You should see Figure A-1.

B. Beard, *Beginning SQL Server R Services*, DOI 10.1007/978-1-4842-2298-0_10

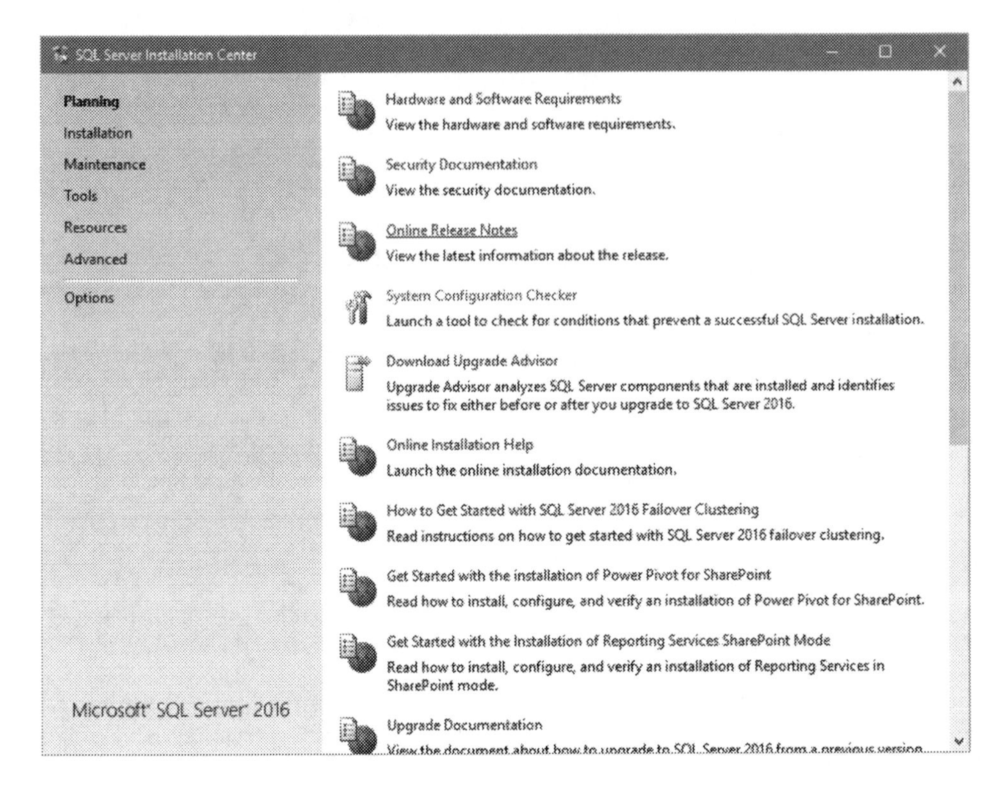

Figure A-1. *Initial SQL Server 2016 installation screen*

If you see the screen asking to make changes to your system, go ahead and click **Yes**.

Figure A-1 shows the first screen that you should see when you start installation.

This screen should look pretty familiar to you. Click the **Installation** link on the left to see what is shown in Figure A-2.

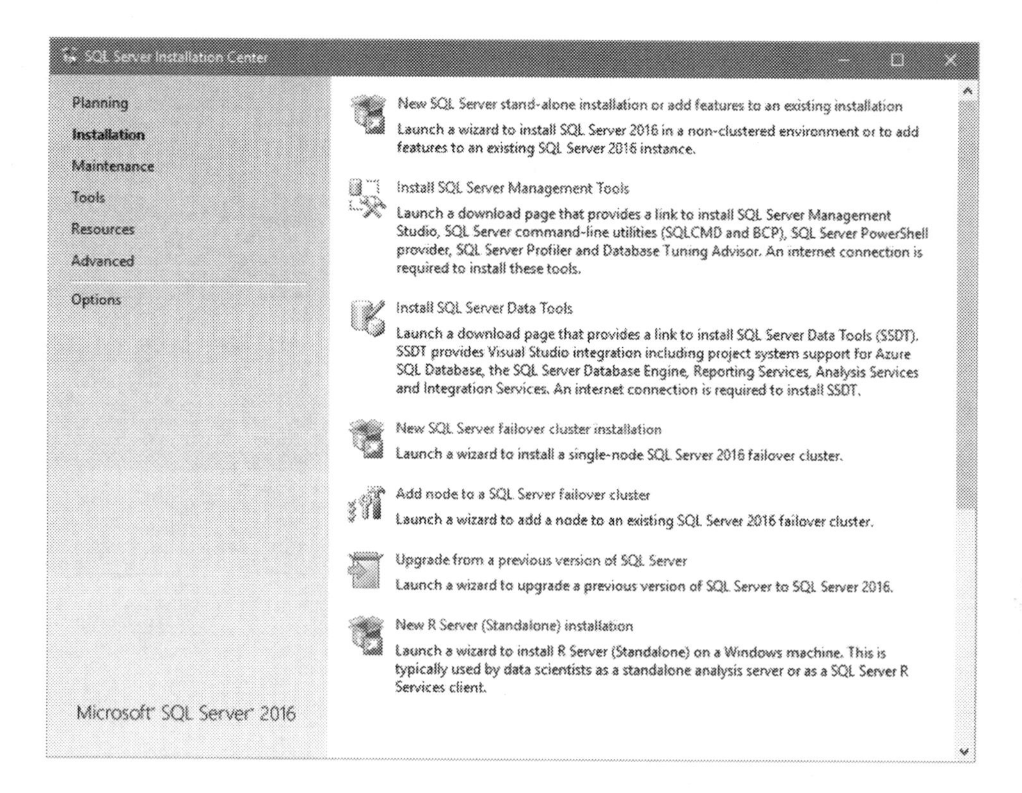

Figure A-2. *SQL Server 2016 Installation options*

Once here, click the **New SQL Server stand-alone installation or add features to an existing installation** link at the top.

Note that the very bottom option is something new. It says **New R Server (Standalone) installation**. You would select this option if you only wanted to install R Server as either a server (standalone, self-contained data analysis server, in other words) or a client (manipulating data from a remote SQL Server R Services installation). Note that you need the SQL Server 2016 services running as well, so this would be to add R services to an existing SQL Server 2016 installation. It cannot be added to previous versions of SQL Server, in other words.

Product Key

Now it's time to enter a product key, or to choose to install the free edition of the product. In other words, if you happen to have a licensed copy of SQL Server, you would have gotten a 25-character license key, so you can enter that here. Otherwise, you can always select a free edition. The purpose of this book is for evaluation purposes, so go ahead and pick Evaluation from the drop-down menu. Figure A-3 shows the screen you see after continuing from Figure A-2 in a stand-alone install process.

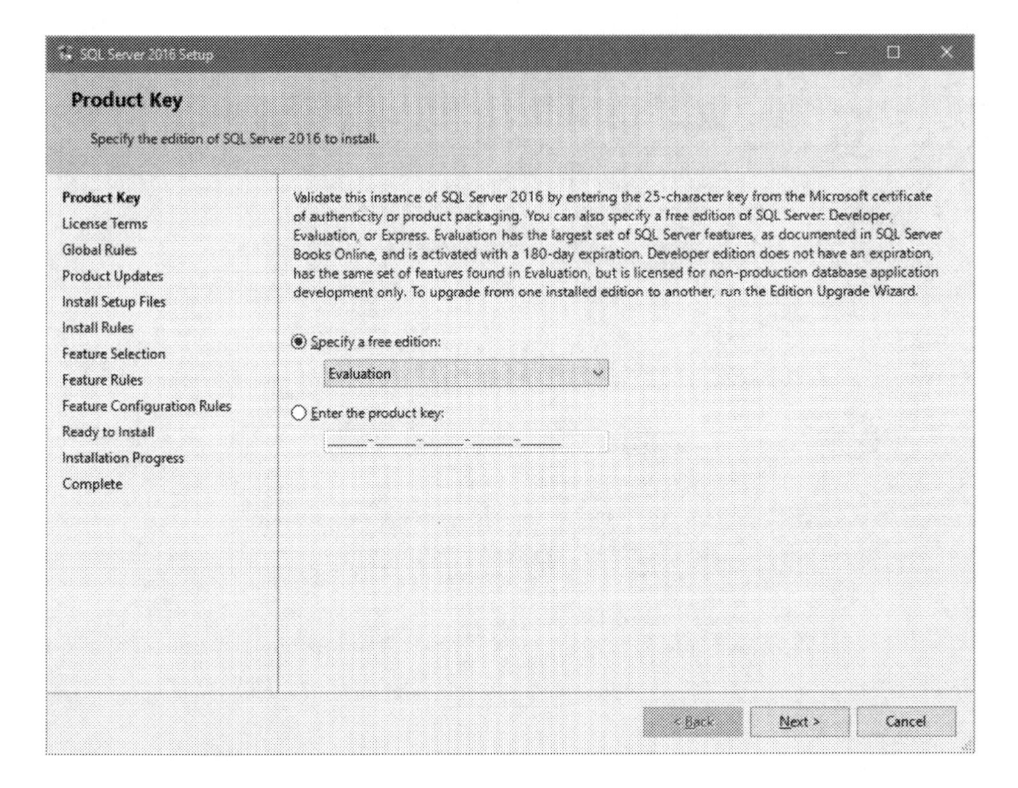

Figure A-3. *Product Key screen*

SQL Server 2016 can be installed in one of three free editions:

- **Evaluation**: A full set of features, basically the Enterprise version of SQL Server 2016, but only good for 180-day spans.
- **Developer**: A full set of features, but cannot be used for production database work.
- **Express**: The smallest, bare-bones installation of SQL Server 2016 that does not expire and can be used for production use.

If you would like to choose an option other than Evaluation, go right ahead. Just understand the implications of choosing that option. For what we need, the Evaluation version is perfect because we will decide certainly within 180 days if this new functionality is something we want to permanently include in our SQL Server installation.

When you have chosen the version you are most comfortable with, click **Next** to continue.

License Terms

The next screen, shown in Figure A-4, simply asks you to accept the license terms.

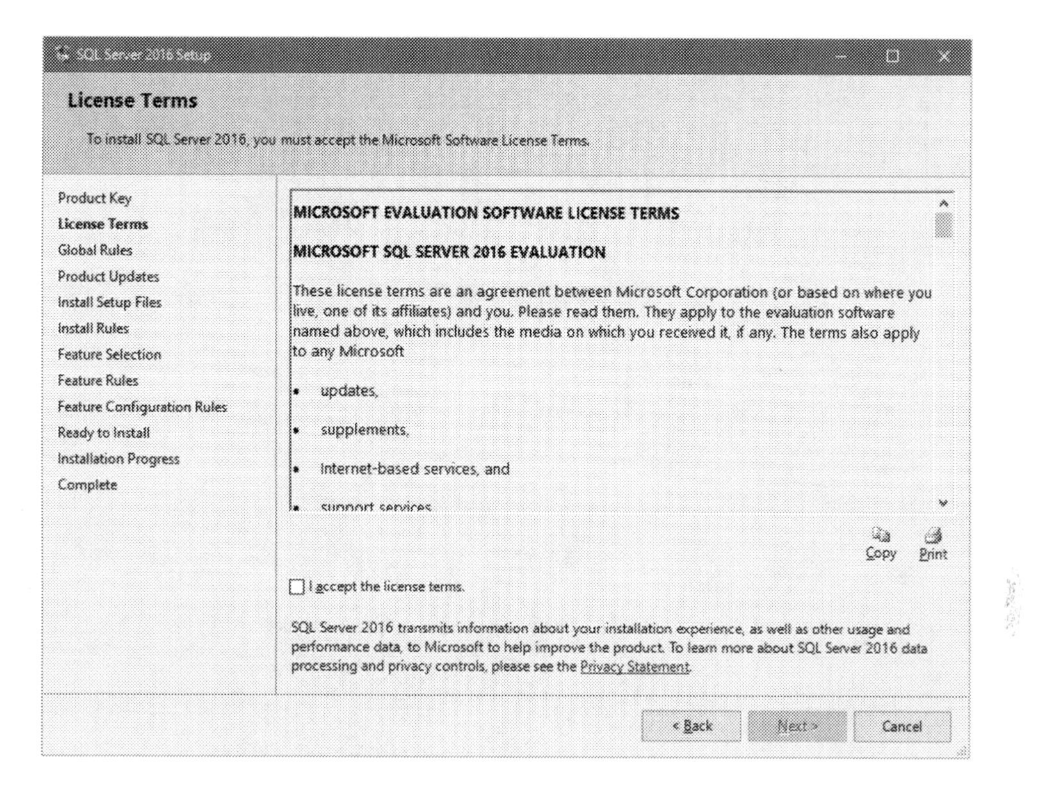

Figure A-4. *License terms*

I honestly have never read this license all the way through and I can't say that I know anyone who has. Obviously, just click the **I accept the license terms** check box and then **Next** to move on.

Install Rules

My screen flashed a few times and eventually ended up at the screen shown in Figure A-5.

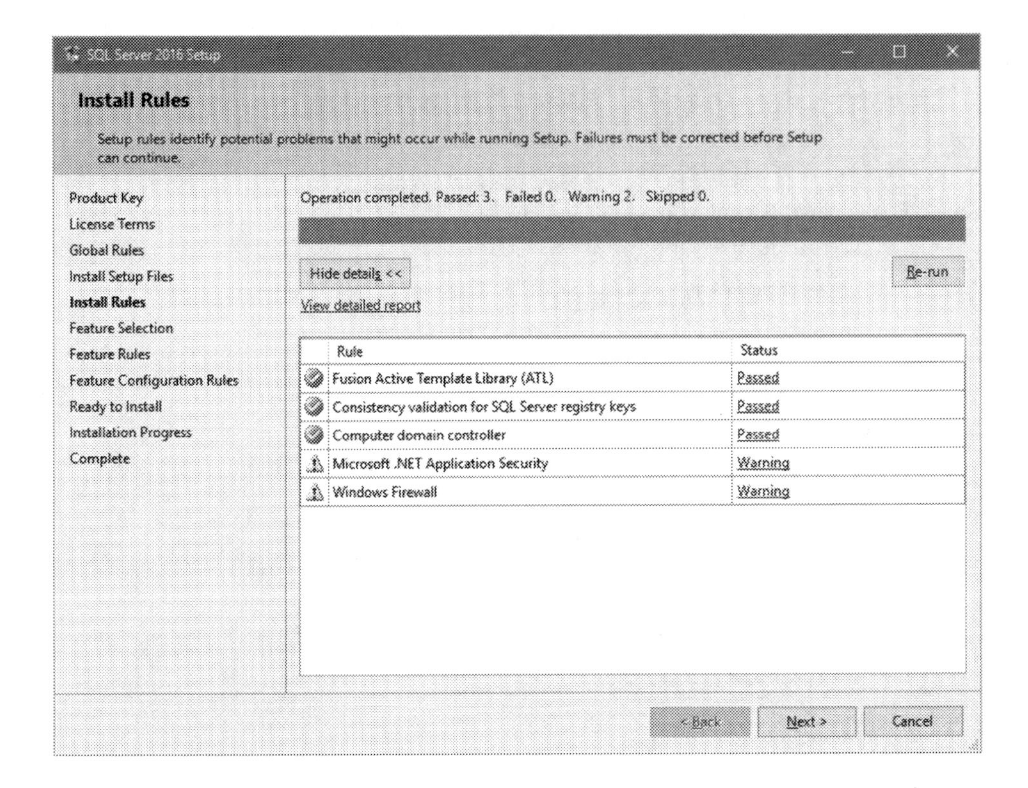

Figure A-5. *Install Rules*

It's worth noting that there could be an update to SQL Server 2016 that gets downloaded and installed during this step; so if a message comes up with that information, go ahead and install it.

So everything looks good except for my .NET application security and firewall rule. This should be fine, so I'm going to click **Next** to continue.

Feature Selection

Now we get to it. Figure A-6 shows the screen we have been waiting for.

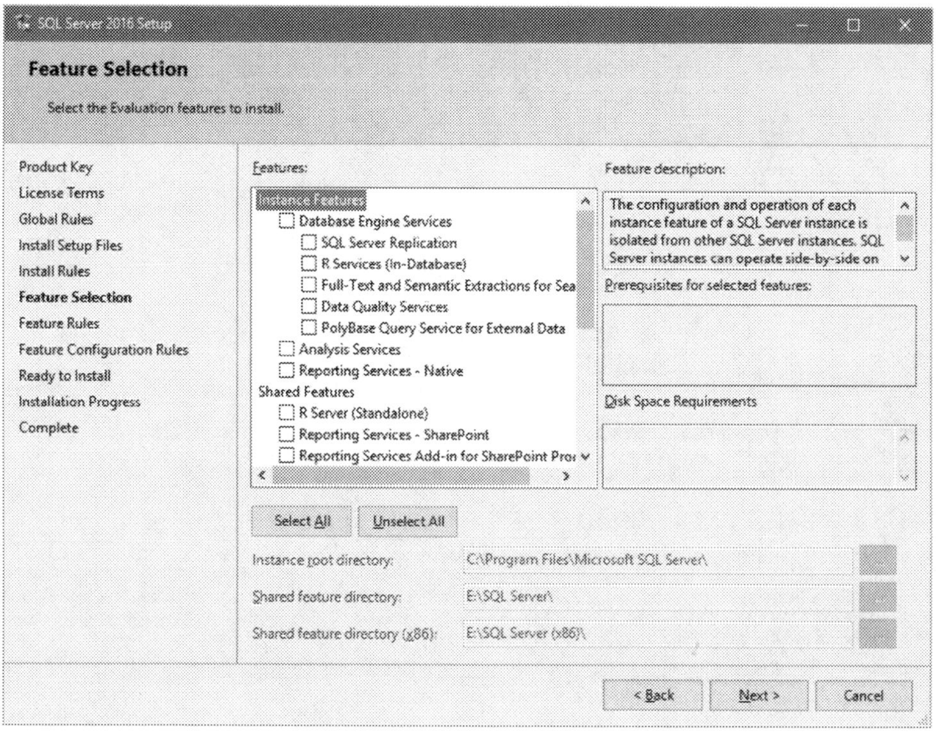

Figure A-6. *Feature Selection*

At this point, we could just press **Select All** and that would be it. If you take a look at the options though, you see what they actually mean. We definitely want to choose **R Services (In-Database)**; otherwise, you can stop reading here. Once we choose that, we see that **Database Engine Services** also gets selected. We also want to select **Reporting Services – Native**, which we'll use in Chapter 7. We don't want to select anything except these three options, though. Figure A-7 shows what you should see selected at this point.

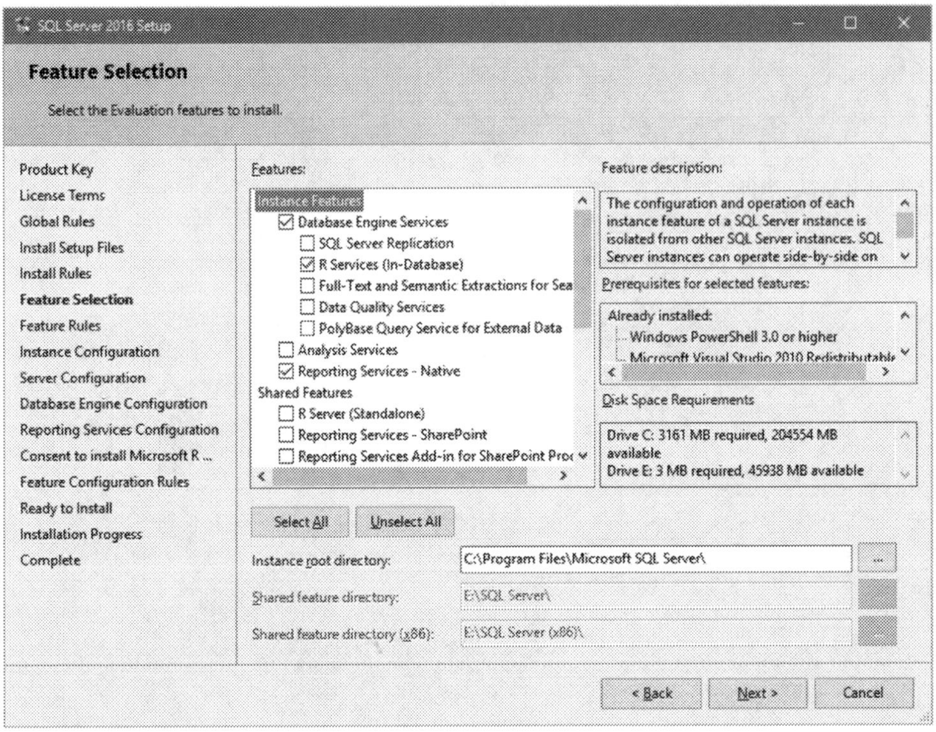

Figure A-7. *Feature Selection with options*

Notice how the bottom shows the default instance root directory and my shared feature directories are pointing to my E drive. That's where I keep my SQL Server stuff for easy retrieval.

Click **Next** here to move on.

Instance Configuration

It takes a second to think about what it wants to do, but eventually, you see the **Instance Configuration** screen shown in Figure A-8.

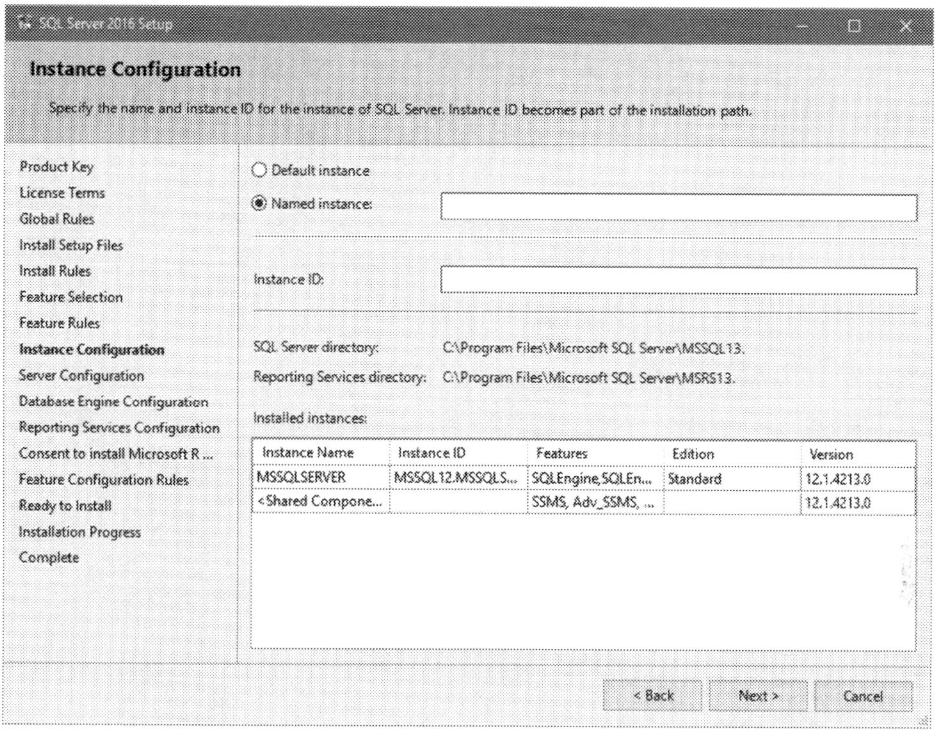

Figure A-8. *Instance Configuration*

At this point, we need to define our new instance. If you look on the **Installed Instances** section, you see that there is already an installed version of SQL Server 2014. We don't want to wipe that out by installing on top of it, so we choose the **Named Instance** option and call it **SQL2016RSVCS** for SQL Server 2016 R Services.

Note that this value is different from the instance name given in Chapter 1. This is to differentiate between the two instances.

Enter that value for the Named Instance field. You see what is shown in Figure A-9.

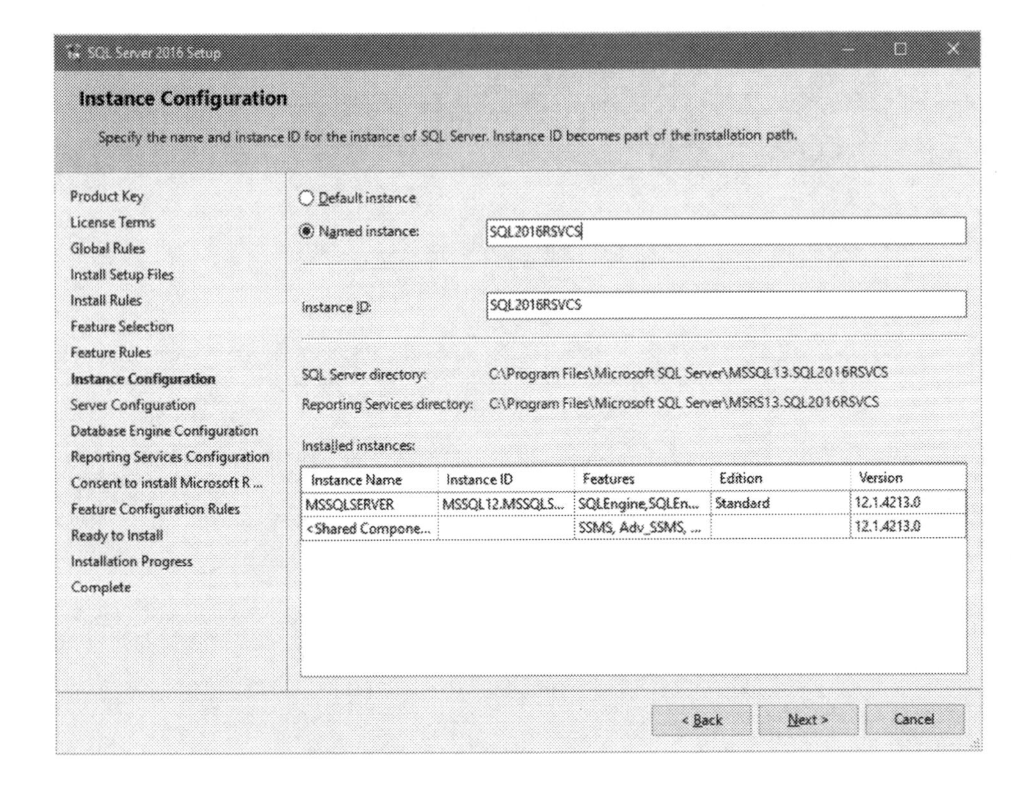

Figure A-9. *Updated Instance Configuration screen*

Pay attention to the **Named Instance** field, the **Instance ID** field, the **SQL Server directory** location, and the **Reporting Services directory** location listed on this screen. Those need to all have SQL2016RSVCS referenced in them. Once you are satisfied that everything is as it should be, click **Next** to continue.

Server Configuration

The next screen is where we define the service accounts and startup types for the services. That screen is shown in Figure A-10.

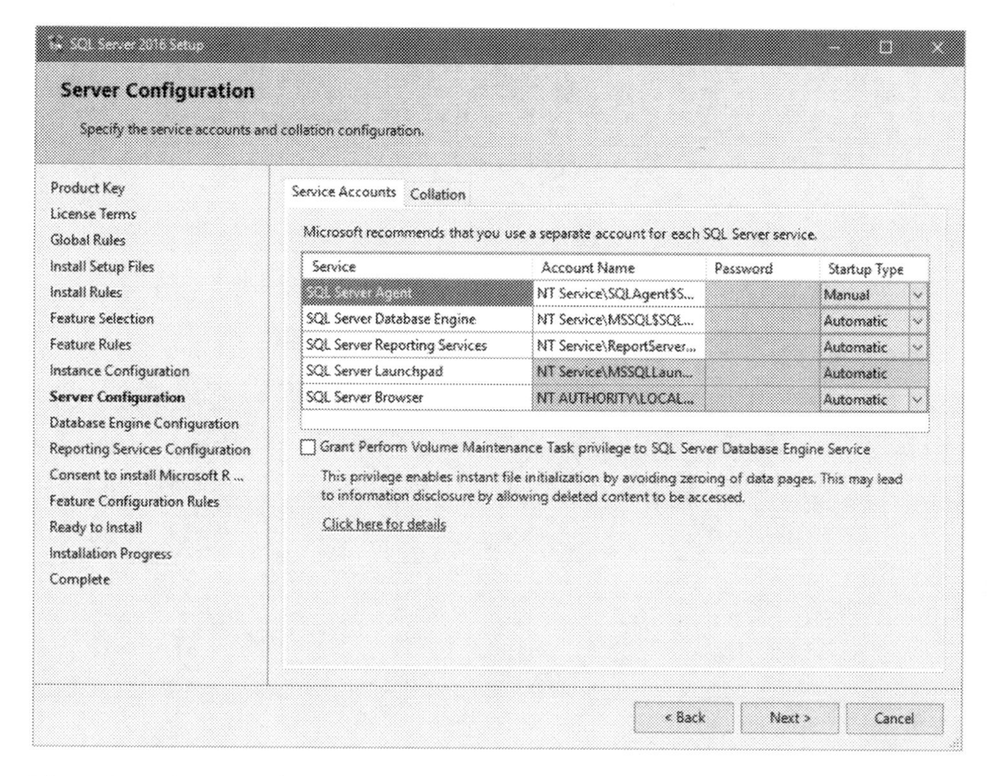

Figure A-10. *Server Configuration*

The following are the service accounts created by SQL Server 2016:

- **NT Service\SQLAgent$SQL2016RSVCS**: Starts and manages the SQL Server Agent service.
- **NT Service\MSSQL$SQL2016RSVCS**: Starts and manages the SQL Server service.
- **NT Service\ReportServer$SQL2016RSVCS**: Starts and manages the Reporting Services service.
- **NT Service\MSSQLLaunchpad$SQL2016RSVCS**: Starts and manages the R Services service.

The SQL Browser service is running under the context of the Local Service, so that isn't a new service account being created. We won't worry about that one, in other words.

These services are the default, but can always be changed to your own service accounts, if you have them. If you don't have your own service accounts, you can keep these suggested service accounts. I know many server administrators that insist on employing the principal of least privilege for services, so if that is the case for your particular environment, then you need to get the service name and login information from the server administrator in order to proceed. Another way you can go about this is to copy these service names and include them in a summary to your system administrator regarding the accounts that were created during installation, so that the system administrator can audit the permissions for this user as needed. It is important to note here that I am referring to a separate individual or entity for "system administrator" that is not a database administrator, but rather the Windows-level administrator. The person in charge of the Operating System level, one step up from the Application layer, in other words.

We only want to change a little bit here; specifically, set the SQL Server Agent service **Startup Type** to **Automatic**. That is the only change we need to make. Figure A-11 shows what you should see at this point.

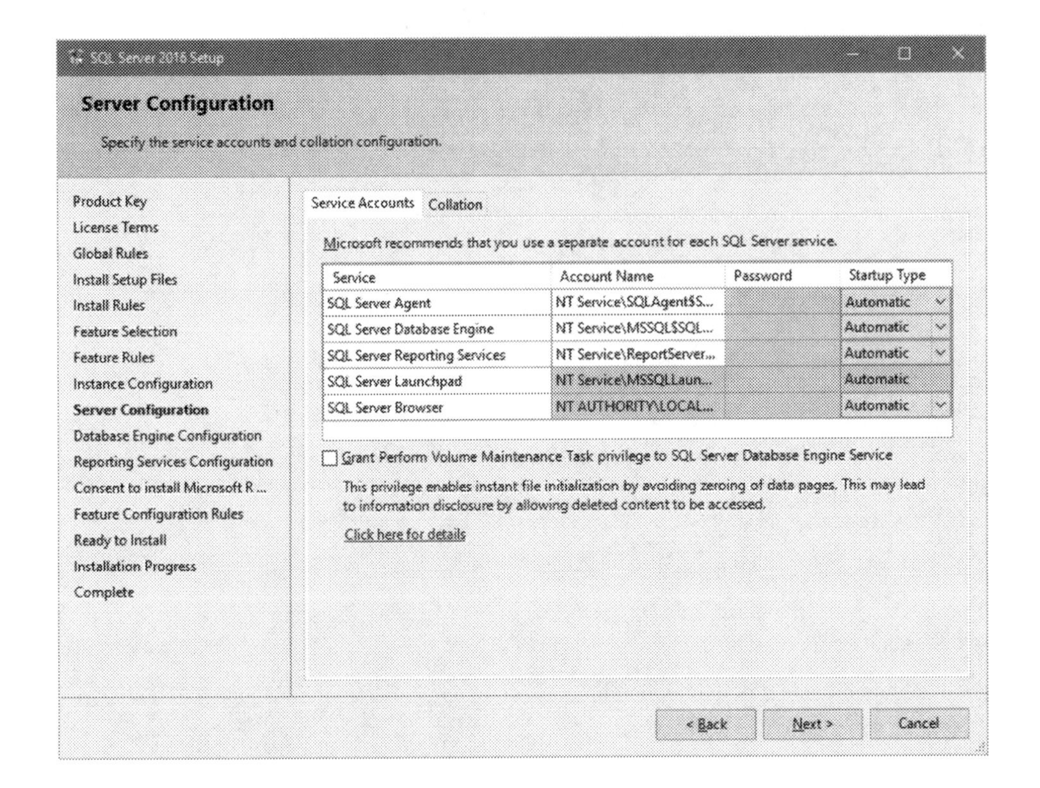

Figure A-11. *Server Configuration updated*

Notice that we cannot set the password for any of these accounts. This is the same as it has been for every installation of SQL Server that I have ever seen. If you were to change the Account Name box from the default to a custom service account name, then the Password box would become active and accept input; otherwise, the password is controlled by SQL Server.

Notice also that there is a new **Grant Perform Volume Maintenance Task privilege to SQL Server Database Engine Service** check box underneath the default services listed. For what we're doing in this book, it's not necessary to check this box. In future installations or for production environments, it would probably be a good idea to enable this.

At this point, all of our services are set to Automatic. Notice that we aren't going to bother with the Collation tab. This should have SQL_Latin1_General_CP1_CI_AS specified in the tab by default. That's it. Go ahead and click **Next** to move on.

Database Engine Configuration

The next screen is the Database Engine Configuration screen shown in Figure A-12. It lets you set options for the engine in four different tabs.

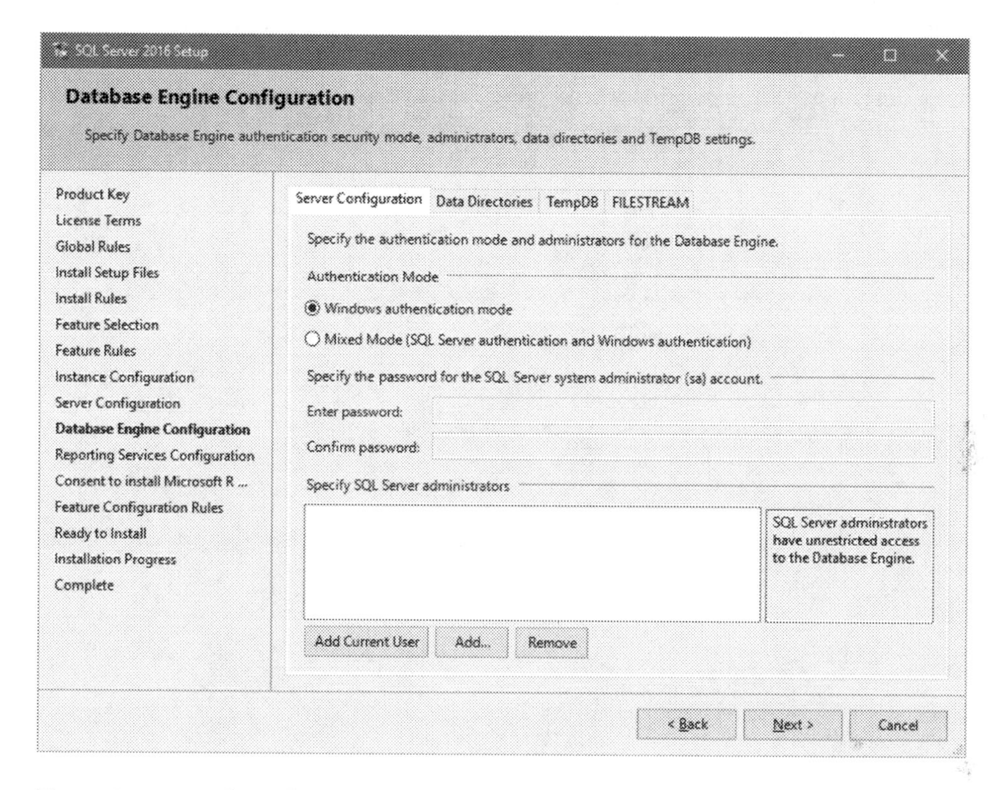

Figure A-12. *Initial Database Engine Configuration screen*

Server Configuration

This tab lets you specify the authentication mode and the administrators for this instance of the database engine. Because this is just for testing and evaluation, I am going to add myself in Windows Authentication Mode as the administrator by clicking the **Add Current User** button at the bottom of the screen with **Windows Authentication Mode** selected. Figure A-13 shows these options selected.

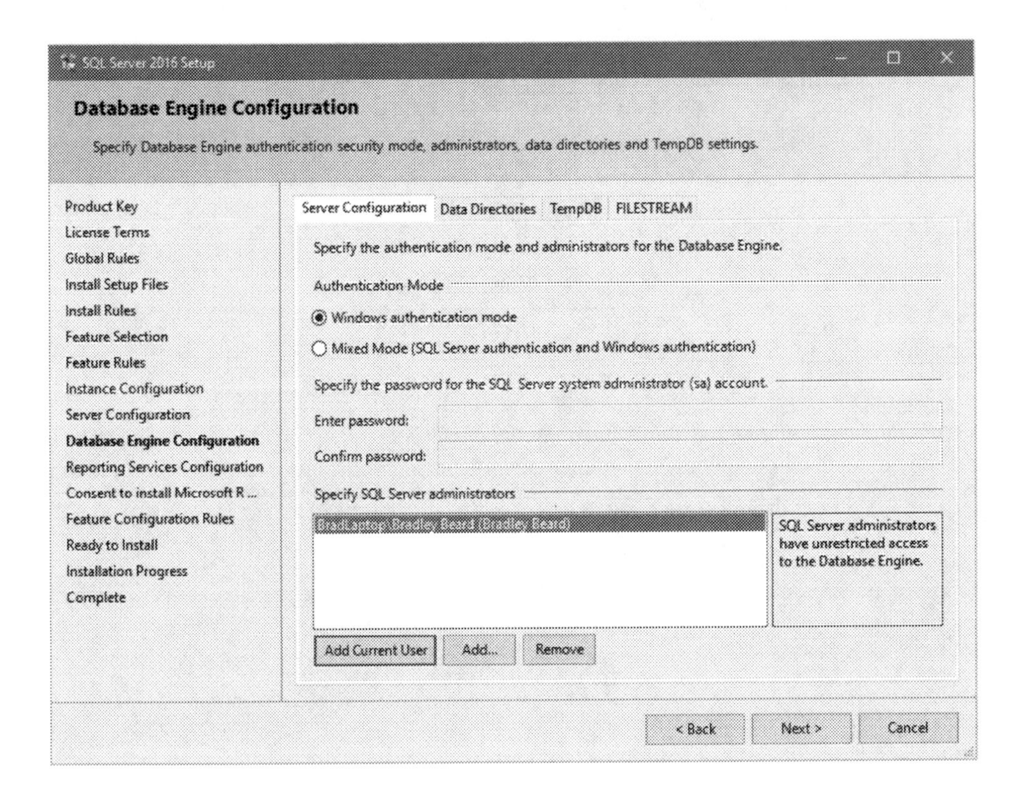

Figure A-13. *Server Configuration tab with options*

Data Directories

Recall the way that I had my file system set up. This is where that comes into play. Figure A-14 shows what this screen looks like initially and Figure A-15 shows what my selected options are. You can leave these however you like, but my personal preference is to not put the files I want in the labyrinth of folders that are the default.

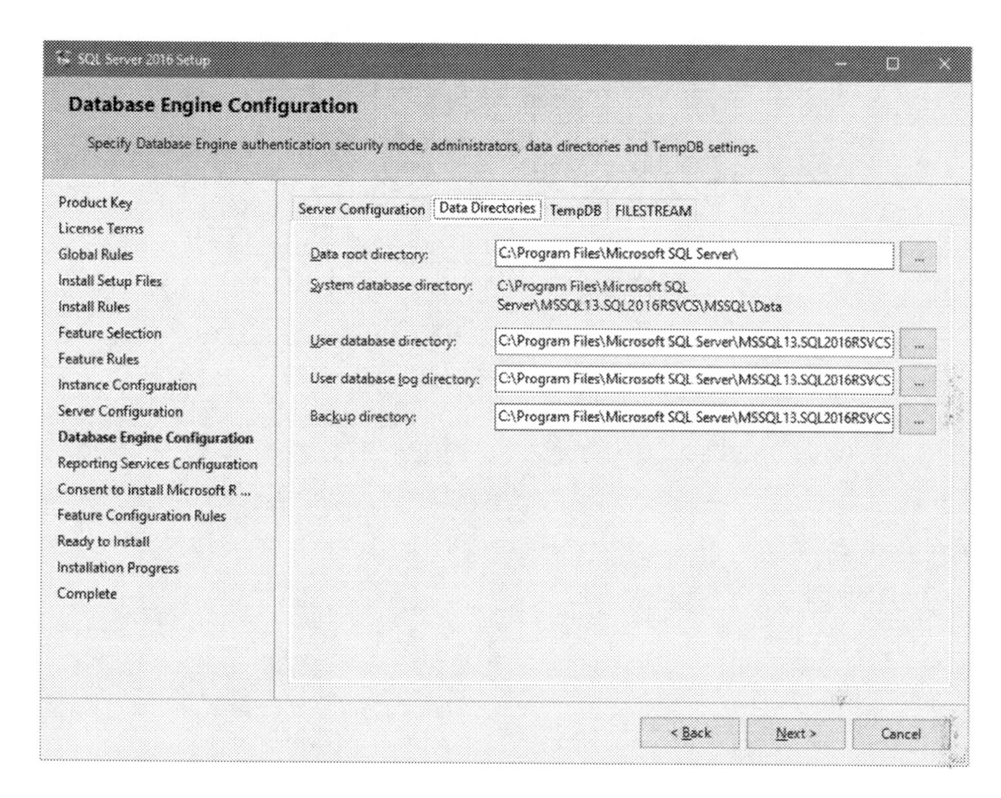

Figure A-14. *Initial Data Directories tab*

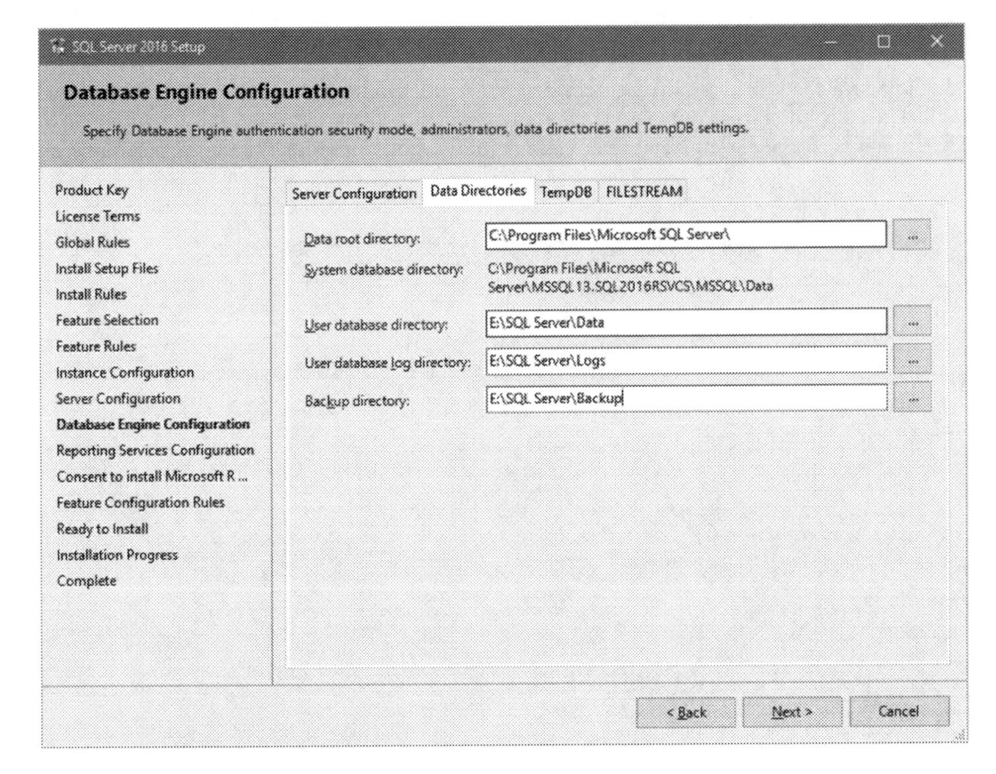

Figure A-15. *Updated Data Directories tab*

TempDB

Usually, I leave this option alone. However, in this case, I set the options to mirror the file system that I have enabled. Figure A-16 shows the default settings and Figure A-17 shows the updated settings.

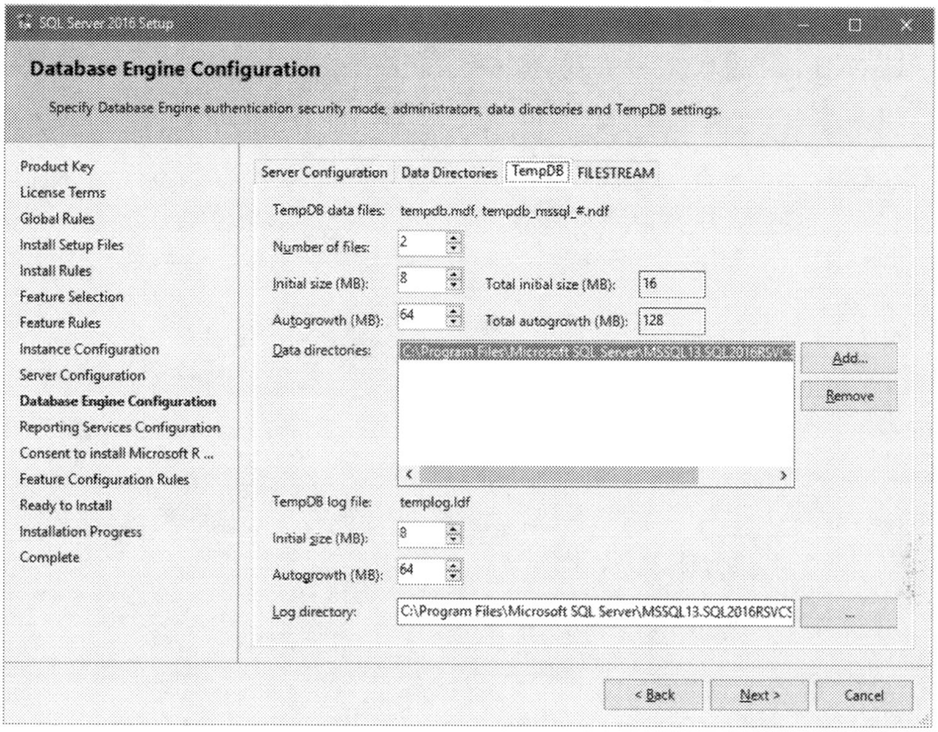

SQL Server 2016 Setup

Database Engine Configuration

Specify Database Engine authentication security mode, administrators, data directories and TempDB settings.

Product Key
License Terms
Global Rules
Install Setup Files
Install Rules
Feature Selection
Feature Rules
Instance Configuration
Server Configuration
Database Engine Configuration
Reporting Services Configuration
Consent to install Microsoft R ...
Feature Configuration Rules
Ready to Install
Installation Progress
Complete

Server Configuration | Data Directories | TempDB | FILESTREAM

TempDB data files: tempdb.mdf, tempdb_mssql_#.ndf
Number of files: 2
Initial size (MB): 8 Total initial size (MB): 16
Autogrowth (MB): 64 Total autogrowth (MB): 128
Data directories: C:\Program Files\Microsoft SQL Server\MSSQL13.SQL2016RSVCS
Add...
Remove

TempDB log file: templog.ldf
Initial size (MB): 8
Autogrowth (MB): 64
Log directory: C:\Program Files\Microsoft SQL Server\MSSQL13.SQL2016RSVCS
...

< Back | Next > | Cancel

Figure A-16. *TempDB default settings*

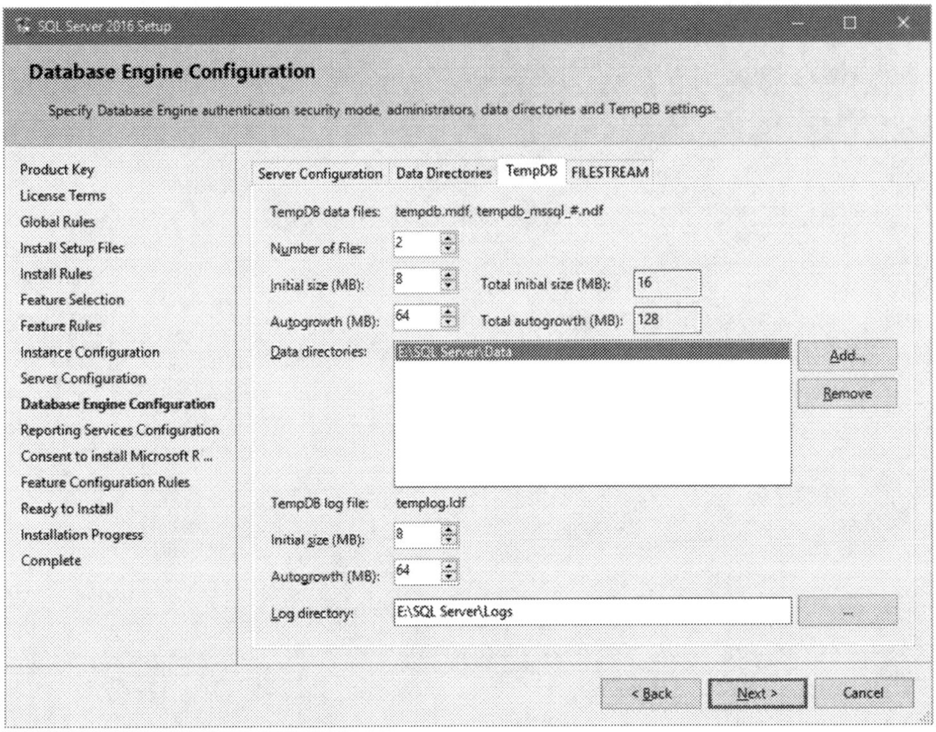

***Figure A-17.** TempDB updated settings*

The changes I made were slight. I first highlighted the existing option in the **Data directories** field and then clicked the **Remove** button. Then I clicked the **Add** button and added E:\SQL Server\Data instead. This location was mirrored in the **Log directory** field, so I changed that to E:\SQL Server\Logs instead. That's it for this tab.

FILESTREAM

Just leave the FILESTREAM tab alone. We won't be using FILESTREAM in this book.

Reporting Services Configuration

Once you've got those tabs all filled in, click **Next**. Figure A-18 shows you what you should see now.

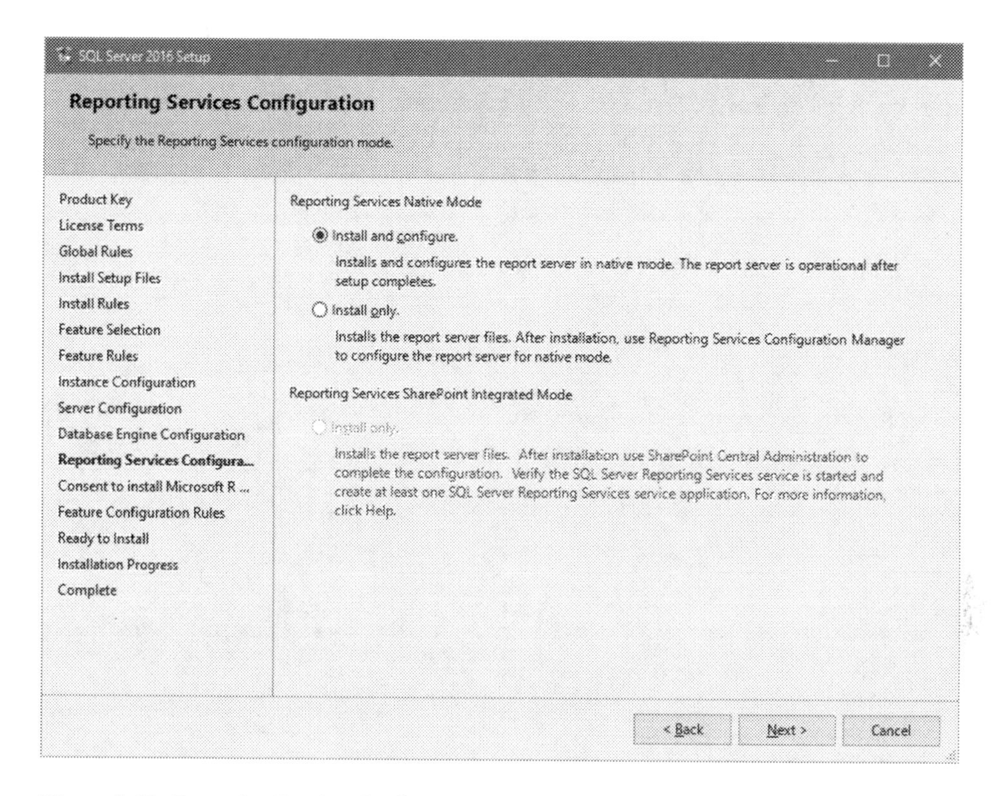

Figure A-18. *Reporting Services Configuration*

This is where we get to configure Reporting Services. We configure this further in Chapter 7 and onward, so we installed it now, using the **Install and configure** option that is selected by default.

One quick thing while we're talking about this; if you go in later and want to Reporting Services because you didn't install it with the database engine, you only have the **Install only** option available to you. The reason for this is because the Reporting Services Configuration Manager must be used to add Reporting Services to an existing database engine instance. Ideally, you should use the *principal of least installation* here, which is a concept that states that, when installing new software, you should only install what is needed and disregard what is not. In this case particularly, that makes perfect sense and is exactly what we are doing my not installing all of SQL Server 2016 right now. In certain other cases, it makes perfect to install the deluxe bells and whistles version of the software, but just keep in mind that this isn't always the case.

Ensure that the **Install and configure** option is selected and click **Next** at the Reporting Services Configuration screen to move on.

Consent to Install Microsoft R Open

You should now see what is shown in Figure A-19.

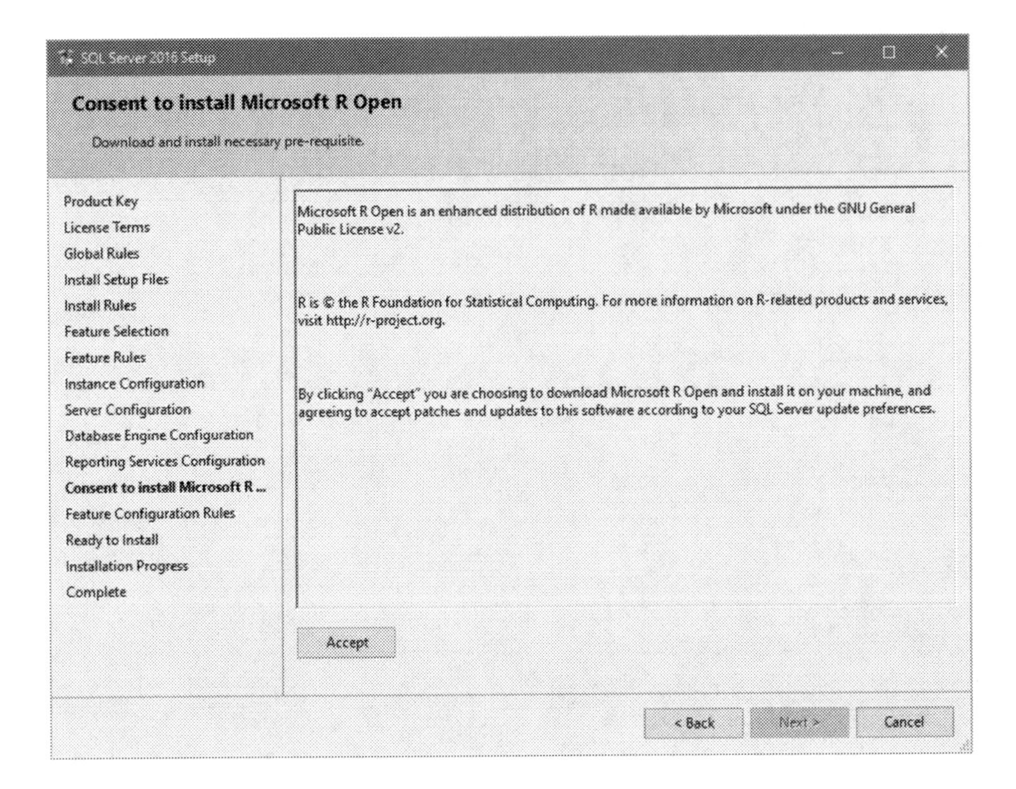

Figure A-19. *Consent to Install Microsoft R Open*

This is pretty cool. Before the full version of SQL Server 2016, you had to download the separate components for R and then install them individually. In this release, you just authorize the download here instead. Click the **Accept** button. The **Next** button becomes enabled. Go ahead and click **Next** to move on.

Ready to Install

Figure A-20 shows the Ready to install screen that you should now see. Read it over and ensure that you match these settings before moving on, if you're going to do the exercises in this book.

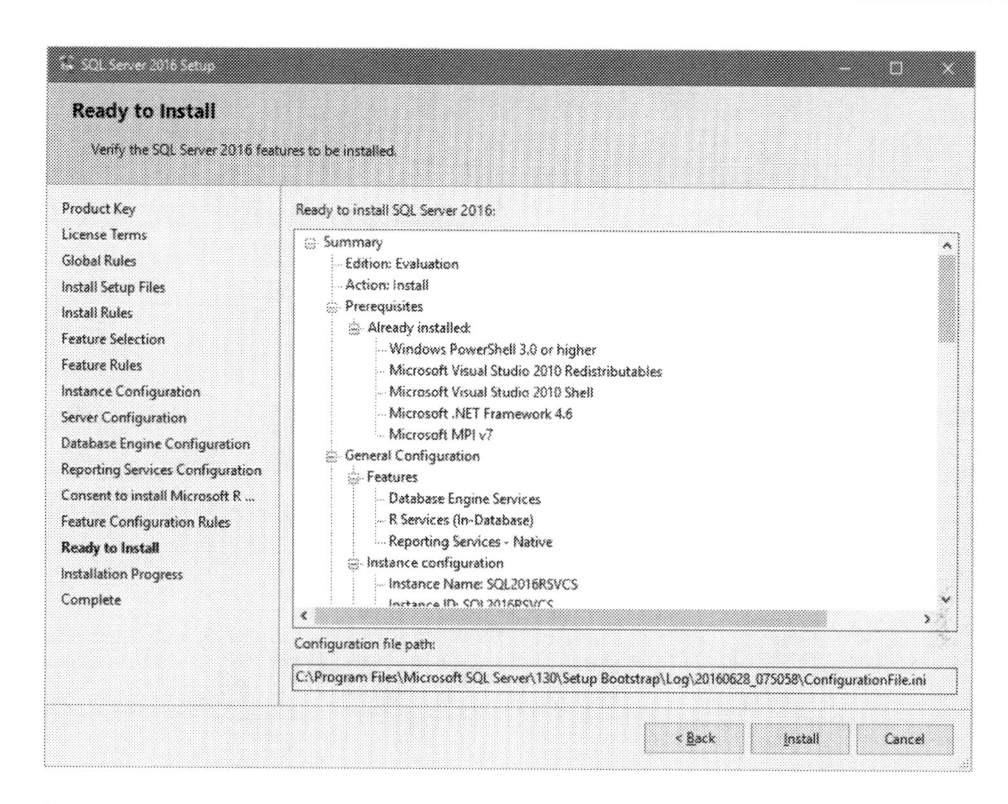

Figure A-20. *Ready to install*

Once you are ready, cross your fingers and click **Install**. Your screen flashes a few times while it is loading and installing what it needs. Figure A-21 shows what you should see when it starts running.

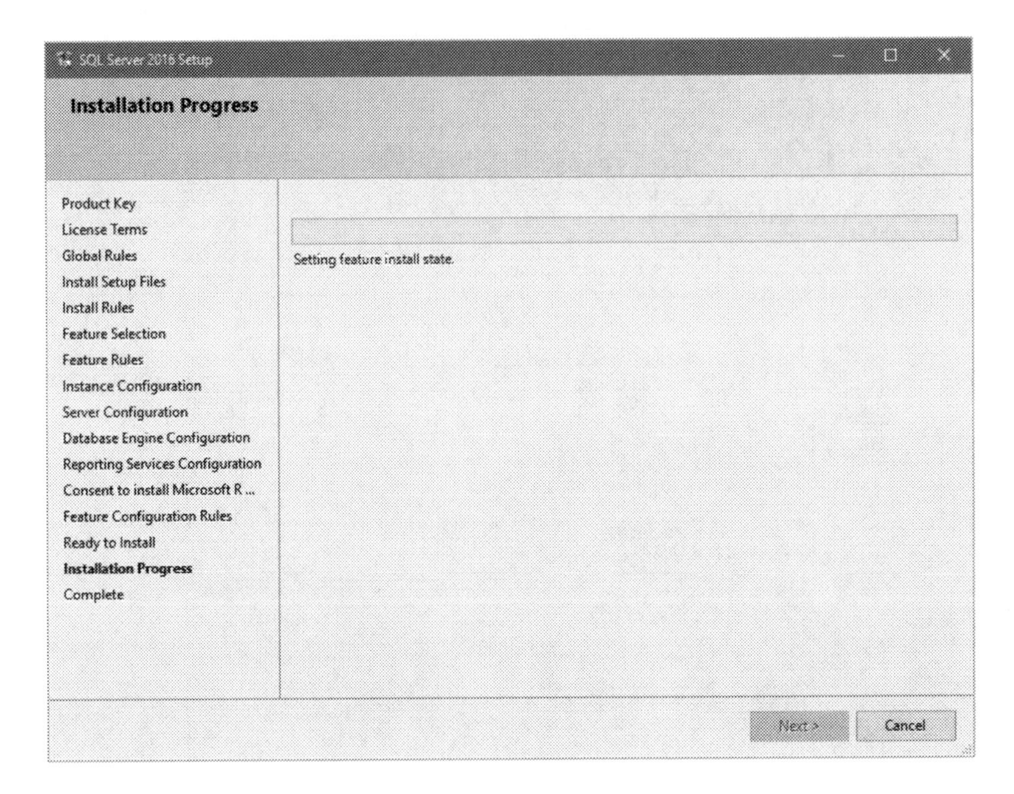

Figure A-21. *Installation Progress*

At this point, it is installing normally. It takes a little while, but eventually finishes with the screen shown in Figure A-22.

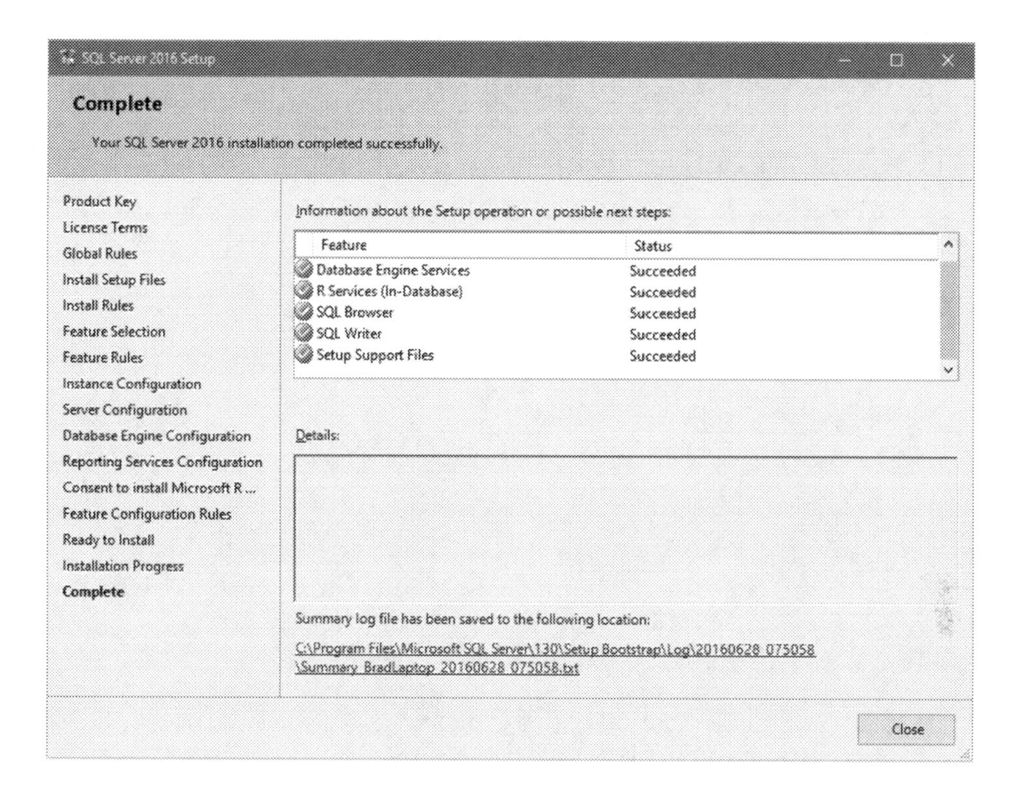

Figure A-22. *Complete*

Complete

This is the screen you really want to see at this point. If you don't see this screen and instead you see an error, then something bad happened. If you followed the directions in this Appendix, then there was probably a conflict with the existing SQL Server 2014 installation.

My installation took about 10 minutes to complete. Scroll down in there to see if everything installed correctly, and then click **Close**.

Congratulations! You have installed SQL Server 2016 R Services. According to Microsoft, we still have a little work to do though.

Open SQL Server Management Studio (yes, the 2014 one) and connect to your newly installed instance. Figure A-23 shows how to connect to the instance.

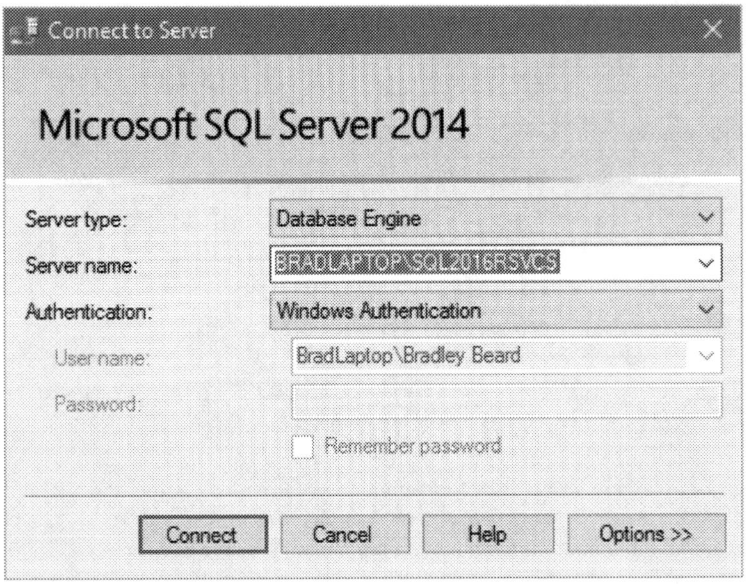

Figure A-23. *Connecting to the new instance*

Recall that I named the new SQL Server 2016 instance SQL2016RSVCS, so that's the instance I am going to connect with. The format for the Server name field is SERVER\INSTANCE, so that's how I have formatted my connection. You can also pull down the menu and navigate to an instance from there. However you are more comfortable is fine, as long as you get there. Click **Connect** to log in to your instance.

The initial screen resembles Figure A-24.

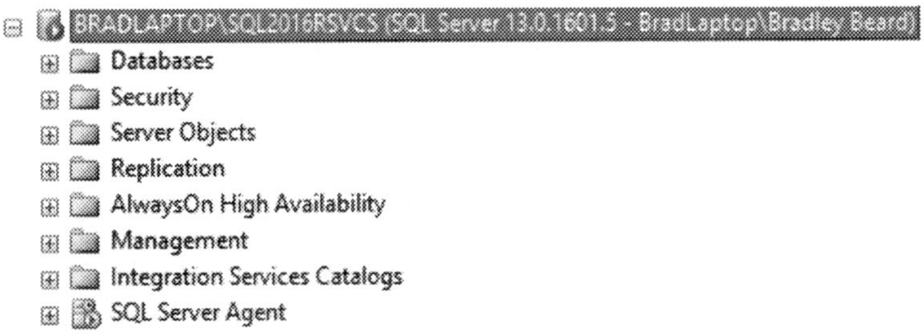

Figure A-24. *SQL Server Management Studio connected to SQL Server 2016 instance*

Pay attention to the named instance and the SQL Server version shown. This means that we have successfully connected to our new instance and we are ready to get going.

Microsoft has put out a post-configuration procedure that we are going to run first. I fully expect for this to be removed and added to the installation in the future, but for now, follow along to complete installation.

Open a New Query window in SQL Server Management Studio and type the following command:

```
Exec sp_configure 'external scripts enabled', 1
Reconfigure with override
```

Figure A-25 shows this action.

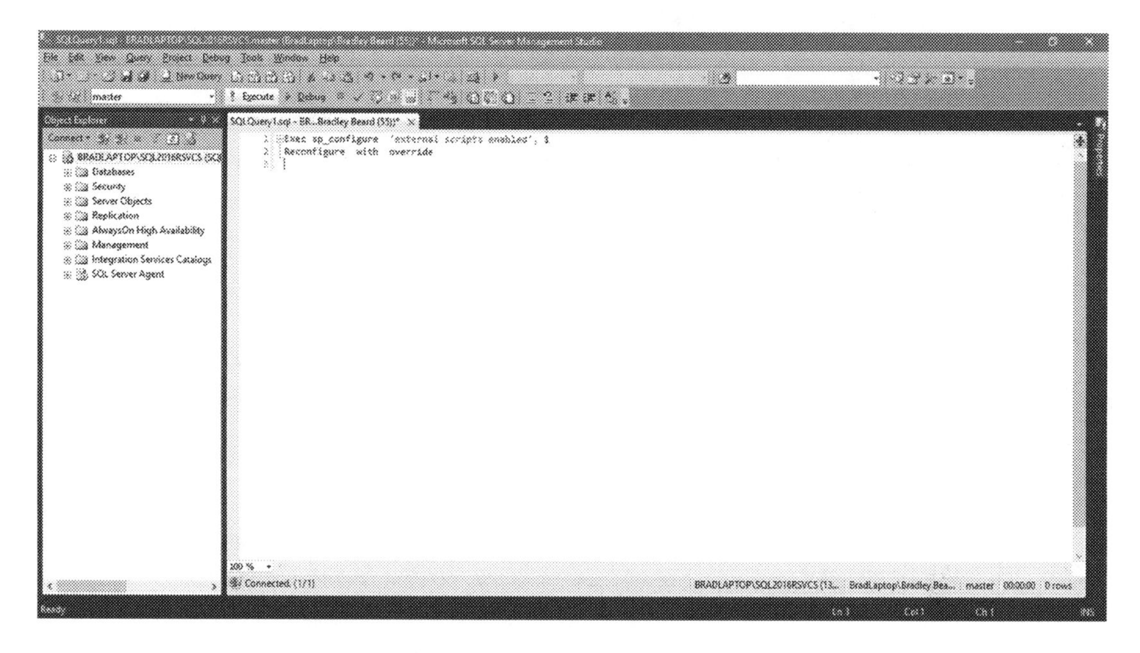

Figure A-25. *Command ready for execution*

Notice that we are executing against the master database.

Run that. You should see Figure A-26, which tells us that the execution was successful.

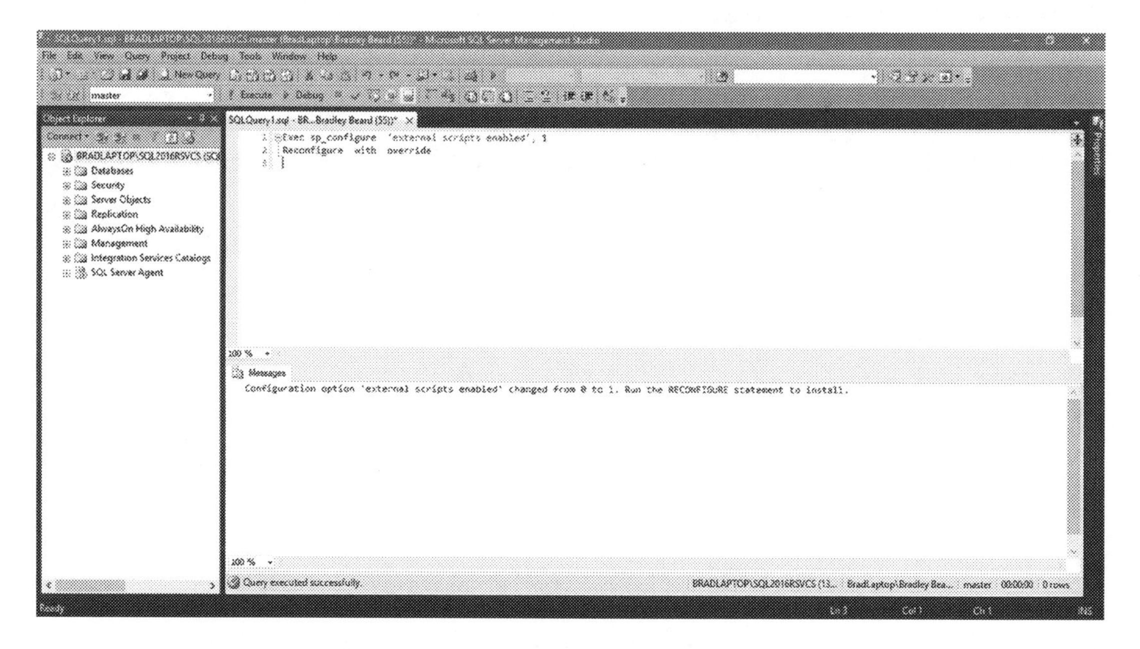

Figure A-26. *Success*

Next, we need to verify that R is indeed running. To do this, Microsoft says to restart the SQL Server instance and run the following script. Restart the instance first. Then open a New Query window and type the following:

```
exec sp_execute_external_script
@language =N'R',
@script=N'OutputDataSet<-InputDataSet',
@input_data_1 =N'select 1 as hello'
with result sets (([hello] int not null));
go
```

For those that haven't yet memorized every single system stored procedure, you won't recognize that `sp_execute_external_script` is a brand-new stored procedure introduced to execute external scripts. This stored procedure can be invoked with the following:

- `@language`: The name of the supported language.
- `@script`: The script executed (you can either type it all in to the stored procedure or reference it as a variable).
- `@input_data_1`: The SQL query you're using to gather data from the database goes here.
- `@input_data_1_name`: The data frame that acts as the result set of the `@input_data_1` query. *This attribute is optional.*
- `@output _data_1_name`: The data frame variable in `@script` that holds the output data. *This attribute is optional.*

Press **F5** to execute the script. The anticipated results are shown in Figure A-27.

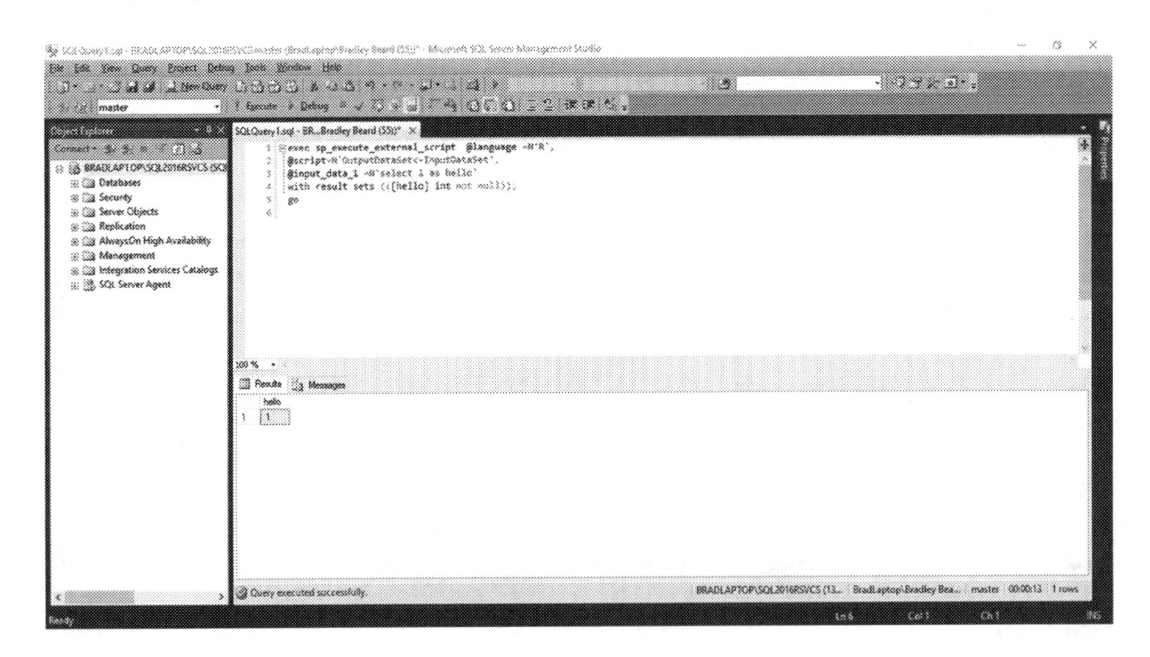

Figure A-27. *R is alive!*

Excellent! R is alive and well, and communicating normally with the SQL Server instance.

Summary

Let's briefly review what we have covered in this Appendix.

1. Installed a SQL Server 2016 named instance on top of an existing SQL Server 2014 default instance.
2. Configured R after installation.
3. Verified that R is installed correctly by running the script specified earlier.

So, essentially, that is how you install SQL Server 2016 on top of SQL Server 2014.

Keep in mind that, as I mentioned earlier, there could still be some conflicts with how the instances are configured. It may be better to just upgrade your instance to 2016, but you should never, ever upgrade a mission critical server just because the latest and greatest version of some application was released. You always want to check and do your homework to make sure that the OS is supported, and that there isn't a huge difference between the versions that you are upgrading. This is the main reason why the primary focus of this book is about installing on a new server, and not one with an existing SQL Server installation.

If you are comfortable with this installation and you are ready to move on, go ahead and start at Chapter 2 now. You should still be able to complete the exercises and examples in this book, but you need to follow those directions to get R Tools for Visual Studio and Reporting Services installed and configured correctly.

APPENDIX B

Software Requirements Document

Project

Beginning SQL Server R Services

Author

Bradley Beard, Lead Developer, We R Pros, LLC

Prepared for

R. Customer, Chief Information Officer, PleaseGiveUs-R-AnalysisData.com

Problem

PleaseGiveUs-R-AnalysisData.com ("Customer") has approached We R Pros, LLC ("Vendor") with a unique request.

Customer has a finite amount of data related to the weather, but Customer does not have the means to analyze this data. Customer would like Vendor to suggest products and services that could be gleaned from the analysis of the data sets.

In essence, Customer does not exactly know what it wants, but knows that it wants something and is relying on the expertise and experience that Vendor has to provide a solution or set of solutions for Customer. The data is quite valuable and is surely hiding quite a few "diamonds in the rough."

Solution

Upon initial inspection of Customer's data, the following information was gathered:

- The data file is 1 Microsoft Excel file.
- The file is approximately 5.5MB.
- The file is exactly 113,333 rows long, minus one header column.

B. Beard, *Beginning SQL Server R Services*, DOI 10.1007/978-1-4842-2298-0_11

- The file consists of the following headers:
  - Year
  - AdjustedMonth
  - AdjustedDay
  - AirportID
  - AdjustedHour
  - Timezone
  - Visibility
  - DryBulbFarenheit
  - DryBulbCelsius
  - DewPointFarenheit
  - DewPointCelsius
  - RelativeHumidity
  - WindSpeed
  - Altimeter

Based on these columns, it was determined that the following reports shall be created:

- Average Wind Speed per AirportID
- Average Temperature per AirportID (°F)

It is noted that many more reports could be gathered, but the customer specifically requested that these reports be created.

Language/Platform

This solution will use SQL Server 2016 to collect and store the data, R to analyze the data, and SQL Server Reporting Services to provide reports based on the data.

Medium

The medium of delivery for reports shall be as requested by the Customer; reports can be delivered in image format, Excel format, or PDF format. The specificity of the format can be determined at a later date, since the delivery method via SQL Server Reporting Services allows for exporting of the reports in a variety of formats.

APPENDIX C

R Plot and Tabular Code for R Tools for Visual Studio

First, a couple of notes about this appendix:

- The full code for the samples in Chapter 5 are here, but you still need to import the data into RTVS from the `Weather_Sample.csv` file in order to actually do anything.
- You may not need to run the `install.packages()` references, but they are there just in case. If you have already installed them, then it won't be necessary to install the packages again.

Average Temperature by Airport ID (Plot)

This code gives the average temperature by airport ID in graphical format. Notice that the `mean()` function contains the `DryBulbFarenheit` attribute, which indicates that this is keying on the temperature field, and not the wind speed field. Also, the presence of the `ggplot2` reference indicates that this is the plot code, and not the tabular code.

```
install.packages("data.table")

library(data.table)

Weather_Sample <- data.table(Weather_Sample)

chart_by_ID <- as.data.frame(Weather_Sample[, mean(DryBulbFarenheit, na.rm = TRUE),
by = AirportID])

install.packages("ggplot2")

library(ggplot2)

ggplot(chart_by_ID, aes(x = AirportID, y = V1)) + geom_point(stat = "identity") + geom_
smooth(method = "lm", formula = y ~ splines::bs(x, 3)) + scale_x_continuous(name = "Airport
ID") + scale_y_continuous(name = "Average Temperature") +
geom_text(aes(label = AirportID), size = 3, vjust = 1.0) +
geom_text(aes(label = round(V1, digits = 2)), size = 3, vjust = 2.0)
```


B. Beard, *Beginning SQL Server R Services*, DOI 10.1007/978-1-4842-2298-0_12

Average Temperature by Airport ID (Tabular)

This code gives the average temperature by airport ID in tabular format. Notice that the `mean()` function contains the `DryBulbFarenheit` attribute, which indicates that this is keying on the temperature field, and not the wind speed field. Also, the absence of the `ggplot2` reference indicates that this is the tabular code, and not the plot code.

```
install.packages("data.table")

library(data.table)

Weather_Sample <- data.table(Weather_Sample)

setkey(Weather_Sample, AirportID)

avg_temperature_by_ID <- as.data.frame(Weather_Sample[, mean(DryBulbFarenheit,
na.rm = TRUE), by = AirportID])

avg_temperature_by_ID
```

Average Wind Speed by Airport ID (Plot)

This code gives the average wind speed by airport ID in graphical format. Notice that the `mean()` function contains the `WindSpeed` attribute, which indicates that this is keying on the wind speed field, and not the temperature field. Also, the presence of the `ggplot2` reference indicates that this is the plot code, and not the tabular code.

```
install.packages("data.table")

library(data.table)

Weather_Sample <- data.table(Weather_Sample)

chart_by_ID <- as.data.frame(Weather_Sample[, mean(WindSpeed, na.rm = TRUE), by =
AirportID])

install.packages("ggplot2")

library(ggplot2)

ggplot(chart_by_ID, aes(x = AirportID, y = V1)) + geom_point(stat = "identity") + geom_
smooth(method = "lm", formula = y ~ splines::bs(x, 3)) + scale_x_continuous(name = "Airport
ID") + scale_y_continuous(name = "Average Wind Speed") +
geom_text(aes(label = AirportID), size = 3, vjust = 1.0) +
geom_text(aes(label = round(V1, digits = 2)), size = 3, vjust = 2.0)
```

Average Wind Speed by Airport ID (Tabular)

This code gives the average wind speed by airport ID in tabular format. Notice that the `mean()` function contains the `WindSpeed` attribute, which indicates that this is keying on the wind speed field, and not the temperature field. Also, the absence of the `ggplot2` reference indicates that this is the tabular code, and not the plot code.

```
install.packages("data.table")

library(data.table)

Weather_Sample <- data.table(Weather_Sample)

setkey(Weather_Sample, AirportID)

# OBJECT NAME            DATA FRAME    DATASET NAME          DATA COLUMN REMOVE N/A    GROUP
BY COLUMN
avg_windspeed_by_ID <- as.data.frame(Weather_Sample[, mean(WindSpeed, na.rm = TRUE), by =
AirportID])

avg_windspeed_by_ID
```

Index

B. Beard, *Beginning SQL Server R Services*, DOI 10.1007/978-1-4842-2298-0

R

S

T

U, V

W, X, Y, Z

Printed in the United States
By Bookmasters